For Peter & Sally
after Kew.

2009

SOUTH AFRICAN BOTANICAL ART

Peeling Back the Petals

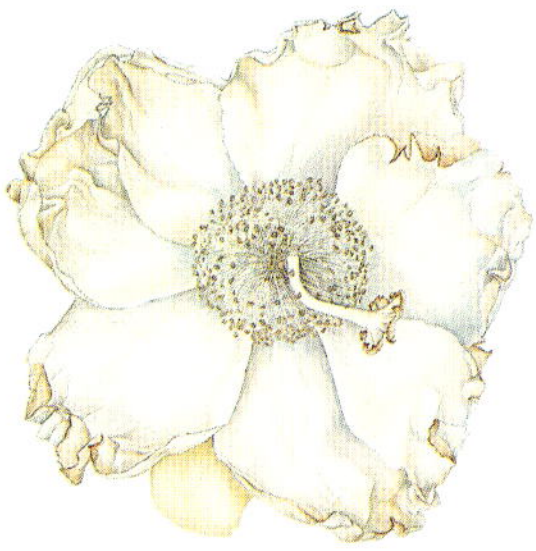

Glynn

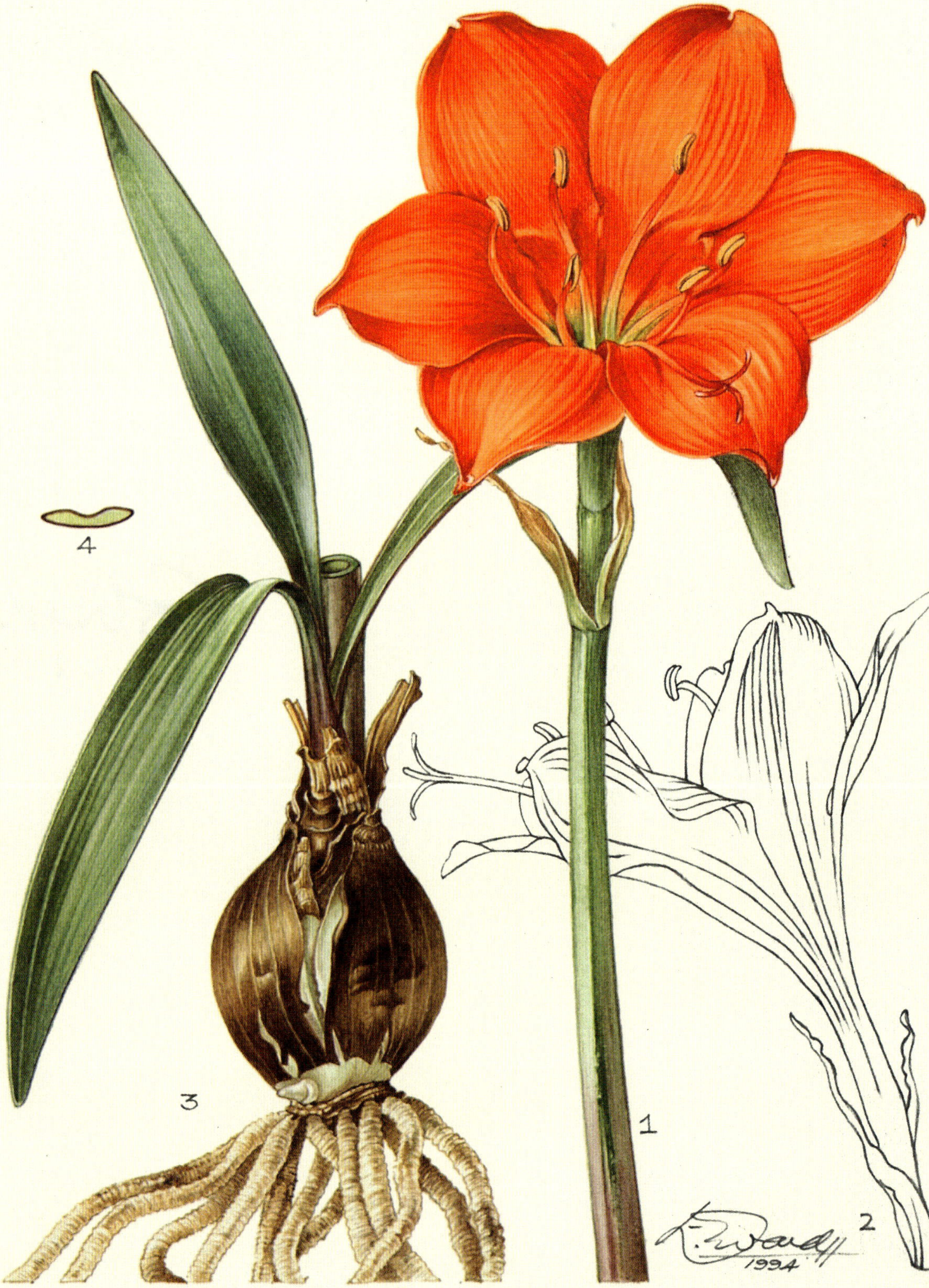

Ellaphie Ward-Hilhorst, *Cyrtanthus flammosus*, 1994, watercolour and pencil:
reproduced in *The Flowering Plants of Africa*, 1995

SOUTH AFRICAN BOTANICAL ART

Peeling Back the Petals

EDITED AND INTRODUCED BY

MARION ARNOLD

AUTHORS

JOHN ROURKE · DEE SNIJMAN

JOHN MANNING & PETER GOLDBLATT

MARION ARNOLD

Endorsed by the Botanical Society of South Africa

FERNWOOD PRESS

IN ASSOCIATION WITH

ART LINK

THE SPONSORS

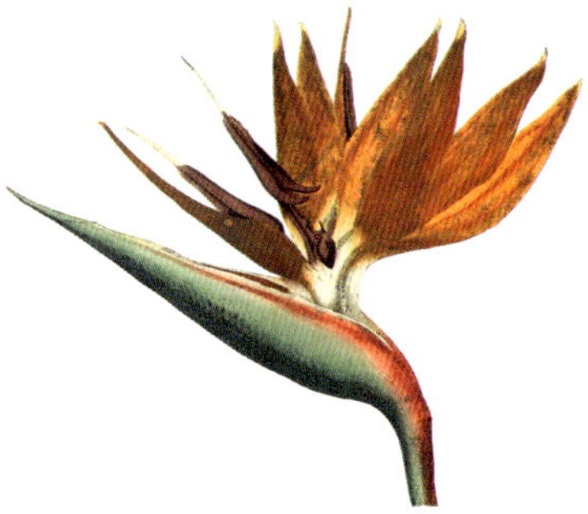

Reduction, Plate 121

Only with generous financial assistance has this publication been possible. All sponsors are warmly and most sincerely thanked for their spirited support and promotion of botanical art in South Africa. Mr Kenneth S. Birch of Johannesburg has made a major contribution. Other sponsors are: FNB (A Division of FirstRand Bank Limited) and the Mia Karsten Fund (National Botanical Institute, Cape Town).

FERNWOOD PRESS

FERNWOOD PRESS
P O BOX 15344
VLAEBERG 8018
SOUTH AFRICA

REGISTRATION NO. 90/04463/07

ART LINK (PTY) LTD
P O BOX 52431
SAXONWOLD 2132
SOUTH AFRICA

FIRST PUBLISHED 2001

Standard Edition ISBN 1-874950-54-7
Collectors' Edition ISBN 1-874950-57-1
Sponsors' Edition ISBN 1-874950-58-X

HOUSE EDITING BY CYNTHIA KEMP, JOHANNESBURG
INDEX BY LEONIE TWENTYMAN JONES, CAPE TOWN
DESIGN & TYPESETTING BY ABDUL AMIEN, CAPE TOWN
PRODUCTION CONTROL BY ABDUL LATIEF (BUNNY) GALLIE, CAPE TOWN
REPRODUCTION BY CAPE IMAGING BUREAU CC, CAPE TOWN
PRINTING AND BINDING BY TIEN WAH PRESS (PTE) LTD, SINGAPORE
BINDING OF COLLECTORS' AND SPONSORS' EDITIONS BY PETER CARSTENS, JOHANNESBURG

CONTENTS

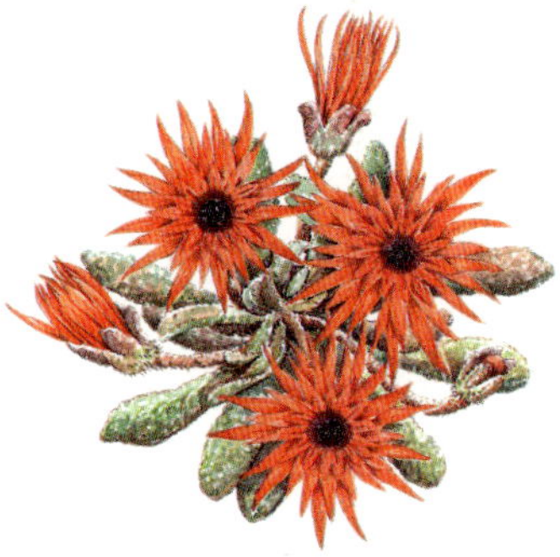

Barbara Jeppe,
Dorotheanthus rourkei,
c. 1970, watercolour

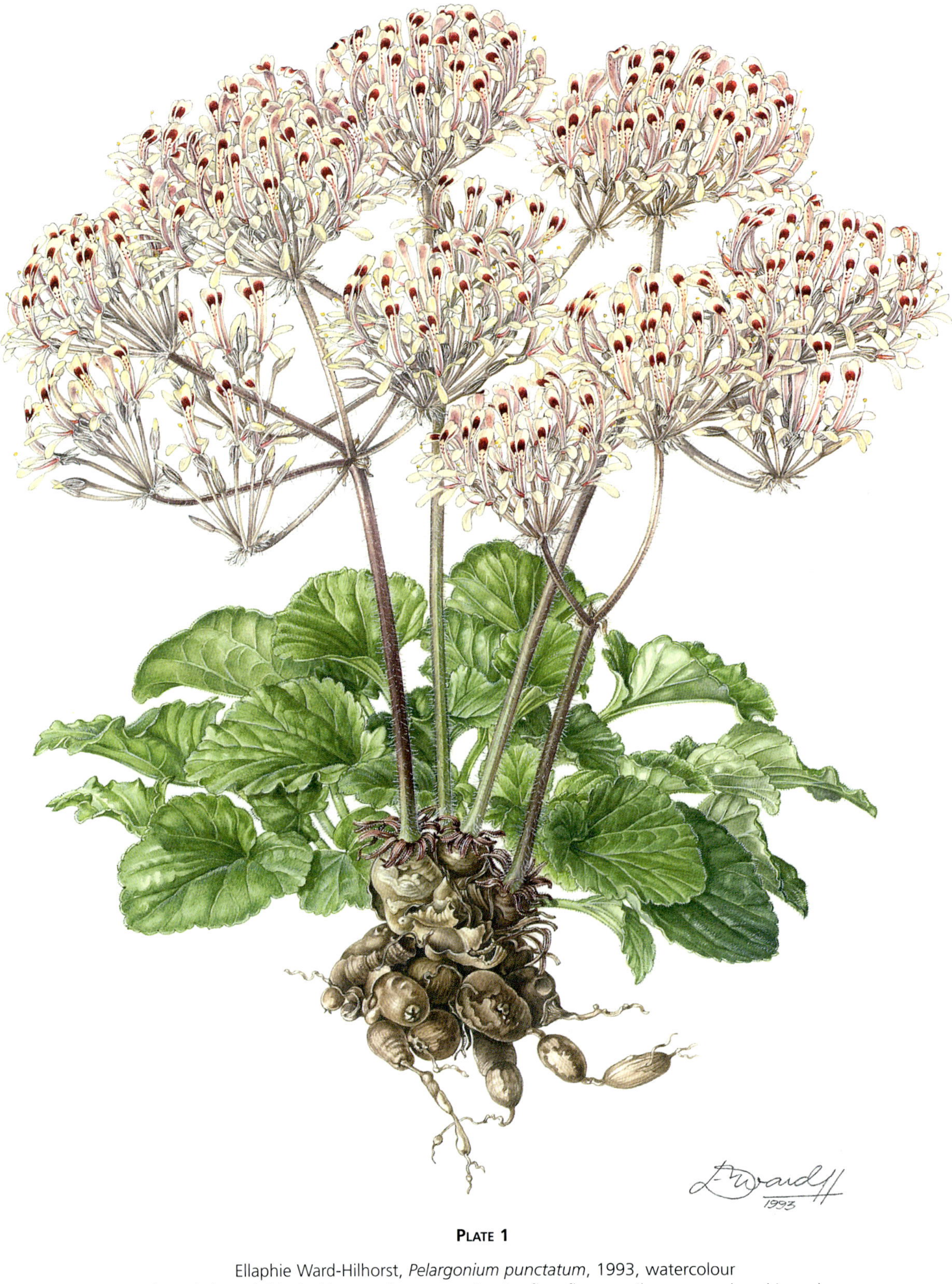

Plate 1

Ellaphie Ward-Hilhorst, *Pelargonium punctatum*, 1993, watercolour
A composite painting of this deciduous tuberous species. Leafless flower spikes are produced in early summer (foreground) while the rosette of leaves appears in winter in response to seasonal rains (background). The specific name 'punctatum' (dotted) refers to the characteristic red dots on the posterior petals.

PREFACE

STEPHAN A. WELZ

In none of the standard recognized handbooks, dictionaries or companions to western European art do we find a reference to 'botanical art' or 'botanical artists'. Even some of the greatest flower painters of all times, such as Nicolas Robert (1614–1685) and Pierre Joseph Redouté (1759–1840), do not warrant a mention – a surprising omission as, from the eighteenth century to this day, by virtue of universal decorative appeal, their work has been copied and reproduced onto every household object ranging from porcelain to curtain fabric to table mats. In his own time, Redouté was heralded as the 'Raphael of the rose' and a hundred years after his death, in the late nineteenth century, his influence was still acknowledged – in the realm of photography. Redouté studied under Gerardus van Spaendonck (1746–1822) who in turn studied under Jan van Huysum (1682–1749), both highly respected artists. Yet why is he, along with other talented botanical artists, ignored by art historians and lexicographers?

The omission can be attributed, in part, to the fact that botanical art is a hybrid genre constituted of two primary discourses – botany as science, and botany as visual art. Its frame of reference embraces not only art and science but also history, philosophy, geography, technology and commerce. Thus botanical art unites function and aesthetics. For example, until the late 1500s, botanical artists were concerned primarily with providing aids for the identification of medicinal plants. Then, in the early 1600s an aesthetic revolution took place and flowers came into their own, seen for their visual qualities.

A direct result of this change was the evolution of the flower garden – an area designed to display floral beauty, to flaunt personal wealth, to exhibit rare plants, and to provide pleasure and recreation. The consequent frenzy of collection and documentation of newly discovered plant species gave birth to the science of plant classification. Sumptuously produced and illustrated florilegia, celebrating private gardens and their botanical treasures followed, commissioned by the wealthy to record detailed selections or collections of flowers in folio or book form. Judged by today's standards, these florilegia served horticultural rather than botanical purposes: they were luxurious expressions of privilege.

Access to florilegia was limited. Usually very few were produced and often they were unique creations – and so it is even to this day. Consequently they remain entombed in museums and private collections, visited only by those who deliberately seek them out. By contrast, non-botanical paintings, particularly family portraits, were displayed with pride in the 1600s. For the educated classes they formed an integral if not essential part of interior decoration. Large portraits of high-ranking friends and relations were hung prominently, and even the frames adorning them assumed great importance.

And so botanical art became the domain of the scientist and bibliophile. It became classified and indexed rather than interpreted, rarely assessed and analysed by art historians or critics. As Marion Arnold notes here in her introduction: 'botanical art has been allowed to belong to the world of science, under-researched by art historians, undervalued by art theorists and critics, uncollected by art institutions, and seldom exhibited in art galleries.' When Goethe was first introduced to the work of Franz Andreas Bauer (1758–1840), regarded by art historian Wilfrid Blunt as the 'greatest botanical artist of all time', he observed that whereas nature was visible, art was hidden. In this pioneering publication, the authors' texts now raise that curtain of concealment. This will surely result in a dramatic reassessment of the role and interpretation of the work of the botanical artist.

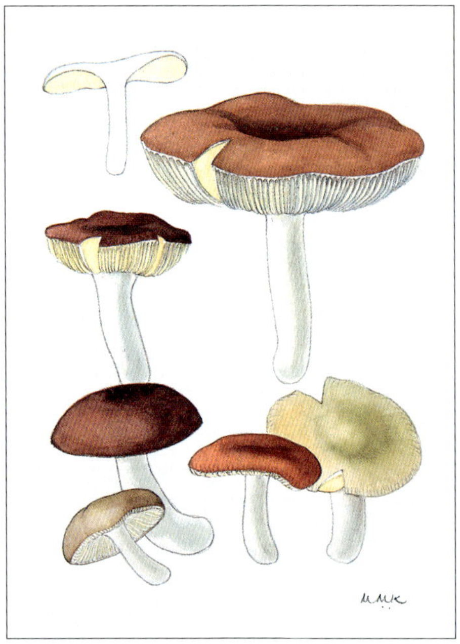

Plate 2

Mary Maytham Kidd, *Russula capensis*, 1950, watercolour and pencil: a native western Cape fungus appearing in early winter

In the extremes of modern parlance and politically correct points of view, the word 'colonialism' is today often used as a synonym for 'pillage' and 'oppression'. In many ways, though, South Africa was fortunate to have been colonized by two of the greatest European colonial powers ever, at a time when they were at the height of their power and wealth. The Dutch in the late 1600s, 1700s and early 1800s, and particularly the English in the 1800s, created an environment conducive to the flowering of the arts and sciences. With the expansion of political spheres of influence and trade-links of these and other European countries, came the exportation into Europe of exotic plant species from the New World. The commodity readily became a conspicuous luxury item, sought after and treasured, not least for its economic importance.

Already by the early 1600s it was perceived that the magnificence and variety of the floral kingdom of the Cape of Good Hope surpassed most other regions in the world. Horticulturalists and botanists were dispatched there by wealthy patrons to search for species which would not only enhance European gardens but also result in financial gain. The motives of these collectors included also the acquisition of personal recognition and prestige, and this led to the publication of some of the finest examples of botanical art. All the plates depicted in Franz Bauer's *Delineations of Exotick Plants Cultivated in the Gardens at Kew* are of ericas collected at the Cape. Similarly, *Coloured Engravings of Heaths*, by Henry C. Andrews (active 1794–1830), of which a large number of plates depict Cape species, exemplified the 'erica-mania' which dominated English horticulture in the first quarter of the 1800s, the cause of which had been the importation of hundreds of new species from the Cape.

The acclaim which southern African plants derived from these publications and the very presence of these visiting botanists, nurserymen, explorers and scientists, benefited the African subcontinent considerably. It stimulated the inhabitants of the Cape to reappraise their natural heritage, inducing in them a sense of pride. Scientists from other disciplines soon began to develop a strong affinity for their surroundings, as seen in the work of the nineteenth-century astronomer, Sir John Herschel, who with his wife, Margaret, set about recording and painting many of the Cape plant species.

All of this prompted and established a tradition and practice, the impact of which is still present and active today. Not only can South Africans be proud of the large number of indigenous flowers and plants painted by some of the world's leading botanical artists over a period of four centuries, but they can pride themselves on some of the finest botanical gardens in the world and of the publication of monumental works such as Rudolf Marloth's *The Flora of South Africa* (1913–1932) and Pole Evans's *The Flowering Plants of Africa* (1921 and ongoing). The large number of botanical works published today, illustrated by South African botanical artists of world renown, bears testimony to the high standards practised in the fields of science and art.

For this status to continue it requires both constant financial patronage and recognition. Just as the large florilegia of yesteryear were produced only with the foresight and resources of astute sponsors – not all wealthy, it must be added, for some died paupers having spent their life savings on grandiose publication ventures – hopefully that tradition will be acknowledged and will continue.

This book examines the aesthetics of botanical art and its place in science, how science has been enriched by the botanical artist and, more importantly, how both artist and scientist are responsible for highlighting – on several levels of awareness – just how crucial plant and animal interdependence is, and how the survival of all living species relies on this coexistence. In recent years plants, particularly those endemic to southern Africa, have become a significant commercial resource for both horticultural and medicinal purposes. In many respects the present situation mirrors that which obtained in the 1600s when the southern African subcontinent lost many of its plants to European horticulturists and hybridizers and European botanical artists grasped the discovery of the new to great commercial advantage. To prevent this from happening again, it is vital that all South Africans heed the contents of this book and that scientists, art historians, art gallery and museum directors, bibliophiles and those responsible for the preservation of our heritage appreciate and respect botanical art for what it is and give it its rightful place in our institutions.

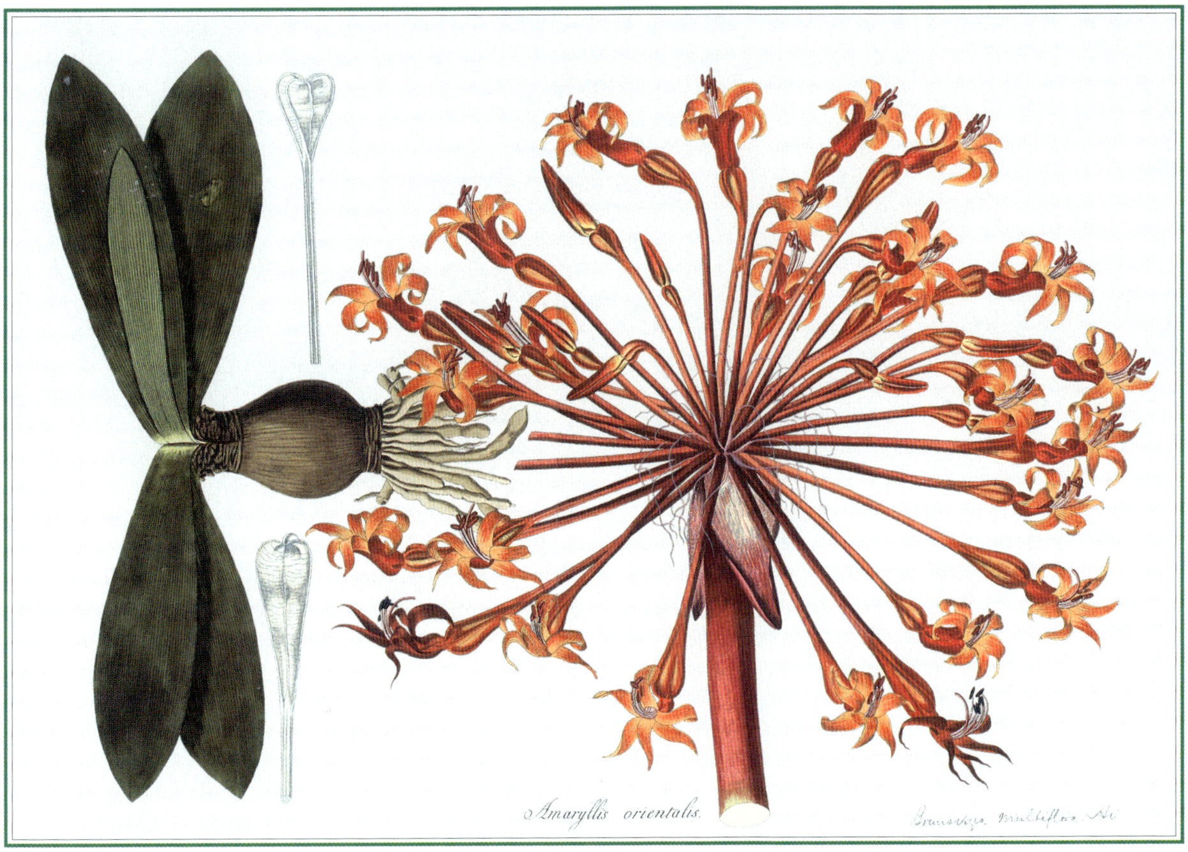

PLATE 3

Brunsvigia orientalis, hand-coloured engraving from an original drawing probably by Franz von Scheidel: Nikolaus Jacquin, *Plantarum Rariorum Horti Caesari Schönbrunnensis Descriptiones et Icones*, vol. 1 (Vienna, 1797). Jacquin's sumptuous production, which appeared in four volumes between 1797 and 1804, illustrated the plants in cultivation at the imperial gardens at Schönbrunn Palace. Many of the plants depicted, mostly for the first time, are southern African, and their publication was largely the fruit of the labours of the gardener, Georg Scholl, who spent twelve years at the Cape collecting for Jacquin.

ACKNOWLEDGEMENTS

Sponsors Kenneth S. Birch of Johannesburg, FNB (A Division of FirstRand Bank Limited) and the Mia Karsten Fund (National Botanical Institute, Cape Town) are thanked for generous financial support (see also imprint page).

Authors Marion Arnold, Peter Goldblatt, John Manning, Dee Snijman and John Rourke are most grateful to the many artists who contributed with enthusiasm to the project, and especially those who allowed them to reproduce their artworks without payment. They also thank the following individuals for their support and assistance: Trevor Arnold, Steve Bales, Auriol Batten, Hanna Botha, Jane Carruthers, Gillian Condy, Susette Foster, Susan Goldswain, Brian Huntley, Patricia Stuart Lorber, Hannelie Snyman, Michael Stevenson, and Piet Vorster. Finally, they thank Cynthia Kemp for her professionalism, editorial and creative skills, and commitment to this project.

The publishers acknowledge Gerhard Schwekendiek, Cape Town, and Heather Hogan, Cape Town, who contributed considerably towards layout, design and typography; artwork for Plate 49 is by Heather Hogan.

ILLUSTRATIONS

Acknowledgement is due to copyright-holders, owners and custodians of artworks and illustrations for permission to reproduce them; in cases where photographic transparencies and other images were generously loaned free of charge, the owners are especially thanked.

The following have permitted inclusion of material here: (see List of Illustrations which provides source details): Fay Anderson (Geary-Cooke), Cape Town (Plates 47, 117); Ann Axelson, Raif Axelson, Cape Town (Plate 65); J-M. Ball, Cape Town (Plate 54); Auriol Batten, East London (Plates 48, 69, 88); Bolus Herbarium, University of Cape Town (Plates 33, 46, 58, 115); Bolus Library, University of Cape Town (Plates 3, 42, 66, 100, 101, 122); The Brenthurst Library, Johannesburg (Plates 16, 19); The Caversham Press, Natal (Plates 109, 110); Central Johannesburg Library (Plate 14); Compton Herbarium, Kirstenbosch, Cape Town (back dustjacket [right], frontispiece, Plates 2, 11, 24, 27, 28, 29, 30, 47, 59, 63, 64, 67, 68, 80, 84, 96, 97, 98, 118, 119); Distillers Corporation Limited, Stellenbosch (Plate 7); First National Bank, Johannesburg (Plates 6, 43, 107, 128); Peter Goldblatt, Missouri, USA (Plate 56); Susan Goldswain, East London (Plates 1, 61, 130); Douglas Goode, Gillitts, Natal (Plates 109, 110); C. Kemp, Sandton (Plates 109, 110); Vivienne Koorland, New York (Plate 127); Count Natale Labia, Cape Town (Plate 103); Library of Parliament, Cape Town (Plate 17); Thalia Lincoln, Cape Town (Plates 43, 107, 124); Local History Museums' Collection, Durban (Plate 105); John Manning, Cape Town (Plates 56, 81, 83, 86, 87, 92, 95); Marie McCrae Collection, Randburg (Plate 114); Liz McMahon, Dunbar, Scotland (Plates 6, 128); Michaelis Collection, Cape Town (Plate 102); MuseuMAfricA, Johannesburg (Plate 20); National Botanical Institute, Pretoria (front dustjacket, half-title, frontispiece, Plates 5, 9, 18, 23, 26, 35, 36, 37, 38, 39, 40, 44, 45, 50, 51, 53, 55, 60, 62, 70, 71, 72, 73, 74, 75, 76, 79, 82, 85, 89, 90, 91, 93, 94, 99, 108, 116, 121, 125); National Museum (Rock Art Department), Bloemfontein (Plate 8); Chris Nel, Johannesburg (Plate 4); Colin Paterson-Jones, Cape Town (Plates 77, 78); Barbara Pike, Johannesburg (Plates 4, 131); Johannes Clemens Porer, Cape Town (Plate 127); private collection, Johannesburg (Plate 131); Rembrandt van Rijn Art Foundation, Stellenbosch (Plate 103); Rijksherbarium (Leiden Library), The Netherlands (Plate 15); John Rourke, Cape Town (contents page, Plates 34, 65, 129, 130); Royal Botanic Gardens, Kew, UK (Plates 22, 106); Sasol Collection, Johannesburg (back dustjacket [left], Plate 126); Saunders Archives, Tongaat (Plate 32); D.A. Snijman, Cape Town (Plate 61); South African Library, Cape Town (Plate 13); South African Museum, Cape Town (Plate 12); Spoornet Collection: The Blue Train (Plate 57); Helmut Starcke, Johannesburg (back dustjacket [left], Plate 126); Michael Stevenson, Cape Town (Plates 104, 120); Vicki Thomas, Cape Town (Plates 41, 54); Trinity College Library, Dublin, Ireland (Plate 10); University of Cape Town Archives (Plates 21, 25); University of Stellenbosch, J.S. Gericke Library (Plates 31, 52, 111, 112, 113, 123).

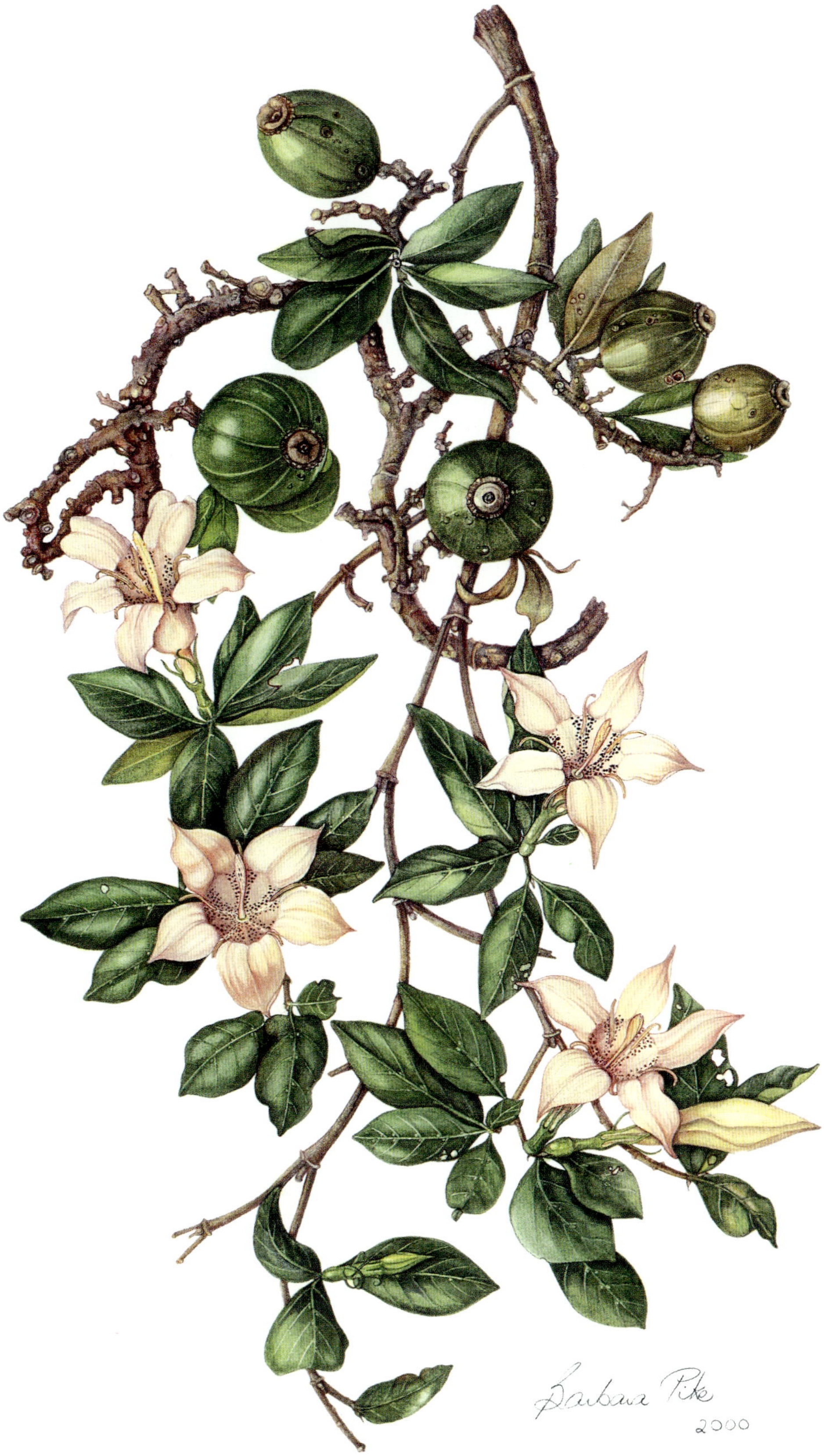

Plate 4

Barbara Pike, *Rothmannia capensis*, 2000, watercolour, 57 cm x 38 cm

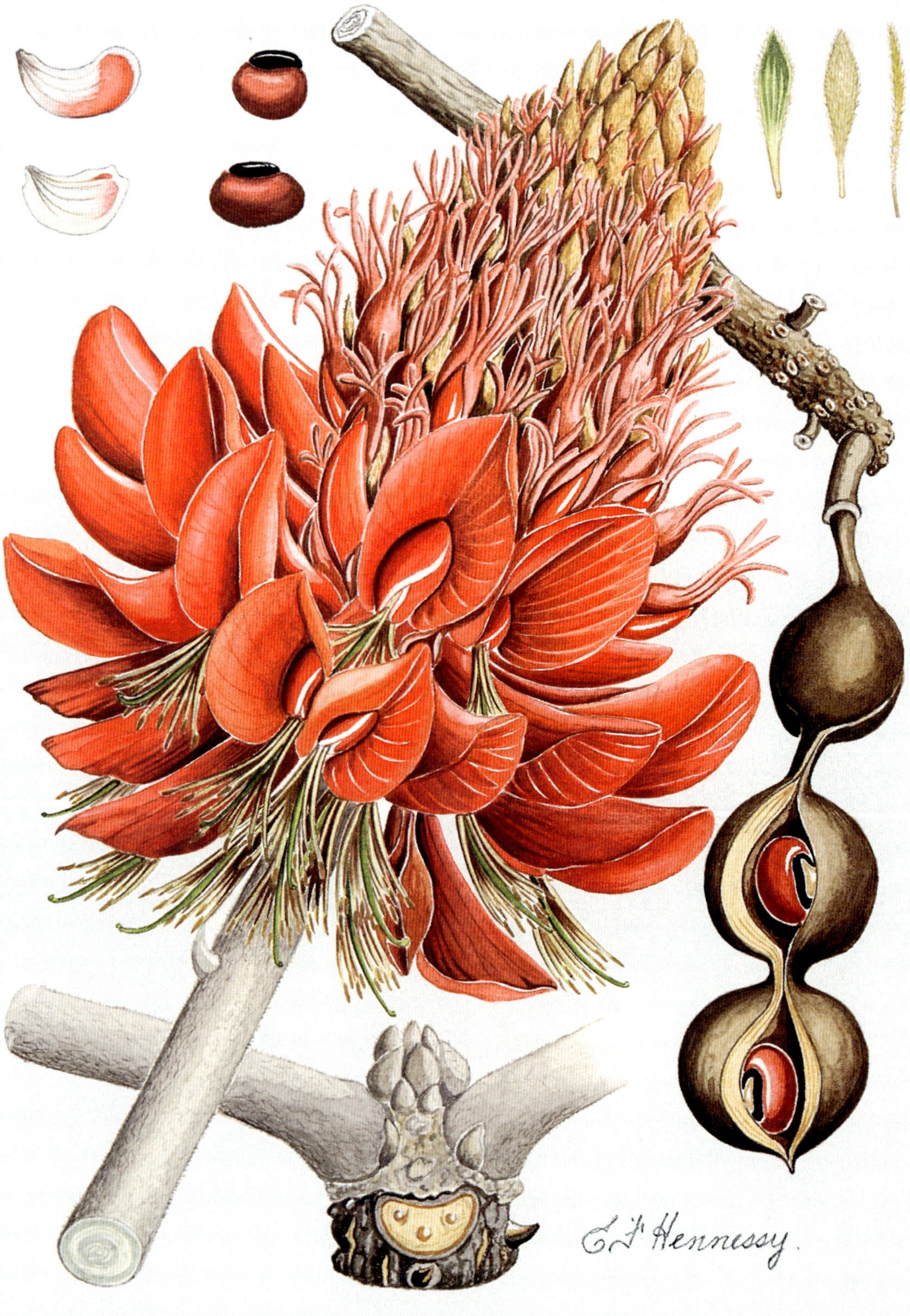

Plate 5

Esmé Hennessy, *Erythrina latissima*, 1973, watercolour: reproduced in *The Flowering Plants of Africa*, 1976

LIST OF ILLUSTRATIONS

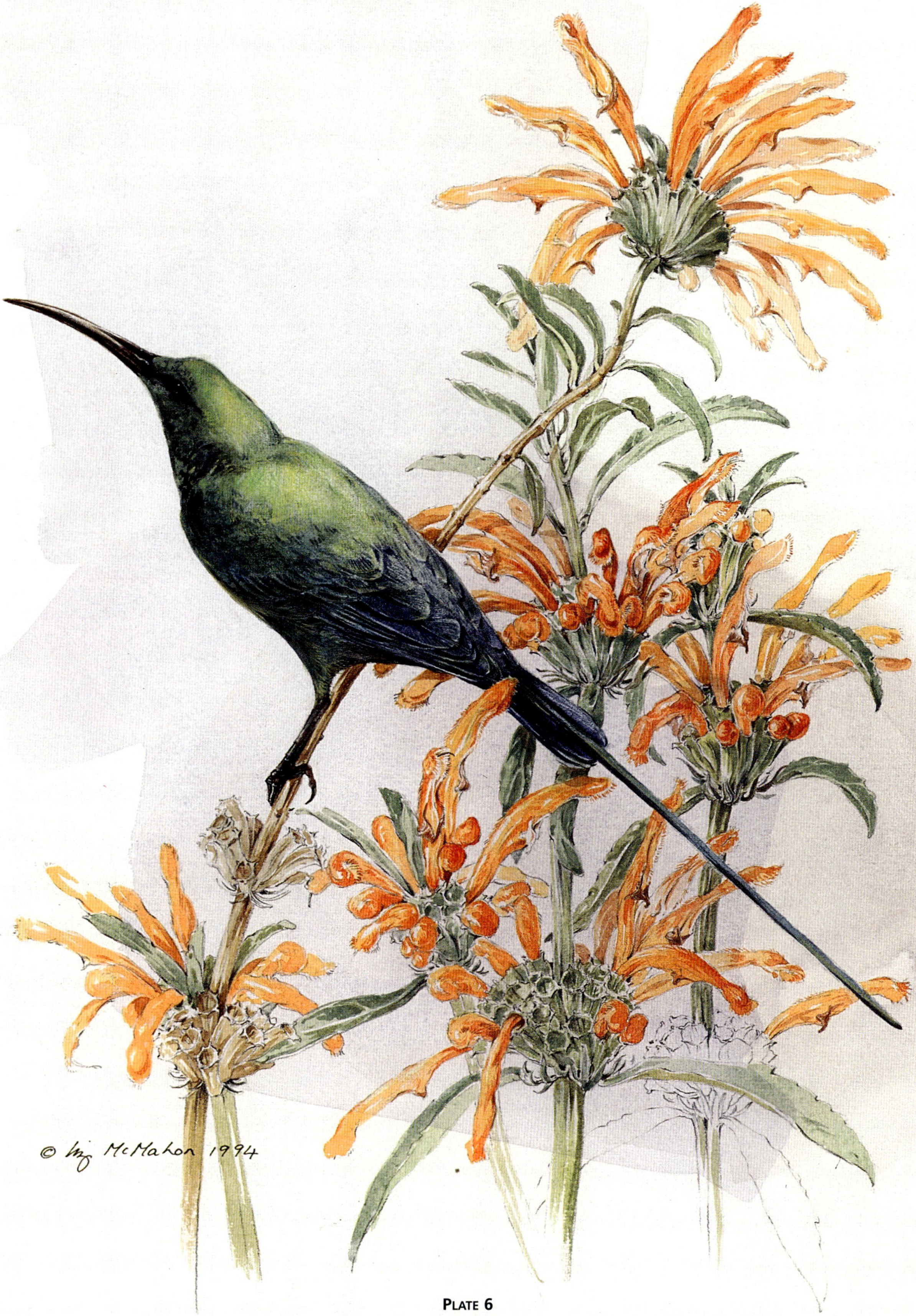

PLATE 6

Liz McMahon, Male malachite sunbird, *Nectarinia famosa*, on wild dagga, *Leonotis leonurus*, 1994, watercolour and pencil; reproduced in Michael Fraser, *Between Two Shores: Flora and Fauna of the Cape of Good Hope*, illustrated by Liz McMahon (Cape Town, 1994)

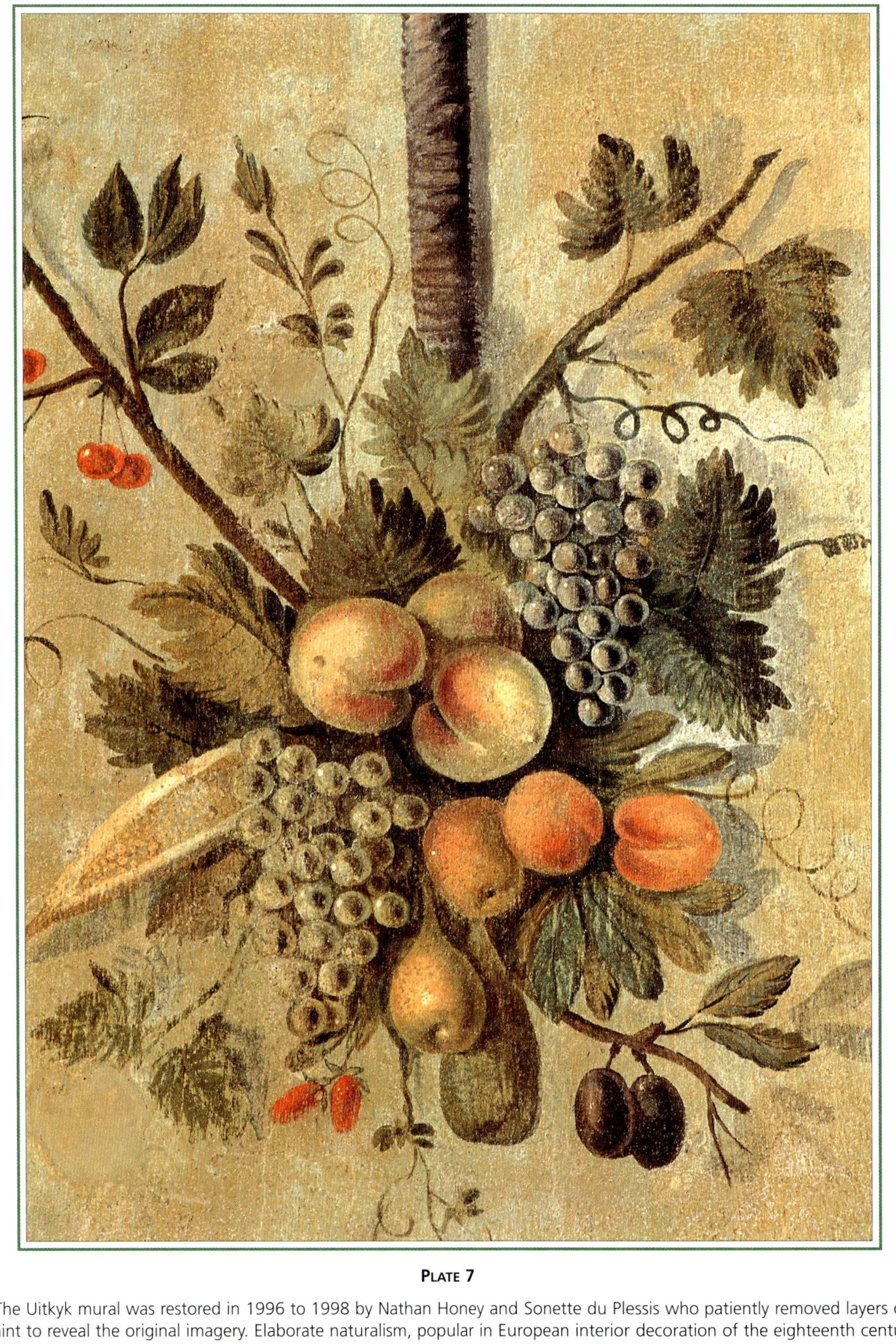

PLATE 7

The Uitkyk mural was restored in 1996 to 1998 by Nathan Honey and Sonette du Plessis who patiently removed layers of paint to reveal the original imagery. Elaborate naturalism, popular in European interior decoration of the eighteenth century, is apparent in realistically rendered flowers and fruits and illusionist shadows seemingly cast on the wall. The artist remains unknown.

INTRODUCTION

MARION ARNOLD

Southern African plant portraiture can be summarized in three images – a bulb depicted on a rock shelter, fruit and flowers painted on a plastered wall, and a detailed engraving of a shrub printed on paper. These representations indicate the complex interactions between seeing and interpreting, an elaborate process of translating and transforming personal vision into signs and objects that are understood differently by different communities.

Although plants have been portrayed in Africa for thousands of years – dating back to ancient Egypt – the extant plant portraits found at the southern end of the continent are of relatively recent origin. Despite the fact that the early indigenous inhabitants of southern Africa – the San (or Bushmen) – were hunter-gatherers, their surviving paintings include few renderings of plants. One such image, (Plate 8) now in rather poor condition with paint flaking off the bulb shape, was painted on a shallow sandstone shelter at Thaba Bosiu in western Lesotho; archaeological evidence suggests that the painting was probably created in the eighteenth century.[1]

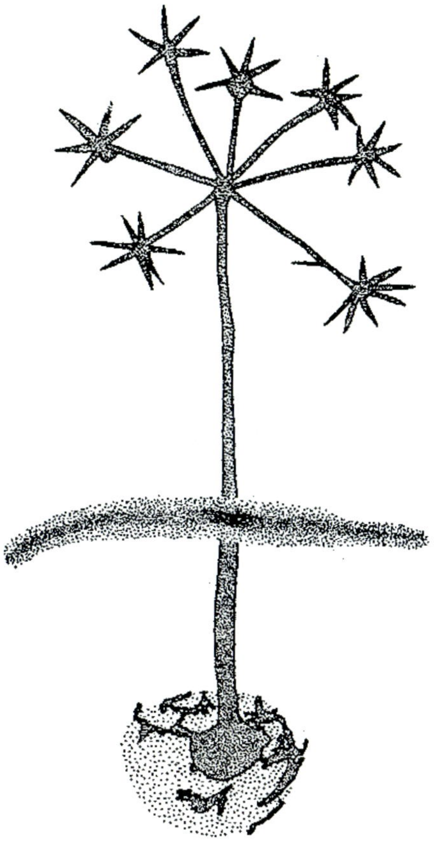

PLATE 8

From the structure of the cryptically rendered bulb and inflorescence, it has been deduced that this plant is a *Brunsvigia*, possibly *B. radulosa*. *(Redrawing of the Rock Art Department, National Museum, South Africa)*

By this time a European presence was well established in southern Africa and, at Uitkyk wine estate near Stellenbosch, an unknown artist decorated the farmhouse entrance hall with two paintings featuring flowers and fruit (Plate 7). Designed by Louis Michel Thibault (1750–1815), the Uitkyk manor house was completed in 1788, so the mural was probably created towards the end of the eighteenth century. The artist, who was obviously conversant with the contrivances of western decorative techniques, demonstrates familiarity with Cape flowers, fruits, birds and insects.

Towards the end of the eighteenth century, when Cape homes reflected the aesthetic tastes of Europe, scientific curiosity motivated naturalists to build collections of African plants. Francis Masson (1741–1806), first sent to the Cape in 1772 from the Royal Botanic Gardens at Kew in England, collected specimens which

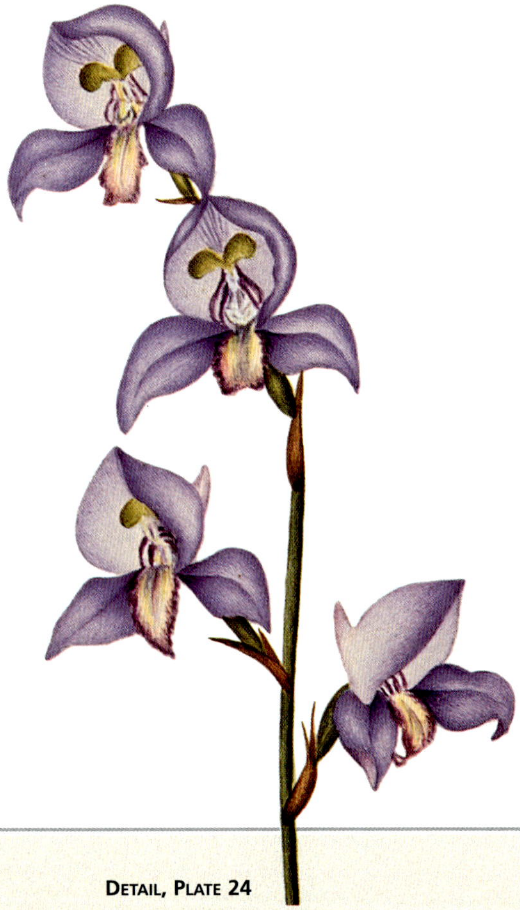

Detail, Plate 24

Just as people have personalities, so do plants. Just as people have personal and social histories, so do plants. A portrait, like a biography, describes likeness through facts, and evokes personality through manipulation of facts. It is a mistake to assume that portraits are 'truth' or facsimiles of 'reality'; information observable in human faces, or flowering plants, is translated into images. As interpretations, images portray – but cannot replicate – reality and in becoming portraits, they explore identity.

he dispatched to England where they were drawn by Franz (Francis) Bauer (1758–1840), a specialist botanical artist. Closely observed and elegantly rendered, Bauer's minutely detailed, structural studies of ericas were published in *Delineations of Exotick Plants* (Plate 9). As Aiton mentions in the preface to this beautiful book, 'each figure is intended to answer itself every question a Botanist can wish to ask, respecting the structure of the plant it represents'.[2]

The anonymous Bushman, the unknown European, and Franz Bauer created uniquely different images but each, in his own way, testifies to the roles played by plants in his own society. The San artist lived in a highly ritualized community, which utilized plants for medicinal and hallucinogenic reasons, and it is probable that the amaryllid bulb drawing relates to one or both of these functions. The Uitkyk painter's *trompe l'oeil* depiction of nature speaks of a need to enhance a gracious home with decoration, and Bauer's study provides visual evidence of discoveries made by Western travellers as they explored southern Africa and extended the frontiers of natural science. Regardless of their reasons for depicting plants, these three eighteenth-century artists create plant portraits.

A portrait is a likeness. It is an image that describes the appearance of a model with sufficient accuracy to facilitate identification, but a convincing portrait is concerned with more than salient external characteristics: it manifests personality. This is something elusive, unseen when first glancing at a subject, but which emerges after prolonged contemplation, and artists have to find ways of associating it with observable physical characteristics that can be described visually.

Plant portraiture is influenced by the fact that it is a hybrid art located within two theoretical discourses – the science of botany, and the visual arts. This gives plant portraits complicated identities and establishes different contexts where meaning operates. Whilst science claims to be concerned with intellectual objectivity and empirical reality, art aims to move the mind through the imagination and senses. Although commonly conceived as being in opposition, the aims of art and science need not be polarized as a conflict between 'objectivity' and 'subjectivity' and, in plant imagery, art meets science.

Both art and botany stake claims on plant representations. The terminology they employ – 'botanical art' and 'flower painting' – carries different connotations, and these terms, often used imprecisely, are neither clearly descriptive nor merely descriptive. Broadly speaking, botanical art is assumed to belong to Science, while

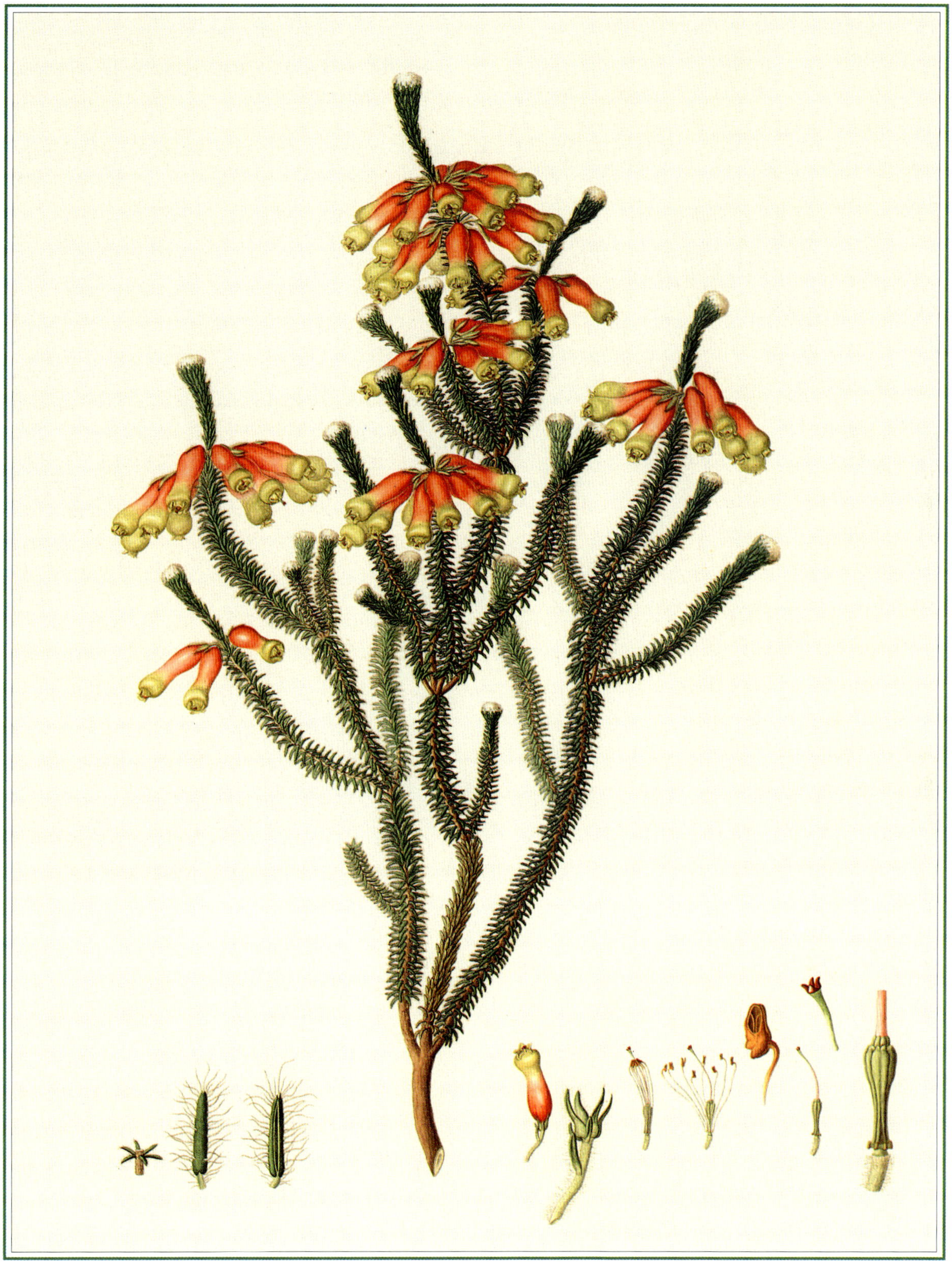

PLATE 9

Erica massonii demonstrates Franz Bauer's sophisticated botanical knowledge and his accomplished mastery of naturalism (hand-coloured engraving).

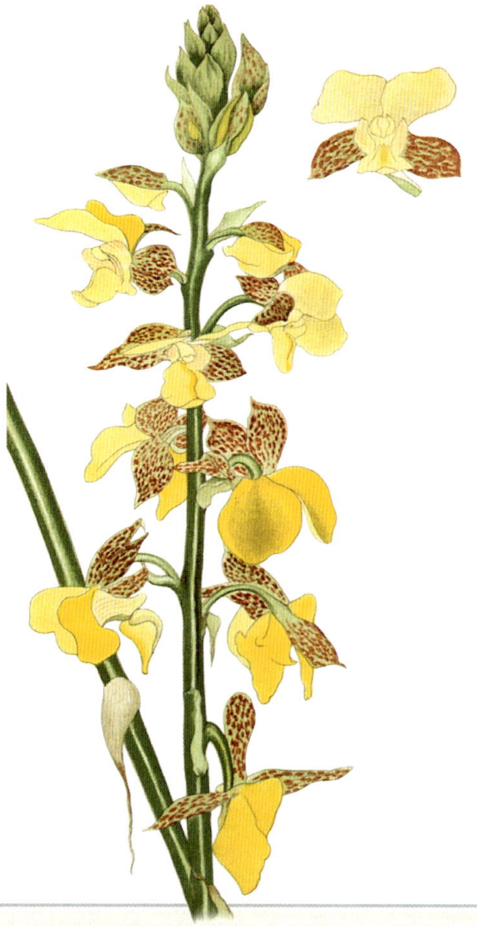

DETAIL, PLATE 26

It is within botany and art that one must metaphorically 'peel back the petals' to explore plant portraiture, and discover the meanings ascribed to the genres of 'botanical art' and 'flower painting'.

flower painting is understood as a genre within Art. The implications of the terminology extend beyond the parameters of subject matter, into the worlds of science and art with their own internal power relations, hierarchies, values, methods and expectations about the production and reception of visual imagery.

Artists and botanists, poets and gardeners see, respond to, and use flowers in different ways but usually the first things they notice are the petals. Coloured, modified leaves, petals are the most visually ostentatious parts of flowers, and – as well as attracting insect pollinators – they have, throughout history, stimulated artistic and scientific enquiries which contribute meaning to plant portraiture.

Representation is a complex procedure including all the processes through which meaning is made and circulated in society. Images acquire significance from their relationship to the reality of the world. The desire to describe the appearance of the world – people, landscape, plants, animals and objects – motivates artists in many societies, but expressive form is culturally determined. The San artist's stylization and Bauer's naturalism are learned conventions. These coded renderings of plants communicate information to viewers who 'read', or interpret, the lines, shapes, tones and colours and connect them with the physical world of appearances.

Within plant representation, changes in convention are related to shifts in scientific thinking. In the seventeenth century, flowering plants were arranged according to the number and structure of their petals but, in the eighteenth century, the great Swedish botanist, Carl Linnaeus (Linné) (1707–1778), devised a new system of naming, ranking, and classifying organisms. Linnaeus, naturalist and professor of medicine and botany, also developed binomial nomenclature (a classification system whereby one name was given for the genus and another for the species, in Latin), as part of his taxonomic scheme (see Plate 49). Petals, according to Linnaeus, do nothing for the cause of plant identity and sexuality. They serve only, he proclaims grandiloquently, 'as the bridal bed which the creator has so gloriously prepared, adorned with such precious bed-curtains, and perfumed with so many sweet scents in order that the bridegroom and bride may therein celebrate their nuptial with the greater solemnity'.[3]

Abandoning the importance of the bed curtains and focusing his attention on the mysteries of 'the bridal bed', Linnaeus envisaged the reproductive process as the key to plant classification. His sexual system was based on the number, form and arrangement of plant reproductive organs, the plant's class being determined by its stamens (male organs) and its order by its pistils

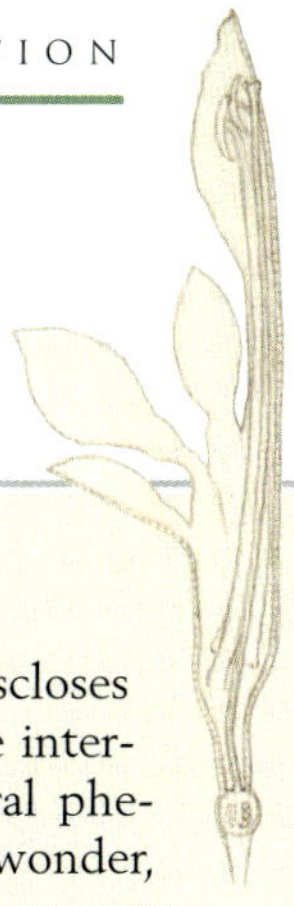

Detail, Plate 71

The exercise of 'peeling back the petals' discloses many ways of seeing and understanding the interdependence of biological, social and cultural phenomena whilst responding, with a sense of wonder, to the beauty and mystery of flowers. This process acquired particular significance at the millennium. The millennium – an arbitrary, man-made date – offered humankind the opportunity to consider history in something other than the dimensions of mortal life-spans.

(female organs). In literally 'peeling back the petals' to pay attention to plant structure, Linnaeus propelled botany into the age of systematically organized knowledge and gave scientific meaning to structures which had long enjoyed attention for their aesthetic appeal.

Theories, derived from world views pertinent to particular societies, are subject to change, and the great change in understanding nature was promoted in the nineteenth century by Charles Darwin (1809–1882). Linnaeus had postulated a static classificatory system, but Darwin completely altered evolutionary concepts by proposing continuous evolution. In his influential publication, *On the Origin of Species by Means of Natural Selection*,[4] he argued for the change of biological species over time, and for natural variations passed on by heredity. In his conceptual scheme, the evolution of species is related to changing environmental conditions, and is a gradual and continuous process.

Darwin's views contributed to the formulation of modern systematics and to a late twentieth-century holistic, ecological paradigm which favours synergy. In science and philosophy, nature is no longer seen as being comprised of separate parts, but as a series of interrelationships that can be understood from many perspectives. However, for a truly ecological view of fundamental interconnectedness to be effectual in the twenty-first century, new ways of understanding life have to be initiated. If humankind persists in abusing nature in the name of 'development' and clings to the belief that man-made, patriarchal systems should dictate social structures, the future looks bleak. Social relationships simply cannot be divorced from environmental relationships any more than art can be separated from science.

Artists and scientists see nature differently: we see what we seek, and our interpretations of plant imagery indicate many levels of engaging with images as aesthetic or scientific documents. When words alone fail to focus attention, the silent eloquence of visual form moves the mind in powerful ways. But if our ideas and images are to raise levels of consciousness they have to reach new audiences. If art addresses only those interested in art, and science speaks exclusively to scientists, the converted merely confirm their premises or prejudices. As has been observed, 'scientists are dangerously ignorant about what arts people really do and arts people are dangerously ignorant about the realities of science'.[5] Interconnectedness is not just a useful term to summarize the ecological paradigm; applied to self-contained disciplines – and to the perception that art and science are in opposition – it challenges artificial parameters, enabling debates to cross barriers of specialization and probe shared philosophical concerns.

Plant portraiture offers a unique opportunity to explore different interpretations of natural forms, and in four independent essays we position plant imagery in historical, artistic and scientific frames of reference. We are concerned with people, ideas, and images, artistic and scientific information, history and knowledge at the beginning of the twenty-first century. By presenting various viewpoints about plant images, we demonstrate that creative thinking and imagination characterize art and science, and that both intuition and logic contribute to persuasive discourse. We offer alternative, but complementary ways of studying images and making deductions that thread together issues about the representation and meaning of biological diversity. Our common starting point is the plant image.

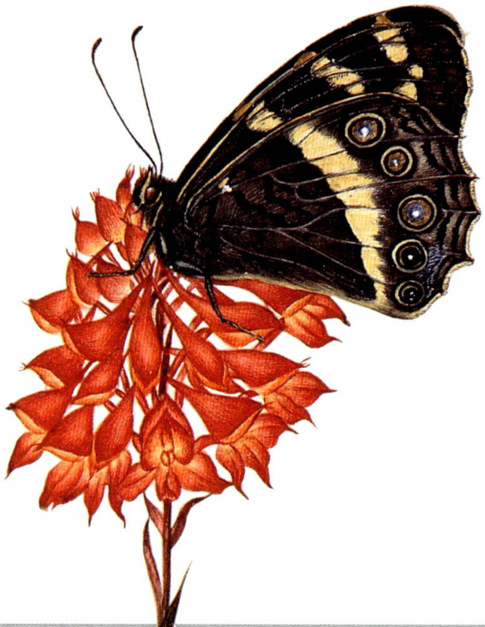

DETAIL, PLATE 52

Our planet has evolved from an angry combustion of gases hostile to life, to a world of land, air and water – elements which sustain plants and animals, including *Homo sapiens*, who arrived late in the earth's long history. To conceive of the future, we need to reflect not only on the past, but also on our being-in-the-world and our place in nature.

DETAIL, PLATE 58

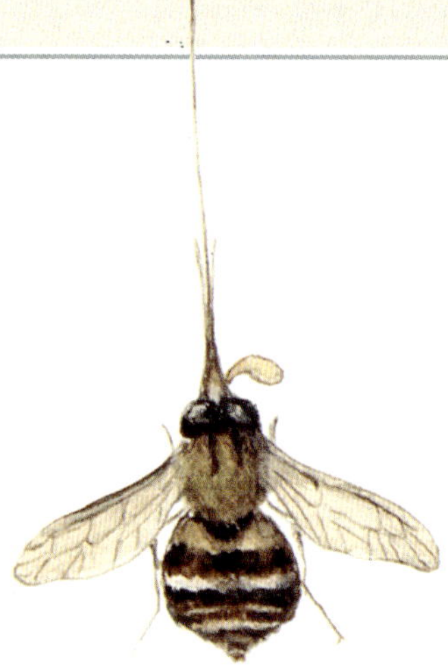

John Rourke provides an overview of the artists who have made a major contribution to the environmental and botanical history of southern Africa. He points out that even though curiosity and practical considerations were strong motivational forces in the desire to create recognizable images, the design, form, colour and texture of southern African plants play a significant part in stimulating visual creativity. Tracing the development of southern African botanical art from anonymous draughtsmen to contemporary professionals, he interweaves the stories of some remarkable people with an account of the publications that celebrate our rich flora.

He raises the interesting point that in the nineteenth and twentieth centuries southern African botanical art has been dominated by women artists. Nineteenth-century upper- and middle-class women were encouraged to number sketching and painting among their accomplishments, and plant study was deemed suitable for the 'lesser' sex, but this does not fully explain the preponderance of botanical women artists in the twentieth century. Although Rourke correctly observes that contemporary art does not generate the professional income male artists desire, the issue is complex, involving more than economics.

It is, as I discuss in my essay, about the hierarchical power structure of the art world. Within High Art the study of humankind – particularly of man – is deemed more significant than the study of nature, and flower painting, often created by women artists, has been stigmatized, and dismissed as being trivial and decorative. Furthermore, botanical art has been allowed to belong to the world of science, 'under-researched by art historians, undervalued by art theorists and critics, uncollected by art institutions, and seldom exhibited in art galleries'.[6]

Rourke's narrative of people – anonymous or named, 'birds of passage' who 'came, collected, painted and departed', or artists who lived and worked here – is complemented by Dee Snijman's stories of the plants themselves. Writing as a systematic biologist she reminds us that discovery (at least 290 plant species were described in southern Africa between 1994 and 1998) is tempered by destruction (58 named southern African plants have recently become extinct and many more are under threat). She focuses on the botanical preoccupations that influence plant representations, a process of 'relentless exploration, discovery and documentation' and, in charting the history of botanical discoveries, demonstrates the significance of producing accurate written and visual descriptions which lead to plant classification.

So sensitive a subject in South African sociopolitical history, classification is non-judgmental in botany. It enables botanists to chart major plant families, genera and species and, as a result, to understand evolutionary relationships. These relationships are explored by John Manning and Peter Goldblatt with particular reference to pollination biology. They assert that for plants, 'form is the key to function', and survival and reproduction depend on the relationship plants enjoy with their environments or ecology.

Observing that flowers are essentially functional structures, they remind us that 'aesthetic considerations are meaningful only to humanity and completely irrelevant to the plant'. In a closely argued discussion they demonstrate that a myriad of insects, animals and birds contribute to plant survival, lured to the reproductive organs of flowers by nectar, pollen, scent or visual deception. Processes of adaptation and change are long and slow, and hence Manning and Goldblatt conclude that 'Botanical portraits are, ultimately, snapshots in the life of a species'.

In my essay I suggest that because plant portraits are a hybrid art lodged both in science and in art, they are defined by expectations from both disciplines and are modified by context and conventions, language, processes, media, techniques and styles. All these issues contribute to the production of art objects that explore plant likenesses and personalities within the picture format, while conveying societal attitudes to image production, aesthetic tastes and values, and scientific achievements.

Undervalued by the art community for complex historical reasons, images of nature in general and plant studies in particular, have been allowed to serve science. But in fact, at the beginning of the twenty-first century with academic intertextuality effecting new fusions of knowledge, plant portraiture is well positioned to respond to new interpretations. In so doing it can be introduced into an ecological world view.

We must respect delicately balanced interactions between planets, stars and space, land structures and weather patterns, plants and pollinators, animals and food sources, people and the environment. In a century that will, in all likelihood, be dominated by information technology, we cannot afford to neglect the environment or to continue wilful interference in the natural relationships that sustain life itself, and expect that humankind will survive.

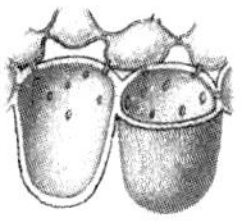

1 J. Loubser and P.C. Zietsman, 'Rock painting of postulated *Brunsvigia* sp. (Amaryllidaceae) at Thaba Bosiu, western Lesotho', *South African Journal of Science*, 90, 1994, p. 612.

2 *Delineations of Exotick Plants Cultivated in the Gardens at Kew . . . the Botanical Characters Displayed According to the Linnean System . . .*, preface by William Aiton (London, 1796–1803).

3 Wilfrid Blunt and William T. Stearn, *The Art of Botanical Illustration* (Woodbridge, Suffolk, 1994), 2 ed., p. 34.

4 (London, 1859).

5 Arthur Marwick, 'Some conclusions: what have we done and why have we tried to do it?' in Martin Pollock, ed., *Common Denominators in Art and Science: the Proceedings of a Discussion Conference Held Under the Auspices of the School of Epistemics, University of Edinburgh, November 1981* (Aberdeen, 1983), p. 188.

6 See chapter below by Marion Arnold, 'Petals and stigmas: reflections on plant portraiture', subsection 'Art and the mark of disgrace'.

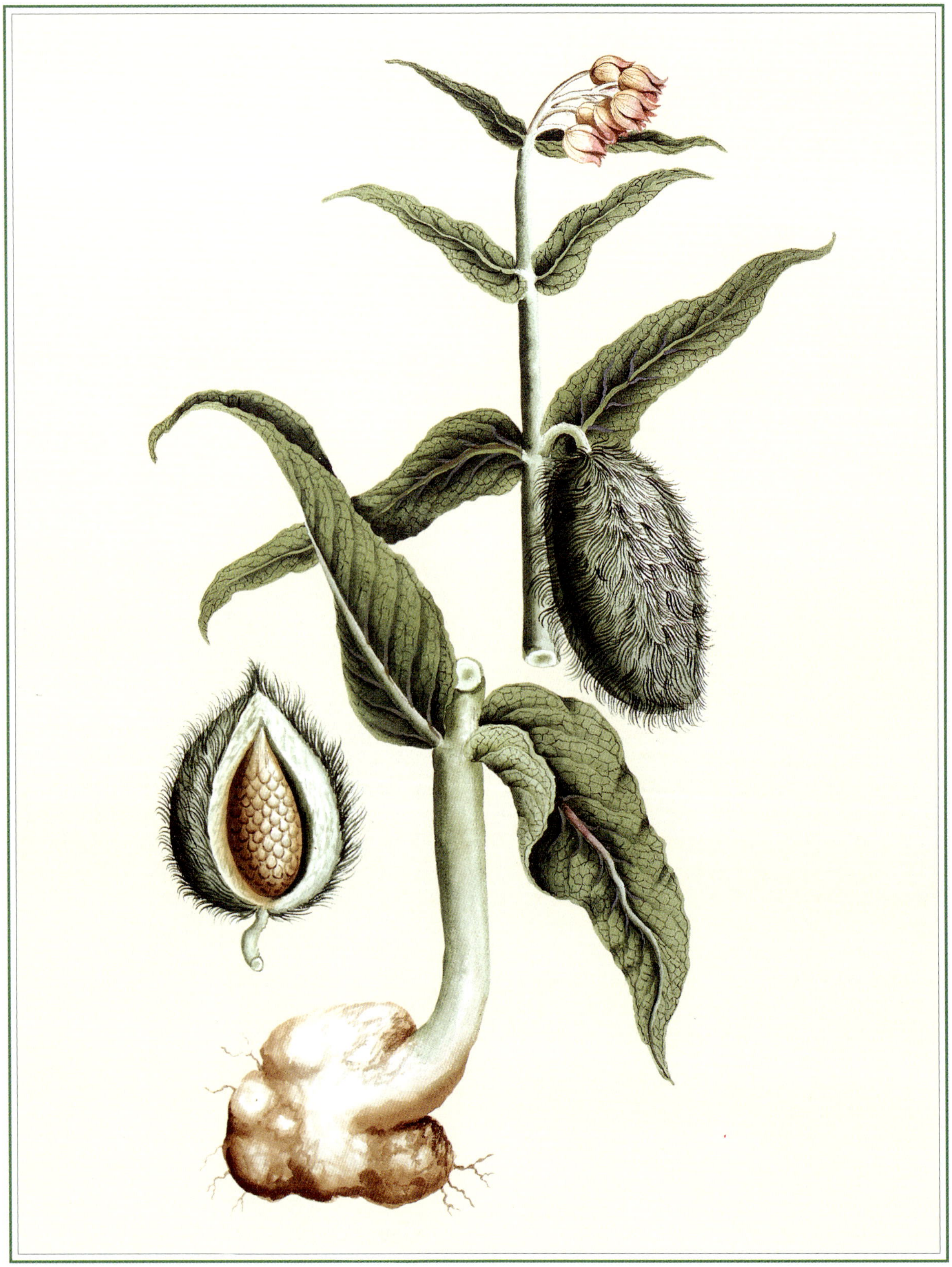

PLATE 10

Asclepias cancellata, watercolour by Hendrik Claudius,
from the manuscript collection of Simon van der Stel's journey to Namaqualand in 1685

BEAUTY IN TRUTH

BOTANICAL ART IN SOUTHERN AFRICA
A BRIEF HISTORICAL PERSPECTIVE

JOHN ROURKE

'Beauty is truth, truth beauty' – that is all Ye know on earth, and all ye need to know
John Keats: *Ode on a Grecian Urn*

In his elegant publication *The Art of the Plant World*, which reviews the world's great published botanical art from 1481 to the 1980s, Martyn Rix made the following significant remark: 'It is perhaps also relevant to mention here that the country which has produced the largest number of well illustrated botanical books in the last two decades is South Africa, which has two great advantages: a uniquely rich native flora and the industries of gold and diamond mining to provide a clientele able to afford the books'.[1]

This was a courageous admission, especially since it was made during a time of intense sanctions against South Africa when the acknowledgement of anything good coming out of the country was almost unthinkable. Yet it highlighted the remarkable fact that by world standards, southern Africa is endowed with an unusually large community of competent publishing botanical artists. How this tradition has slowly developed, matured and finally blossomed, is an interesting story that closely mirrors Western man's relationship with the environment in this part of the African sub-continent over almost four hundred years.

Since the late seventeenth century southern African plants cultivated in Europe have been illustrated by some of the world's greatest botanical artists: George Dionysius Ehret, Pierre Joseph Redouté, Franz Bauer and Nikolaas Jacquin. Greenhouse-grown specimens were the subjects of their virtuoso pieces, which were created using the best available materials, papers and paints, executed in the comfortable atmosphere of sympathetic surroundings, and generally bolstered by generous patronage. Resident artists at the Cape had rather different conditions under which to work. It is not the purpose of this review to discuss the images of southern African flora created by these great European masters, but rather to consider the struggles of resident artists in this country from the earliest times till the post Second World War period, when southern African botanical artists began to make an impact on the Western world.

What then were the stimuli that drove artists to paint flowers in a remote colonial outpost like the Cape, even in the initial years of the Dutch East India Company's rule from the mid 1600s? Certainly there were practical considerations in creating recognizable images of these wondrous forms of indigenous plant life that confronted the first settlers in Table Valley. Some species were edible, others poisonous, while several had useful medicinal properties – all sound reasons for painting them. Scientific curiosity was also a major factor as was that most fickle of human foibles – fashion. In seventeenth-century Europe it was socially desirable to possess collections of natural history specimens or portfolios of paintings depicting plants and animals. While in vogue, there was a strong demand for fine botanical paintings, but this was a passing phase.

Yet there is no doubt in my mind that over and above all these considerations it is the very

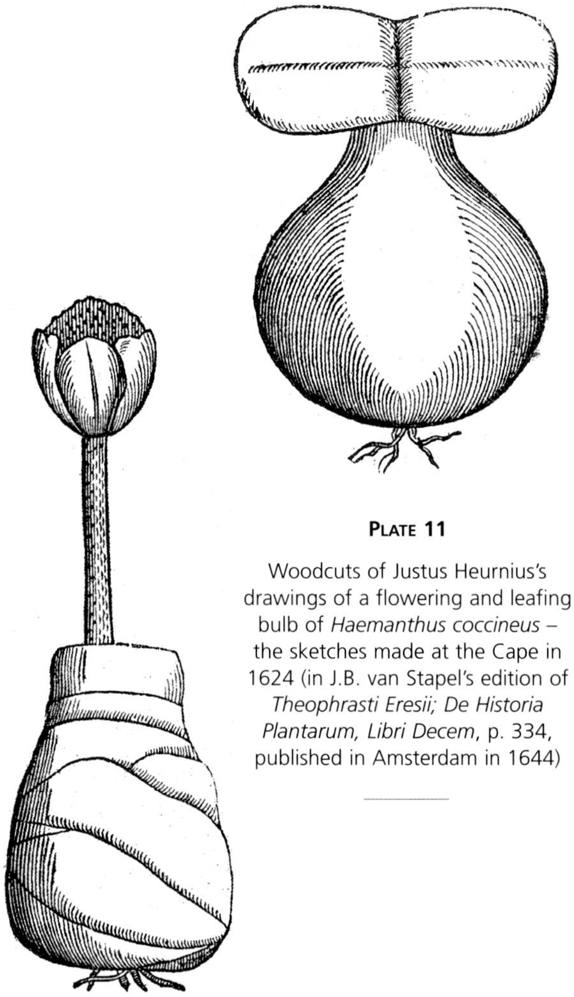

Plate 11

Woodcuts of Justus Heurnius's drawings of a flowering and leafing bulb of *Haemanthus coccineus* – the sketches made at the Cape in 1624 (in J.B. van Stapel's edition of *Theophrasti Eresii; De Historia Plantarum, Libri Decem*, p. 334, published in Amsterdam in 1644)

nature of the plants themselves that has served as the most powerful sustaining stimulus to artists in southern Africa from the early seventeenth century to the present. The challenges of design, form, colour, texture and diversity remain as strong as they ever were and the Cape flora provided all of these stimuli in generous abundance.

Anonymous draughtsmen

Long before western Europeans arrived in 1652 to found a settlement in the shadow of Table Mountain, the plant life of this distant realm was being described and illustrated in learned publications. Botanical curiosities in the form of bulbs, dried specimens or seeds had begun to filter back to Europe from the Cape of Good Hope as early as 1597 when an unknown mariner gathered a dried seedhead of *Protea neriifolia* (Plate 42) which was later illustrated in Carolus Clusius's (1526–1609) great work *Exoticorum Libri Decem.*[2] Predictably, the artist who produced the beautiful wood engraving is not recorded.

However, Justus Heurnius (1587–*c.*1653) is probably the first person who could be described as a resident botanical artist at the Cape, having spent most of April 1624, and almost certainly part of that May, sketching on the shores of Table Bay before proceeding to Batavia. His simple drawings published in 1644 as eleven woodcuts, depict ten native plant species. Interestingly, *Haemanthus coccineus* was featured twice, showing two developmental stages in its growth: one, a leafless flowering bulb which must have been completed within days of arriving in early April, the other a bulb showing well-formed emergent leaves (the post-flowering stage) which must have been drawn at the end of his stay, probably in May 1624, for the leaves to have reached such an advanced stage of growth. Though quaintly naive, Heurnius's drawings are almost all identifiable to species level and have the enormous added interest that they were the first southern African plants drawn under field conditions.

Soon after the establishment of a permanent Dutch settlement at the Cape in 1652, small expeditions were intermittently sent out to investigate the natural resources of the country. But it was not until 1685, under the direction of the second commander at the Cape, Simon van der Stel (1639–1712), that a serious attempt was made to survey and document plant and animal life in the interior. Van der Stel's expedition to search for copper deposits in Namaqualand took place between August 1685 and January 1686 and among the sixty-two accompanying persons was an artist, generally believed to be Hendrik Claudius (*c.*1655–*c.*1697). A collection of seventy-one of what are probably Claudius's coloured drawings depicting landscapes, people, animals and some forty-three different plants, now in the library of Trinity College, Dublin, provides us with vivid insights into the expedition's achievements. Other collections of original paintings relating to this expedition and similarly attributed to Claudius, although different in style and artistic quality, are in the South African Museum, Cape Town, the South African Library, Cape Town, and several other places.[3]

There has been much published debate as to the identity of the artists in question as it is very evident that Claudius permitted his originals to be copied by less-talented hands.[4] Nevertheless, it seems likely that the unsigned Trinity College collection has the strongest claim to be Claudius's own work. What is very clear, however, is that artists of this period were deemed to be so lowly in the bureaucratic hierarchy of the Dutch East India Company that their names were rarely if ever mentioned in official documents – they were in effect reduced to the level of anonymous draughtsmen.

Somewhat later, in the early years of the eighteenth century, another fine collection of paintings of Cape plants was assembled by Nicolaas Witsen (1641–1717), a director of the Dutch East India Company from 1693 and onetime burgomaster of Amsterdam. Witsen's extensive collection, estimated to contain at least fifteen hundred paintings of plants, came to be known as the '*Codex Witsenii*'.[5] These paintings are chiefly of plants found in the immediate environs of Cape Town, the Cape Peninsula and adjacent areas. Regrettably, no information survives as to the identity of the artist or artists who created this extraordinary record. The collection eventually passed through the hands of three eminent Dutch academics, Caspar Commelin (1668–1731), Johannes Burman (1707–1779) – for whom it provided the source material from which most of the engravings of his classic work *Rariorum Africanarum Plantarum* [6] were produced – and eventually to Burman's son, Nicolaas Laurens (1733–1793) from whose estate the collection was ultimately dispersed by auction in 1800 with the component parts apparently being widely scattered.

The South African Museum's volume of this historic material attributed to or copied from Claudius, actually bears Witsen's signature and an inscription indicating it was made at the Cape in 1692 – just before Claudius is believed to have left for Holland.[7] This is clearly part of the *Codex Witsenii*, while another volume *Icones Plantarum et Animalium*, now in the Central Johannesburg Library – though not annotated by Witsen – is also thought to have a good claim to being part of the *Codex Witsenii*. Most of these paintings are especially fine and very sensitively executed. They are the work of several

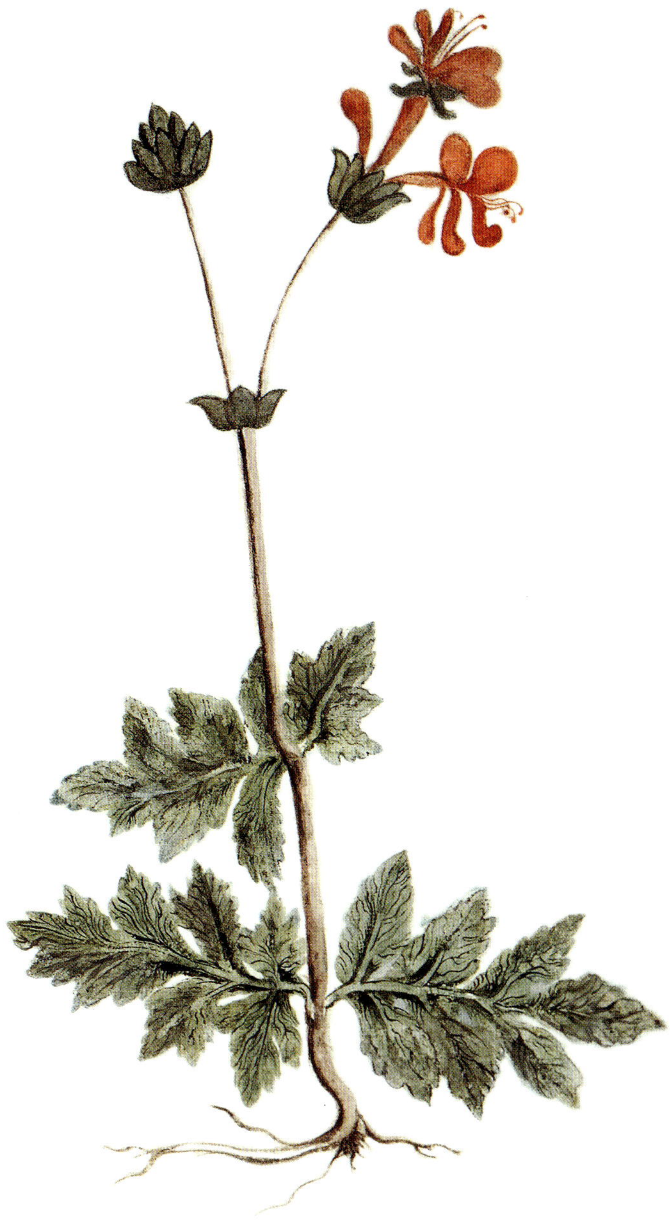

PLATE 12

This depiction of *Pelargonium fulgidum* is from the *Codex Witsenii* in the South African Museum, Cape Town. Although attributed to Hendrik Claudius, it is probably a late seventeenth-century copy by an inferior artist.

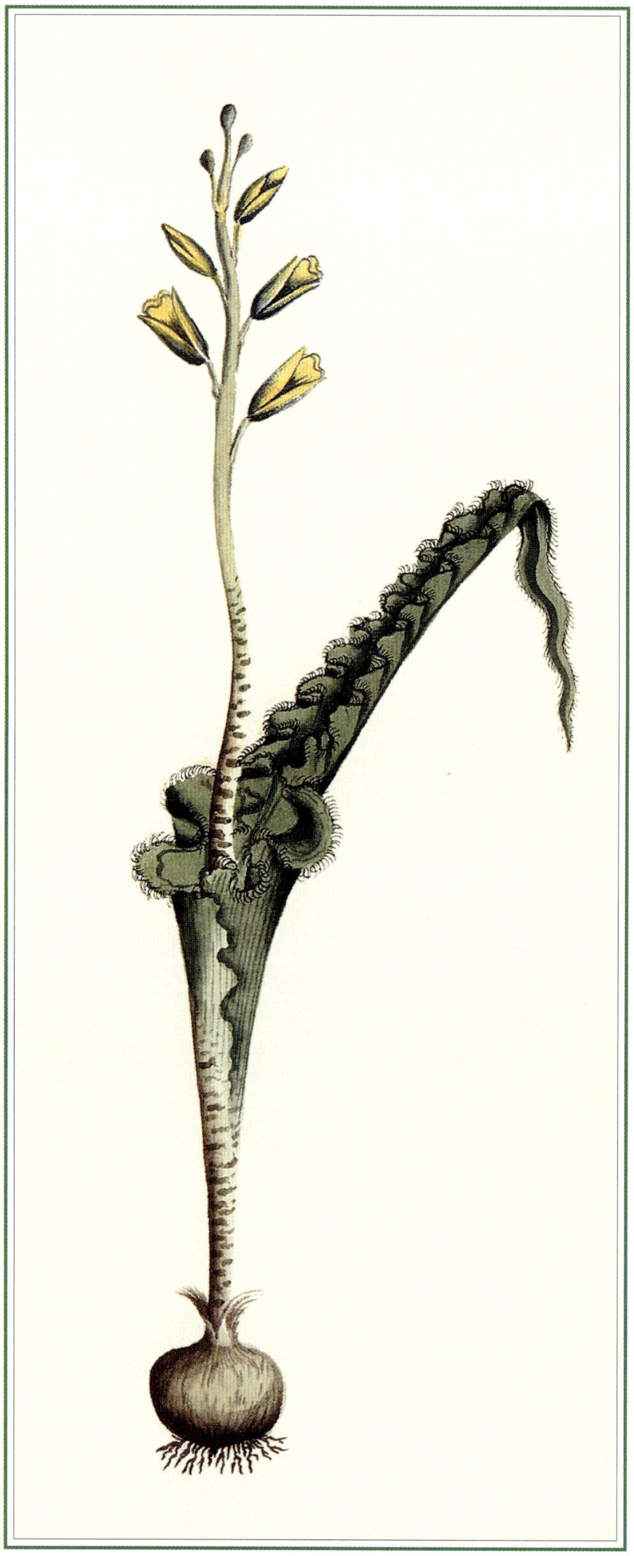

PLATE 13

Lachenalia hirta, late seventeenth century, watercolour, attributed to Hendrik Claudius but probably a copy by an inferior artist

meticulous artists with the majority having a freshness about them as if their subjects had just been picked from the veld. Yet again the identities of the artists remain shrouded in mystery.

Another fine florilegium of this period bearing Johannes Burman's signature of ownership, is in the Mary Gunn Library, National Botanical Institute, Pretoria. Here an unacknowledged, unknown artist responsible for the 148 paintings it contains, has faithfully depicted mainly bulbous species, some showing characteristic leaf damage caused by insects and grazing animals. Such clues point unmistakably to this work having been created at the Cape and not from specimens grown in Europe.[8]

While some of the indigenous plants growing on the shores of Table Bay – like grasses (Poaceae) and daisies (Asteraceae) – could easily be associated with comparable European plants, others like members of the Proteaceae were wholly unfamiliar to Western eyes. They made an immediate impact on all who saw them, so it is not surprising that Jan Hartog (*c.*1663–1722), the Dutch East India Company's master gardener from 1697 to 1715, sent a collection of boldly executed, unsigned watercolours depicting some twenty-three different species to Herman Boerhaave (1668–1739) in Leiden. Boerhaave, considered one of the leading Dutch scholars of the day, had copper engravings prepared from these plates. These he duly published (uncoloured) in 1720. The original watercolours passed through various hands including those of Adriaan van Royen, Boerhaave's successor, and were eventually deposited in the Rijksherbarium at Leiden.[9]

Again we must ask the question: who was the mysterious artist whose work depicts Cape proteas with such strength and vigour? It was certainly not Hartog, who could not draw. Hendrik Bernard Oldenland (1663–1697), Hartog's predecessor as master gardener is a possibility as he is known to have had some artistic abilities, but that is pure speculation. The fact remains that yet again the artist, whoever he was, was not deemed important enough even to receive mention.

This strange pattern of failing to acknowledge the artists who produced these early

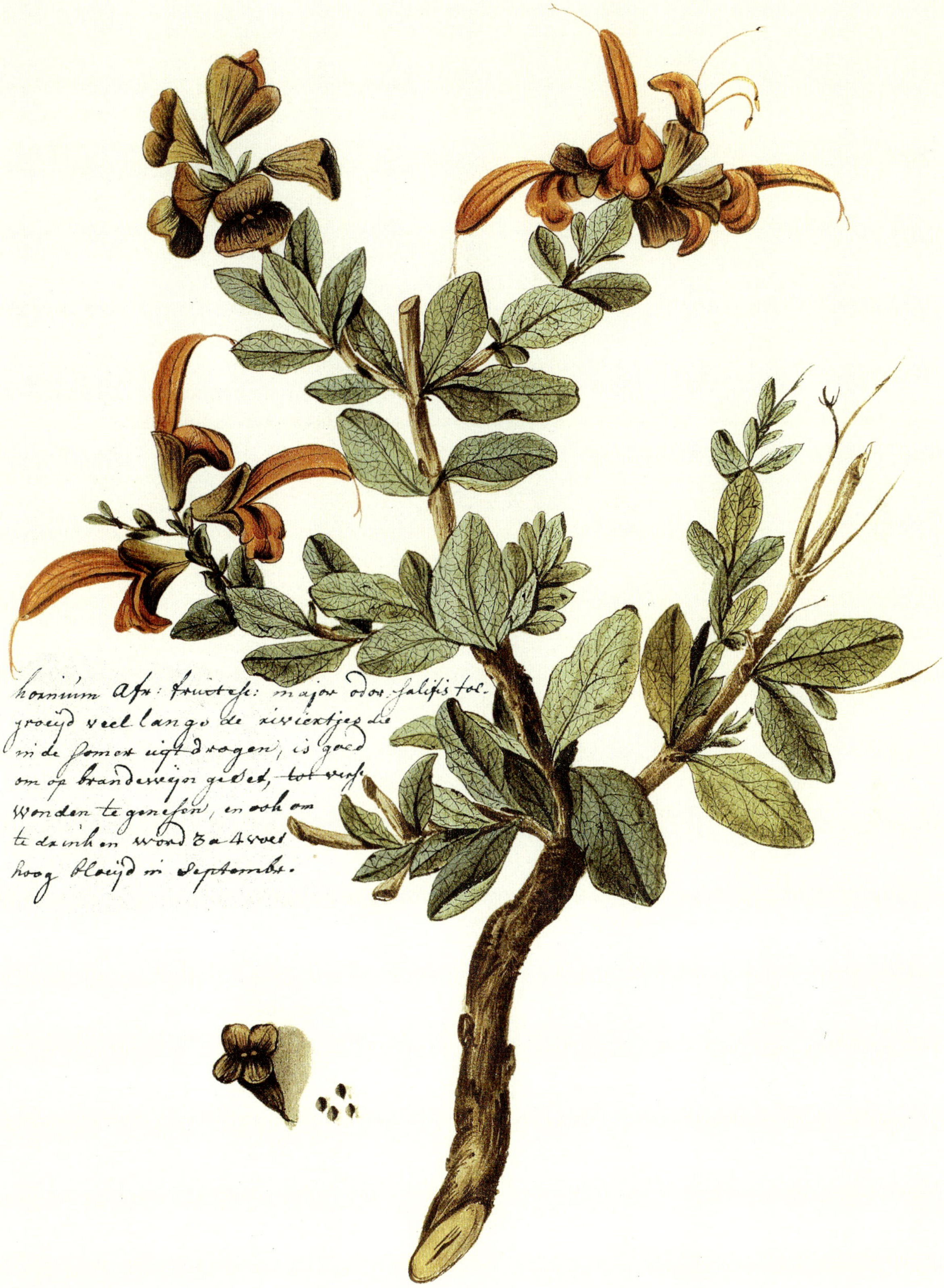

PLATE 14

Salvia aurea, c. 1690, body colour, artist unknown, in *Icones Plantarum et Animalium*, Central Johannesburg Library

images of southern African flora persisted throughout the eighteenth century with one notable exception – that of Robert Jacob Gordon (1743–1795). Colonel Gordon, commander of the Dutch garrison at the Cape from 1780 to 1795, was a man of exceptional energy and intellect as well as an accomplished artist. Gordon's famous atlas of landscapes and natural history paintings is one of the most important records of late eighteenth-century exploration in southern Africa both in content and artistic quality. Significantly, however, there seems little doubt that it was Gordon's high social standing which ensured that the creative work of his own hand was correctly acknowledged.

Although much of the material in the Gordon collection is undoubtedly Gordon's, he did employ an assistant, soldier-artist Johannes Schumacher (active 1770–1789), as well as another unknown artist whose task it was to undertake mainly botanical paintings. Gordon was accompanied on his epic 1779 journey through Namaqualand to the Orange River by William Paterson (1755–1810), plant collector to the Countess of Strathmore. It was entirely due to Gordon's generosity that Paterson also acquired a fine collection of plant portraits, completed under trying field conditions. Indeed, William Paterson's collection of often equivalent botanical paintings was almost certainly the work of a 'draughtsman of Capt. Gordon' as Paterson described this shadowy figure.[10]

Other more or less contemporaneous travellers like the French zoologist François le Vaillant (1753–1824) and Sir Joseph Banks's (1743–1820) botanical emissary Francis Masson (1741–1805) also possessed equivalent copies of paintings in the Gordon collection. They too were executed no doubt by some lowly anonymous painter. Social status was clearly a prerequisite for an artist's recognition in the strict hierarchical order that prevailed throughout this period of Dutch administration. Herein, perhaps, lies the answer to this curious riddle of the unnamed artists – this was an age in which social rank still outweighed creative talent.

Looking back at the sheer volume of botanical paintings produced at the Cape in the 143 years of Dutch East India Company rule, one is astonished at the diligence and effort that went into creating these works. The majority are now in public collections in Europe and southern Africa. A variety of styles can be detected, some naive, some almost grotesque, while others are carefully and sensitively delineated. Yet all of these images seem to possess a coarse, primitive robustness, distinctive of their period, that conveys a powerful impression of the rugged frontier conditions under which they were painted.

Much ink has been spilled in learned publications in attempts to unearth the identity of these artists, to establish which of their works are originals and which are copies – for copying did not seem to have the same connotation of plagiarism in the 1700s as it does today. But despite these profound enquiries it is a sad fact that we will probably never know very much about the men who produced these delightful paintings – humble, largely anonymous servants of the mighty Dutch East India Company.

Birds of passage

As the eighteenth century drew to a close, the Cape experienced political, social and economic changes in the wake of the Dutch East India Company's declining power and Britain's assumption of control of the southern tip of Africa in 1795. Apart from a brief return to Dutch rule between 1803 and 1806, the 1800s ushered in new styles of leadership, especially politically and culturally. For botanical art in southern Africa the end of the eighteenth century was also a watershed of pivotal importance. A vital, powerful and progressive new force manifested itself – the emergence of women as artists. The group was to grow in competence, stature and influence. Indeed, women were to dominate the entire field of southern African botanical art for the next two centuries. Lady Barrow, wife of Sir John Barrow, private secretary to Lord Macartney, Britain's first governor of the Cape, was to lead the way.

Anna Maria Truter (1777–1857) was born at the Cape and married British civil servant John Barrow (later Sir John) in 1799. Little is known about her schooling or art training in eighteenth-century Cape Town, but by the time she left South Africa in 1803 to settle in England, Anna

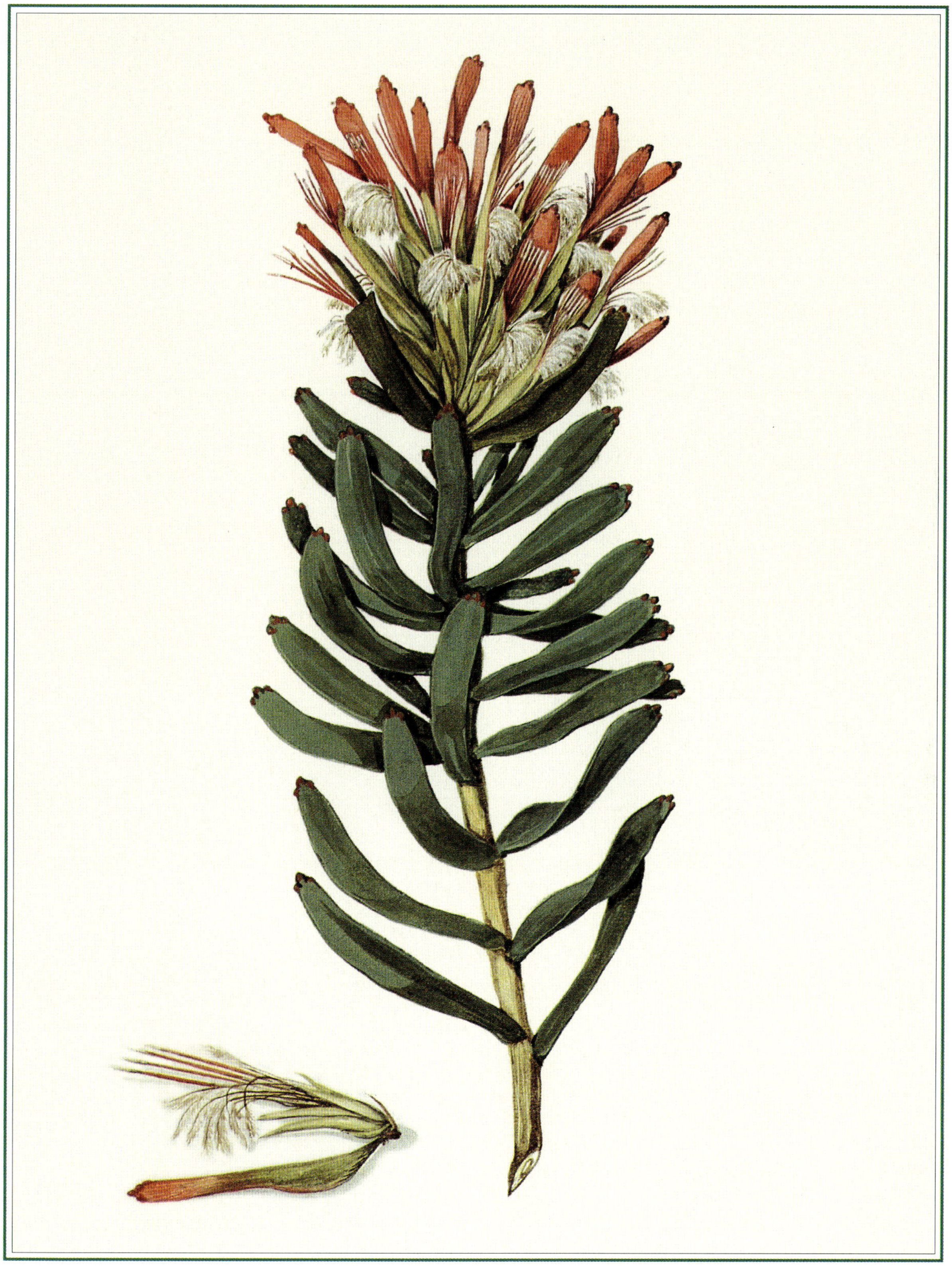

Plate 15

Mimetes cucullatus, c. 1700, watercolour, by artist unknown though possibly H.B. Oldenland

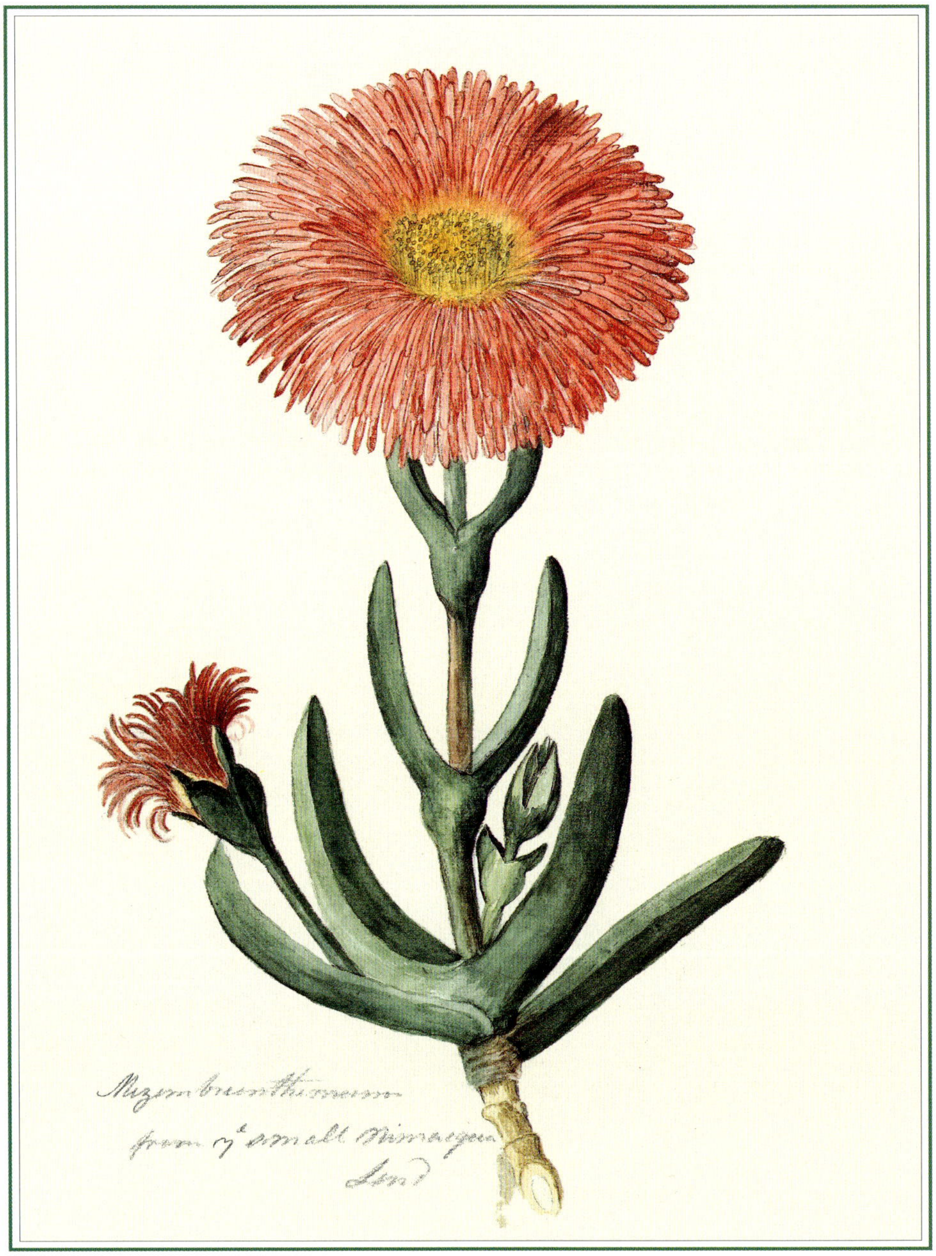

PLATE 16

Carpobrotus sauerae, c. 1779, watercolour, by either Robert Jacob Gordon or copied by one of his unnamed artist assistants

Barrow had completed an album of Cape flower studies as well as an album of Cape scenes. As the first South African-born woman to assemble a portfolio of flower paintings, Lady Barrow occupies a unique position in the history of botanical art in southern Africa.

Those who followed during the first fifty years of British rule were almost invariably birds of passage – sojourners who spent a few months or even a few years in southern Africa before moving on. Often they were the wives or daughters of colonial government officials, military officers, scientists, or 'Indians' – British civil servants stationed in India but temporarily on furlough or invalided to the Cape to escape the torrid Indian climate. Studying painting, music and dancing

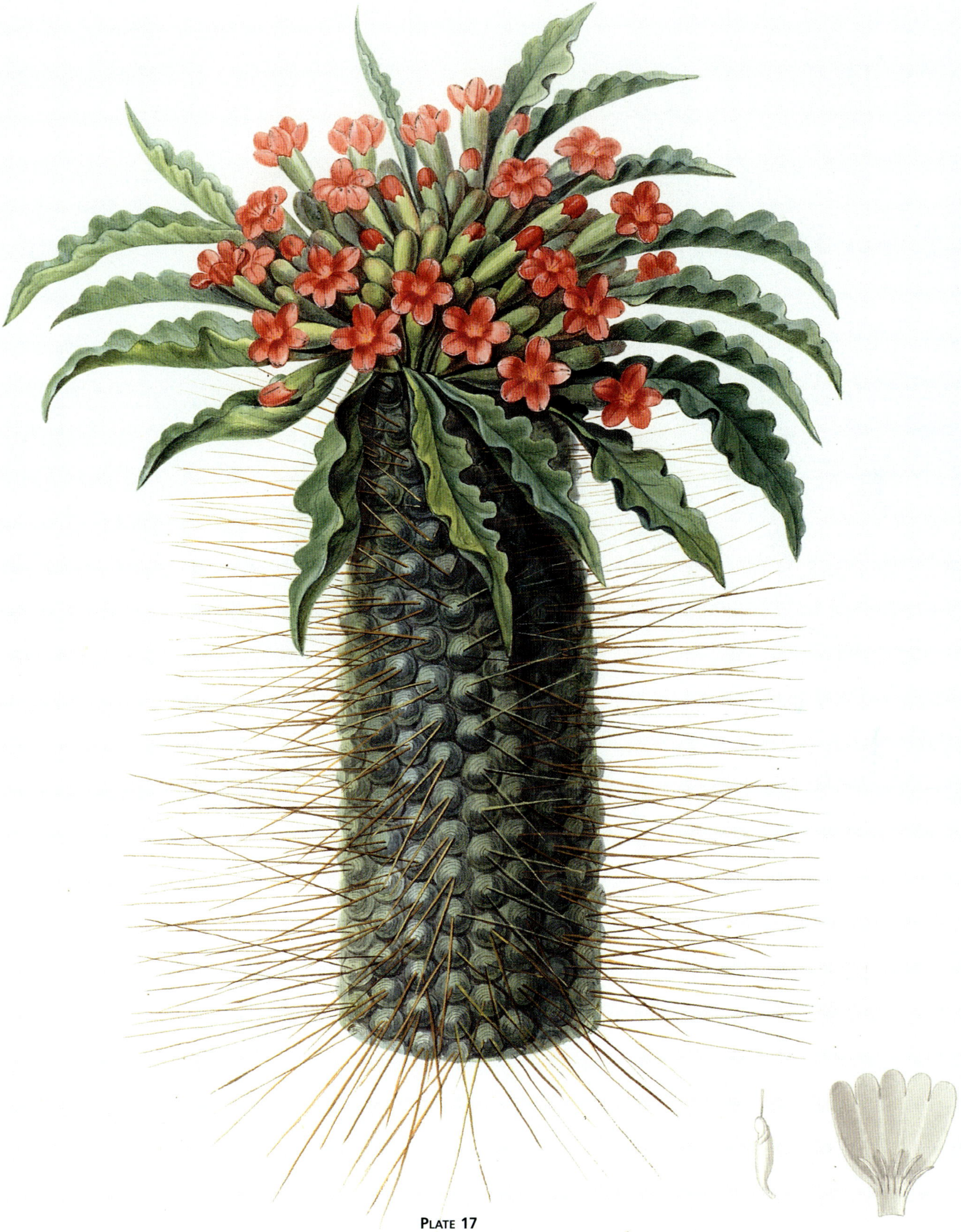

PLATE 17

Pachypodium namaquanum, *c.* 1779, watercolour, by either Robert Jacob Gordon or copied by one of his unnamed artist assistants

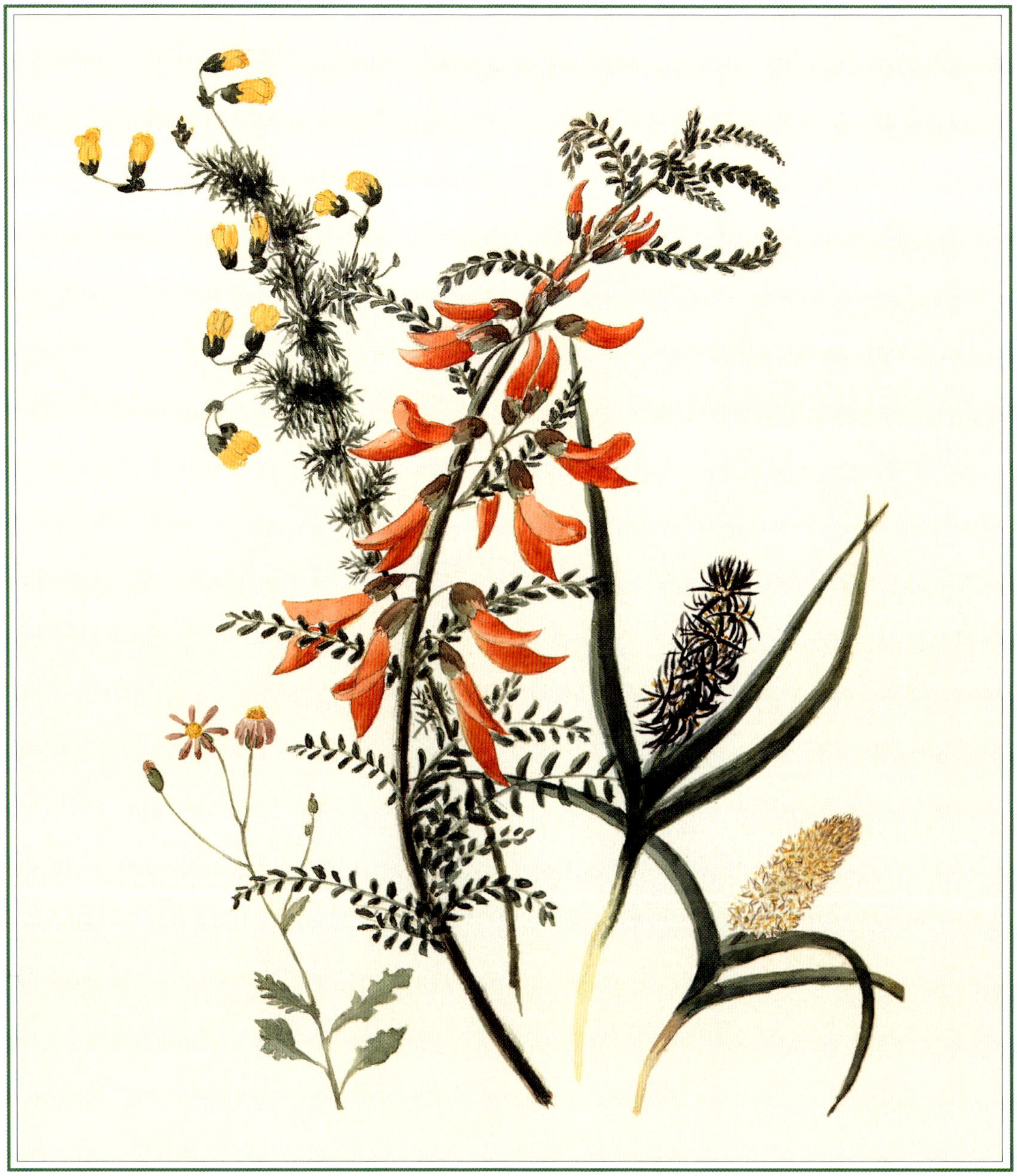

PLATE 13

A selection of Cape flora by Anna Maria Truter (Lady Barrow): *Sutherlandia frutescens*, two species of *Wurmbea*, *Hermannia pinnata* and a small *Senecio*, c. 1800, pencil and watercolour

were among the essential accomplishments expected of every well-educated lady at that time, while the more daring might even venture to study elementary botany. Their portfolios were treasured memories of imperial service or travels in distant lands. Even today, collections of early nineteenth-century botanical paintings appear on the market as family lines die out and the accumulations of generations come under the auctioneer's hammer. So numerous are these collections that it is not possible to comment on more than a few.

Typical of this period were Mrs Margaret Petrie (active 1820–1830) and her daughter (also Margaret), wife and daughter respectively of William Petrie, deputy commissary general.

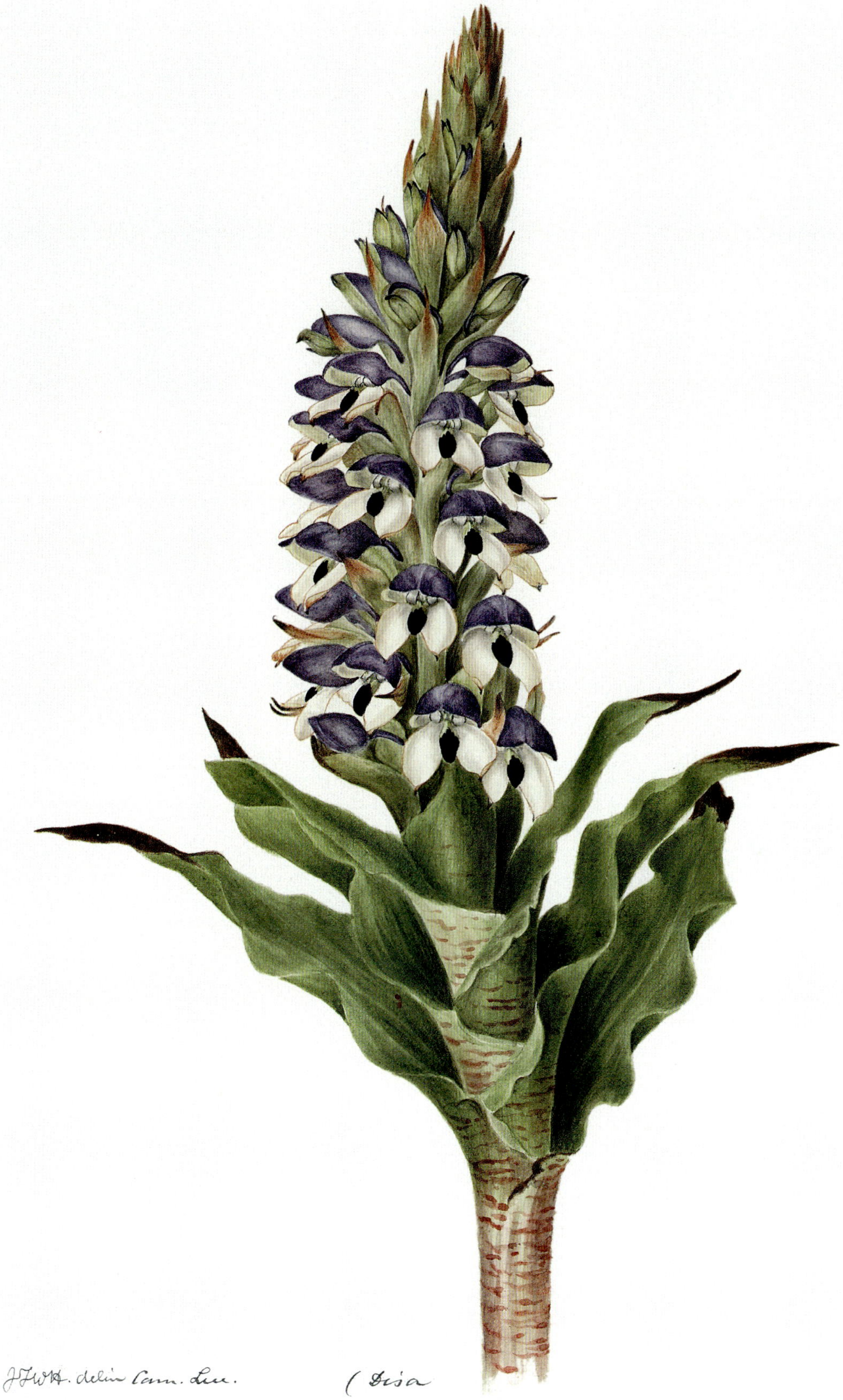

Plate 19

Disa cornuta, November 1835, pencil and watercolour, outlined by John Herschel and painted by Margaret Herschel

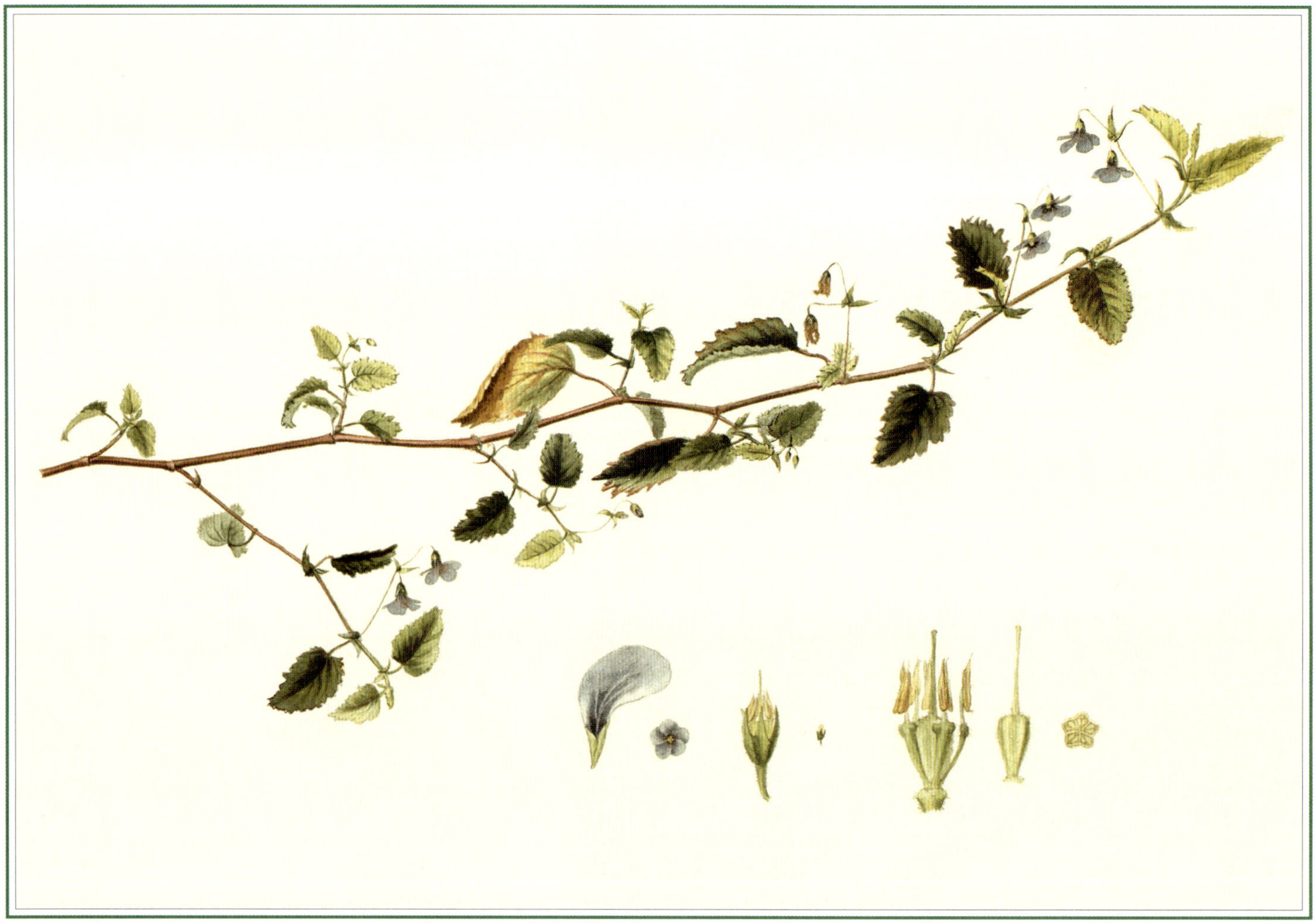

PLATE 20

Mahernia violacea, 19 May 1813, pencil and watercolour, by William John Burchell

There was also Eliza Murray (active 1830s), wife of Dr John Murray, physician to the governor's household, who formed a collection of over a hundred paintings of various Cape heaths. They came, collected, painted and departed.

An unusual combination of talents was the husband-and-wife team of John (1792–1871) and Margaret Herschel (1810–1884) whose four years at the Cape between 1834 and 1838 yielded a portfolio of 131 mainly coloured botanical illustrations. Sir John Herschel, one of Britain's most eminent astronomers selected 'Feldhausen', a fine old estate on the south-east side of Table Mountain as a base where he set up his twenty-foot reflector to survey the southern skies. The couple were captivated by the native plant life around them, to the extent that Margaret occupied most of her limited leisure time in painting it. In these plates the element of artistic interpretation was somewhat downplayed as Sir John was highly skilled in the use of the camera lucida, an optical drawing device that enabled him to produce perfectly accurate outlines of each subject. Margaret then painted in the details. Thus Margaret was responsible for the greater part of these combined efforts. Sir John acknowledged: 'Maggie teaches the children, and draws the pretty flowers in a very good and (now indeed) masterly style – She has improved herself greatly in this pleasing art, and surely if ever there was an inducement for it it is here'.[11]

Although their portfolio of exquisite paintings was originally intended as a personal family record, it achieved far more than other comparable contemporary collections of botanical mementoes. For the most part these paintings lack floral dissections, yet Sir John's strong scientific influence nevertheless ensured a degree of accuracy in the Herschel collection never before achieved by any other resident botanical artist. This exceptional record remained in the possession of the Herschels' descendants in England until 1996 when The Brenthurst Library in Johannesburg acquired and subsequently published the collection.[12]

One of the few male botanical artists active at this time was William John Burchell (1781–1863) who is also generally regarded as the most important early nineteenth-century explorer of southern Africa.[13] During four years of travel into the interior as far as Kuruman, he completed nearly five hundred sketches, many of them carefully portraying plants observed en route. Burchell's botanical drawings are not well known because few were ever completed and even fewer published. Moreover, his work also tended to have a stronger scientific content rather than an artistic one. Embellished with copious dissections and annotations, Burchell's sketches were made to record the morphological characteristics of the species he studied, thus adding extra data to select collections of the huge herbarium of over 8 700 gatherings he amassed while in southern Africa before returning to Britain to publish his discoveries.

Arabella Roupell (1817–1914) must rank among the bolder lady artists of the early Victorian period in southern Africa. She took the almost unprecedented step of having some of her work published – albeit anonymously. The rather coy title of her work, *'Specimens of the Flora of South Africa by a Lady'*,[14] indicates only too clearly that this was still an age when it was deemed somewhat improper for ladies to draw attention to their achievements. Married to Thomas Boone Roupell, an East India Company official and later judge in Madras, the couple spent a period of leave at the Cape in 1843 and 1844 when Arabella was about twenty-seven. No serious scientific motive underpinned the creation of her paintings, and an eminent botanist, William Henry Harvey (1811–1866), in the preface to her book, quoted the artist herself: 'They were made solely for the amusement of leisure hours.'[15] There is something especially fascinating about the Roupell paintings. Delicate, with an almost feminine fragility and scarcely a hint of any brushstrokes, they so perfectly mirror the leisured, genteel times of Arabella's young womanhood.

Still later, between 1882 and 1883, Marianne North (1830–1890) an intrepid Victorian lady of independent means arrived in southern Africa. Her pilgrimage through the western Cape, eastern Cape, Transkei and Natal over a period of some ten months added substantially to the astonishing collection of 848 vibrantly coloured oils – depicting some of the world's most spectacular plant species – which she assembled over a lifetime of travel[16] (see below, Plate 106).

Although Miss North was probably one of the last of the migrating Regency and Victorian botanical artists to pass through southern Africa, she was perhaps the most original. Marianne North will ever be remembered for a wholly new dimension she brought to botanical painting. Not for her the disembodied images and dissected parts detached from their environment and set against a stark white background. Instead, the majority of her plant portraits were framed in natural landscapes with associated vegetation. In short, Marianne North showed plants in their environments. Pollinating birds, insects, butterflies and other creatures important to the ecology of the species were often also included. Here was a departure from the conventional. But it was not entirely new. Robert Thornton had used natural backgrounds in his *Temple of Flora*,[17] a rare work then known only to a select few. Miss North's huge collection, depicting plants in their habitats, since 1882 on permanent public display in the Royal Botanic Gardens at Kew, England, made a powerful statement to a wide audience. An ecological approach had arrived on the scene. It would appear again in the twentieth century but only in the hands of the most confident practitioners.

Visitors to nineteenth-century Cape Town,

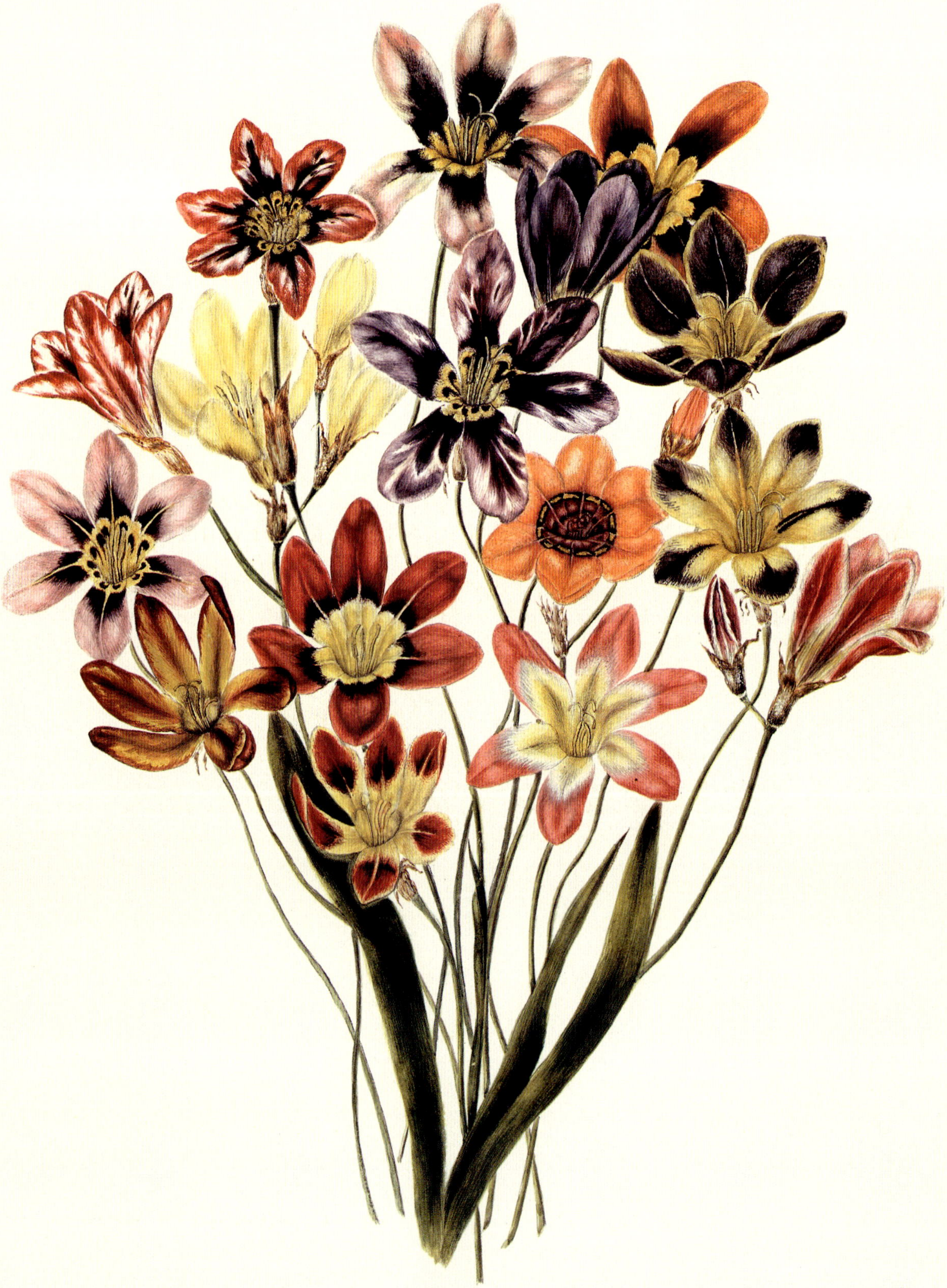

Plate 21

A collection of *Sparaxis* species and hybrids, c. 1843, watercolour, by Arabella Roupell

unable to draw or paint themselves, were able to purchase flower paintings as mementoes, from Jean Villet (1817–1877), a Dealer in Natural Curiosities, at 42 Long Street. Villet (who was apparently self taught) in 1836 joined his father's business which offered 'Drawings of Cape Flowers' among its stock items of saleable merchandise. For those requiring specific species to be illustrated, Jean Villet could also happily oblige, announcing in his published advertisements that 'Plants and Flowers are drawn from nature at a moderate charge'.[18] The business evidently thrived from the early 1830s until the late 1850s, suggesting that numerous customers must have availed themselves of this unusual service. Indeed, Sir John Herschel, noting the visit of a friend in his diary on 15 December 1835, recorded: 'Mrs Hawkins spent the day here. Brought a lot of flower-paintings of Villet.'[19]

None of these paintings, sold through the firm of C.M. Villet and Son, were signed so it is now almost impossible to identify or locate the majority of Jean Villet's works with any certainty. While it is easy to dismiss them as having much the same status as picture postcards, several were commissioned with a scientific purpose in mind. One of these exceptions is a plate of the then newly discovered *Orothamnus zeyheri*, a rare and highly ornamental member of the Proteaceae. In June 1845 a Cape Town medical practitioner, Dr C.W.L. Pappe (1803–1862), sent specimens and drawings of several Cape plants to Sir William Hooker (1785–1865), director of the Royal Botanic Gardens at Kew. Among them was Jean Villet's painting of *O. zeyheri* which was presumably completed the previous year or earlier as it is dated 'July'. Walter H. Fitch (1817–1892), artist and lithographer to *Curtis's Botanical Magazine* adapted Villet's painting to fit the journal's format. The published plate was of course laterally inverted due to the printing process when it was published in 1848.

Unfair as it is to judge Villet's artistic abilities from a single example, it is clear that his original figure lacks the perspective so essential to a meaningful likeness: Fitch's published reworking is a significant improvement. Thus while the Cape Town-born Villet undoubtedly possessed natural artistic talent, the society in which he lived was still so small that it lacked the academic and technical skills to further train, develop or refine his talents. Moreover, its horizons were as yet so truncated that critical comment on artistic matters was virtually non-existent. Nevertheless, Villet was the first of a new breed of resident botanical illustrators – locally born artists or settlers who were firmly rooted in southern Africa. Unlike transitory visitors who carried off both specimens and paintings, these artists were here to stay.

Victorian confidence

Villet was in many ways typical of a new society that was rapidly growing in Regency and early Victorian Cape Town – a society of educated settlers who were putting their roots down and making southern Africa their permanent home. Scientific and educational institutions were being founded: the Royal Observatory (1820), the South African Museum (1825), the South African College (1829), as well as associations of intelligentsia like the South African Institution with its publication *The South African Quarterly Journal*. Later, in 1877, the South African Philosophical Society (forerunner of the Royal Society of South Africa) came into being, while the South African Fine Arts Association (1850) became a congenial home to those with artistic skills. Attention was also focused on botany with the appointment of Dr C.W.L. Pappe as first Cape colonial botanist in 1856.

The superabundance of spectacular plant life on the very outskirts of Cape Town proved a powerful stimulus to many citizens from flower painters to serious scientists intent upon recording, observing, dissecting, measuring and drawing, in their efforts to document the astonishing diversity presented by the Cape flora. With the passage of time a few art teachers began to appear on the scene, like Thomas Thwaits (active 1820–1840), George Henry Ford (active 1820–1840), and Thomas Bowler (1812–1869), and they offered their services to members of the local populace desirous of improving their skills.

Emily Thwaits (*c.* 1860–1906), whose magnificent collection of watercolours is now in the

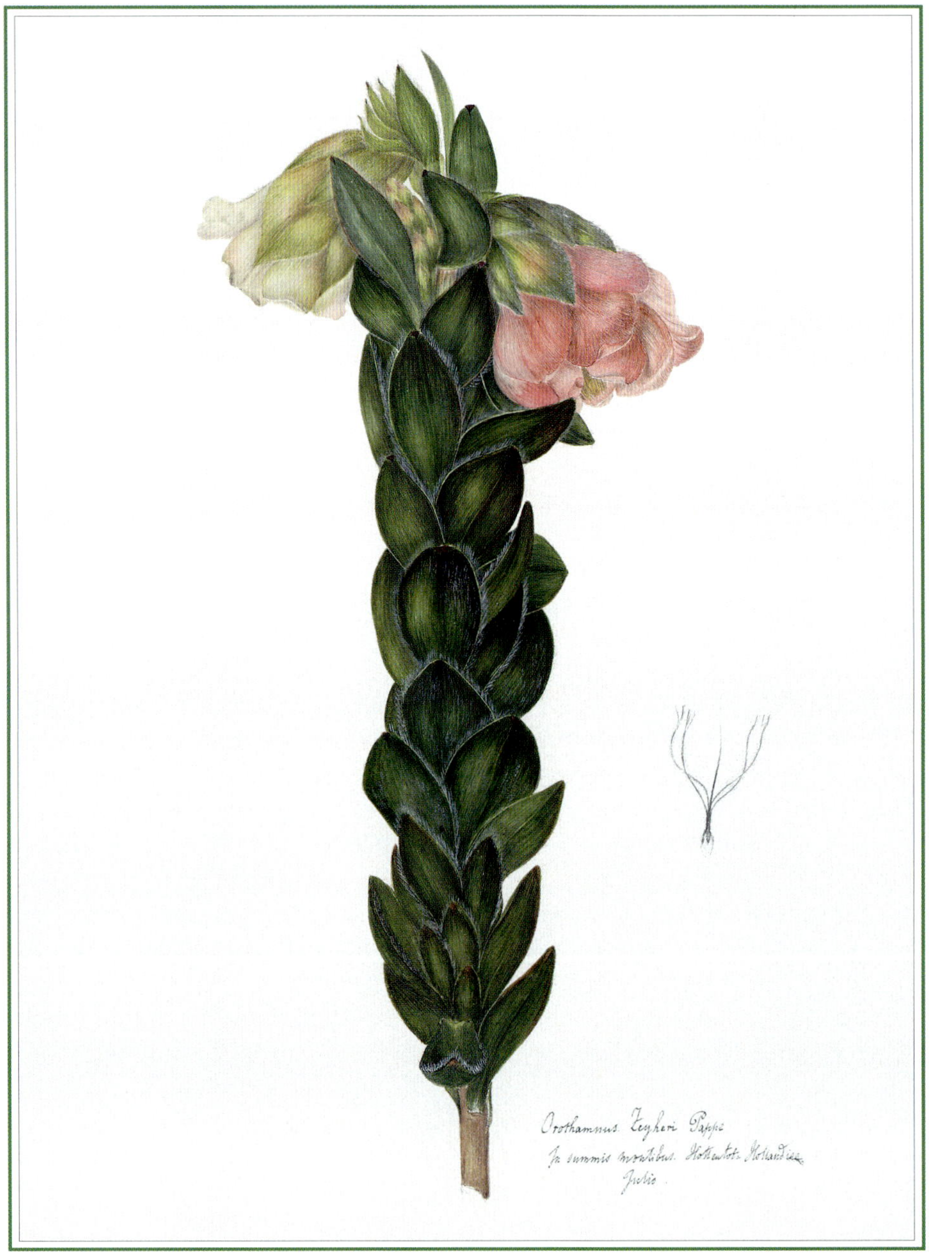

PLATE 22

Orothamnus zeyheri, c. 1844, watercolour, by Jean Villet

Compton Herbarium at Kirstenbosch in Cape Town, was among the most successful of a circle of lady flower-painters who flourished in Cape Town during the high Victorian age. She seems to have inherited her artistic talents from her grandfather, Thomas Thwaits, who had arrived in Cape Town from England to be art master at the Reverend James Beck's School in Roeland Street. Indeed, Emily's botanical paintings were apparently greatly appreciated by her contemporaries, to the extent that in 1880 she was awarded a medal by the South African Fine Arts Association for the best watercolour paintings shown that year. Powerfully concentrated, richly coloured, yet executed with incredibly fine brushstrokes, her paintings are all enhanced with background shadows to produce a visual illusion of depth, thus creating an overall impression of

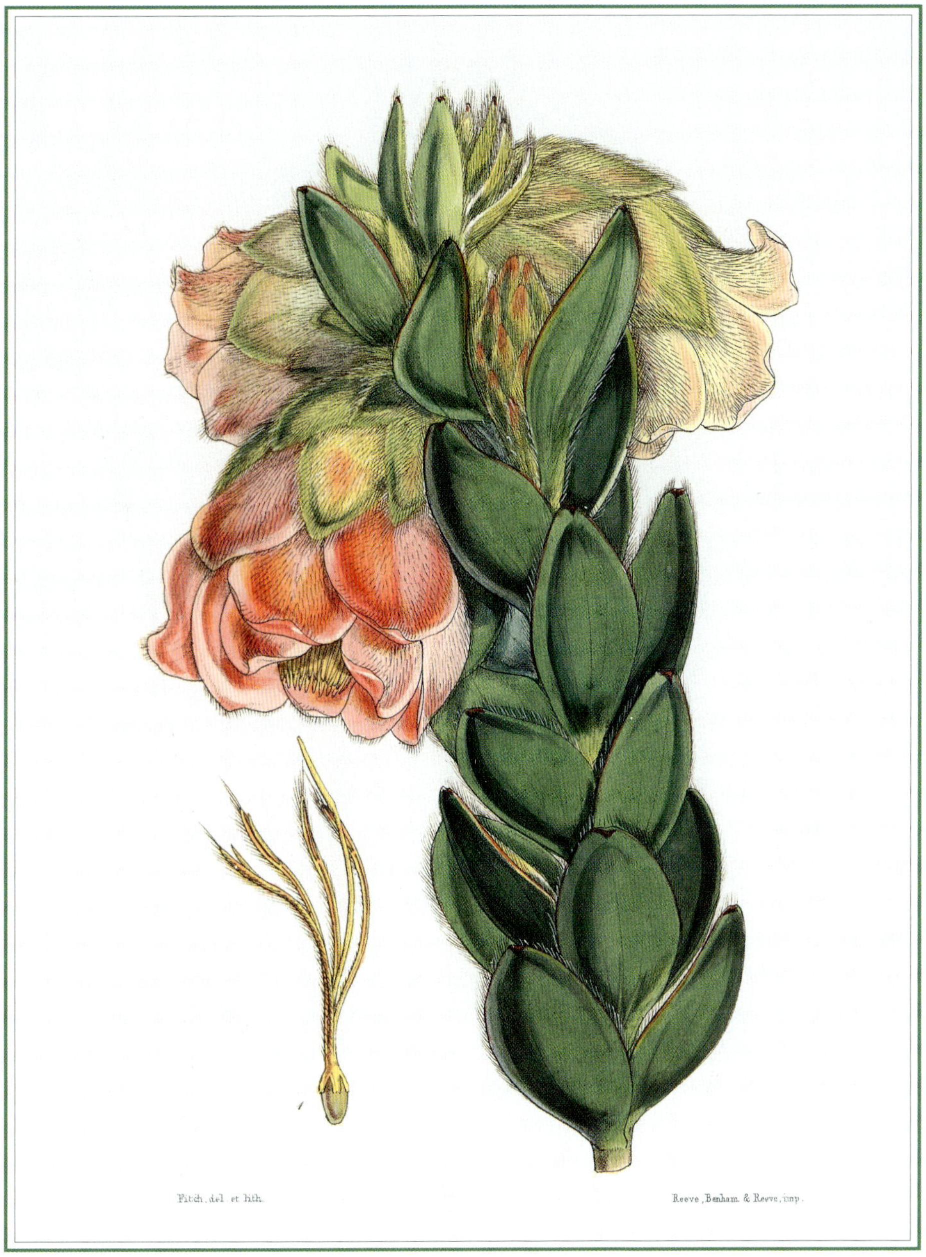

PLATE 23

Orothamnus zeyheri, coloured engraving: Walter H. Fitch's reworking of Villet's watercolour of the plant, as published in *Curtis's Botanical Magazine*, 1848. The image is inverted, the stem shortened and an extra flowerhead has been added.

solid stem, leaf and petal tissue, so unlike the ethereal transparent watercolour washes of other artists. Emily Thwaits was not strictly a botanical artist; she painted for the pure joy of painting, yet with a little botanical guidance her work could have enhanced any taxonomic monograph.

Harry Bolus (1834–1911), stockbroker, botanist and philanthropist who endowed a chair of botany at the South African College, was among the many botanists who must have felt frustrated by his own lack of artistic skills and training. Unable to find an artist to illustrate his classic tomes, *The Orchids of the Cape Peninsula*[20]

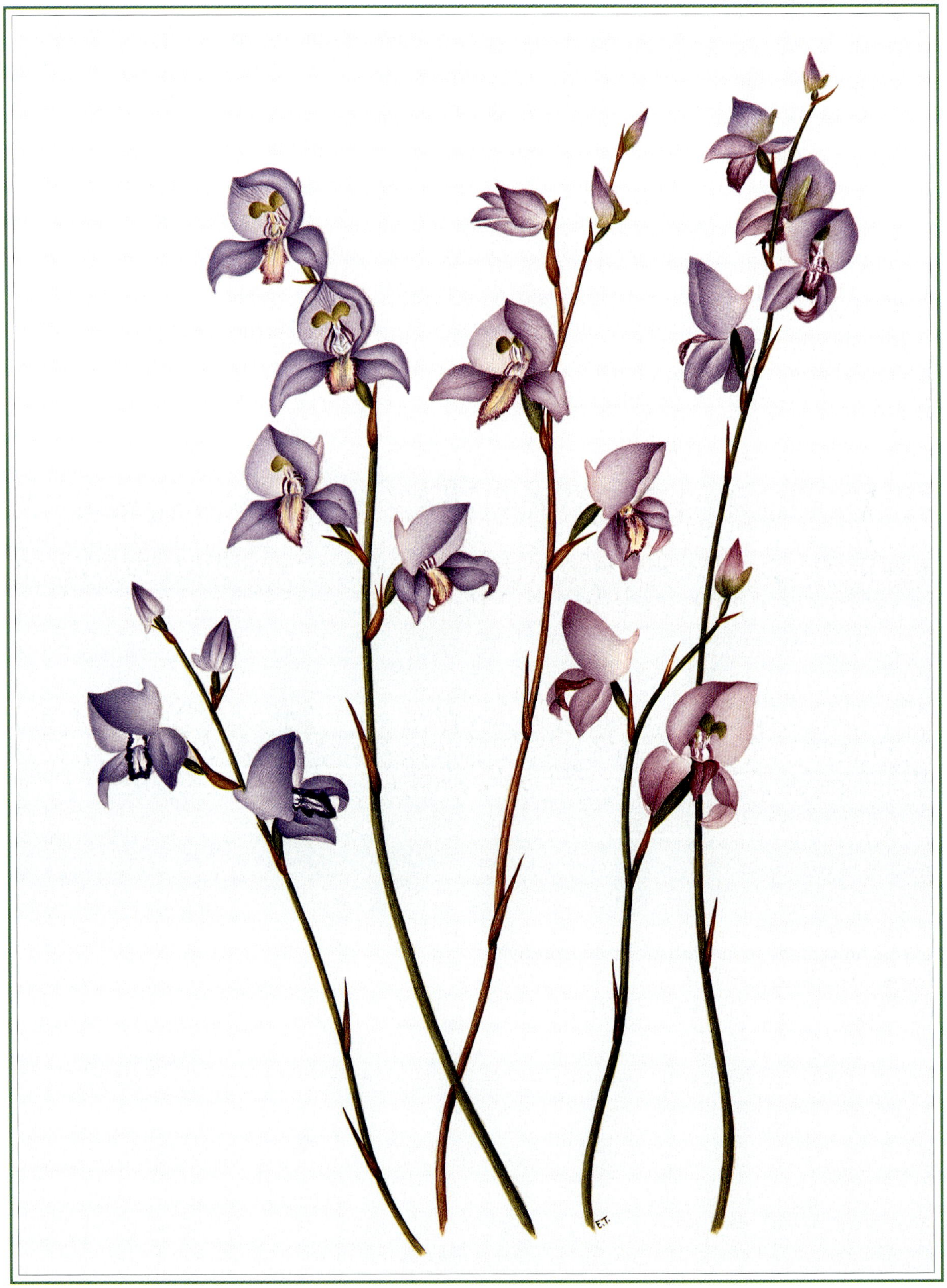

PLATE 24

Disa (Herschelianthe) purpurascens, c. 1880, watercolour, by Emily Thwaits

as well as the three-volume *Icones Ochidearum. Austro-Africanarum Extra-Tropicarum*,[21] Bolus set out to prepare the illustrations himself. His drawings are remarkably accurate in terms of dimensions, morphological details and proportions. However, he struggled with those elusive qualities like depth and perspective, but especially with colour.

Live specimens were used for the great majority of Bolus's plates. It was relatively easy to obtain fresh material of the western Cape species as Bolus lived in Kenilworth, near Cape Town, but obtaining live subjects from the eastern Cape and Natal was another matter in the 1880s. To overcome this problem he used a network of correspondents in remote parts of the country who preserved the freshly collected flowers in bottles of glycerine and then posted them to the great orchidologist. As the glycerine did not leach pigments from the leaves or petals, the specimens arrived in reasonable condition, still showing a fair degree of colour. After washing off the glycerine in tap water – a tedious procedure – Bolus sat down to prepare his portrait.[22] The example shown here (Plate 26) is *Eulophia streptopetala* collected for Harry Bolus by Alice Pegler at Kentani in the Transkei; it is more fully coloured than most of Bolus's work.

Paintings by Bolus represent the most extreme examples of scientific drawings. They are botanically accurate yet flat, lifeless, lacking any depth or perspective and usually with merely the smallest hint of colour – but they are quite sufficient for taxonomic purposes. Absolutely no concession is made to any artistic or interpretive element, yet these compact coloured lithographs fulfil precisely and admirably the purpose for which they were intended.

Alfred Bodkin (1847–1930), mathematics master at the Diocesan College, Rondebosch, Cape Town, was among Harry Bolus's intellectual and climbing companions. Together they delighted in exploring the Cape mountains for new, undescribed plant species. Ada (active 1880s–1890s), Bodkin's wife, was an amateur watercolourist who was soon caught up in the excitement of these botanical forays, painting the rare and curious together with the commonplace species that were brought home. She

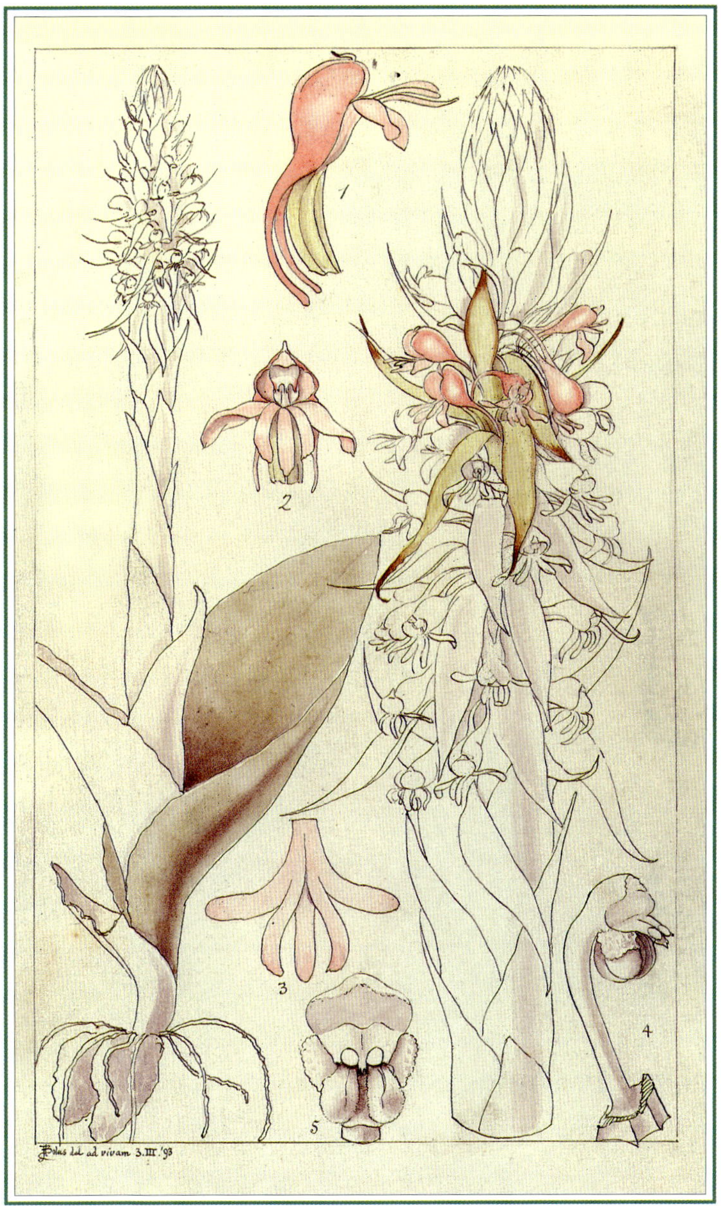

PLATE 25

Satyrium macrophyllum, 3 March 1893, pencil and watercolour, by Harry Bolus

assembled two large portfolios of some four hundred paintings, now housed in the Compton Herbarium at Kirstenbosch. Dated, localized and correctly identified, Ada Bodkin's methodically documented paintings reveal the strong botanical influence of Bolus. Her rendering of a spectacular long-tubed heath, *Erica junonia*, discovered by Bolus and Bodkin in the Koue Bokkeveld during December 1891, reveals the

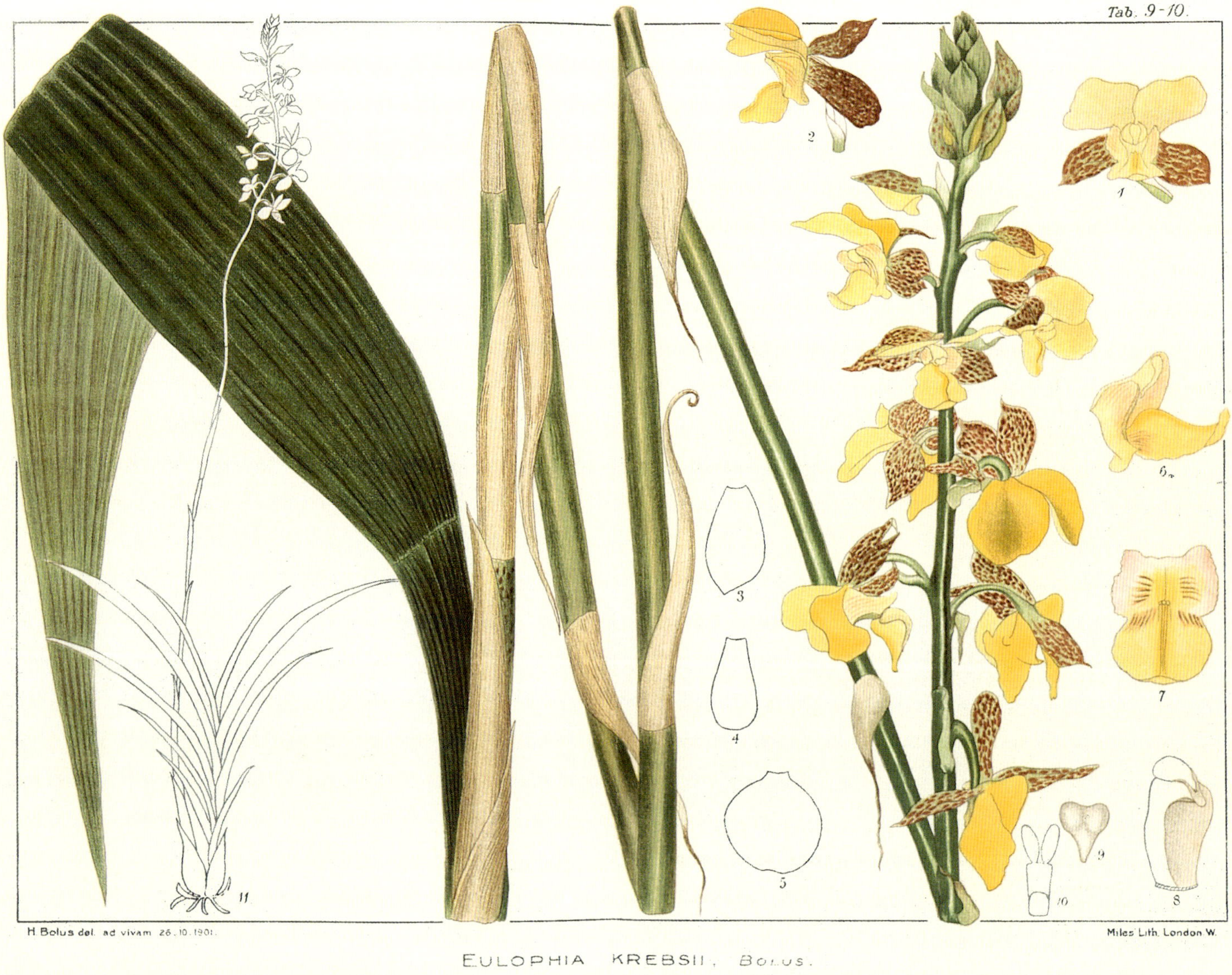

Plate 26

Eulophia streptopetala (= *E. krebsii*), 23 October 1901,
coloured lithograph by Harry Bolus; reproduced in H. Bolus, *Icones Orchidearum. Austro-Africanarum Extra-Tropicarum* (London, 1913)

depth of potential talent in this completely unknown flower painter, leaving one to wonder why she never collaborated with Bolus in the colouring of his orchid plates.

But activities in the Cape did not entirely dominate the scene in the last decades of the nineteenth century. Scientifically and aesthetically, plant life in Natal and the Transvaal was every bit as interesting as the Cape's. Colours were often more vibrant, shapes and forms even more bizarre. It was a time when not only botanists and explorers but even artists themselves were pushing back the frontiers.

Edith Struben (1868–1936) was among the most original of the early Transvaal flower painters.[23] Born in Pretoria in 1868, the daughter of H.W. Struben (one of the earliest gold pioneers on the Witwatersrand), Edith must have revelled in the then lush highveld grasslands in their full summer glory. A portfolio of her childhood paintings now in the Compton Herbarium at Kirstenbosch reveals just how profoundly she was inspired by these flowery meadows long before prospectors trekked across them to pitch the tents that would ultimately grow into present-day Johannesburg. Edith's splendidly naive creations, painted when she was little more than a teenager, are set at the level of a child's eye, as though peering through dense turf. Flowers, beetles, grasses, butterflies, bulbs, grasshoppers – even an occasional spider – are juxtaposed to create an ecological impression of a highveld plant community, quite uninhibited by formal conventions or constraints. Years later, after a period of formal study at art schools in Paris and London, Edith Struben's style matured to that of a more conventional botanical artist, as her *Gardenia thunbergia* (Plate 30) so clearly indicates.

In Natal, 1854 saw the arrival of James Saunders, later a pioneer of the local sugar industry, together with his energetic wife, Katharine (1824–1901). As permanent settlers, the Saunders couple initially had to concentrate on establishing and running a new home, yet within months of arriving Katharine was already at work painting the showy plants in the subtropical bush around 'Tongaat', the Saunders base north of Durban.[24]

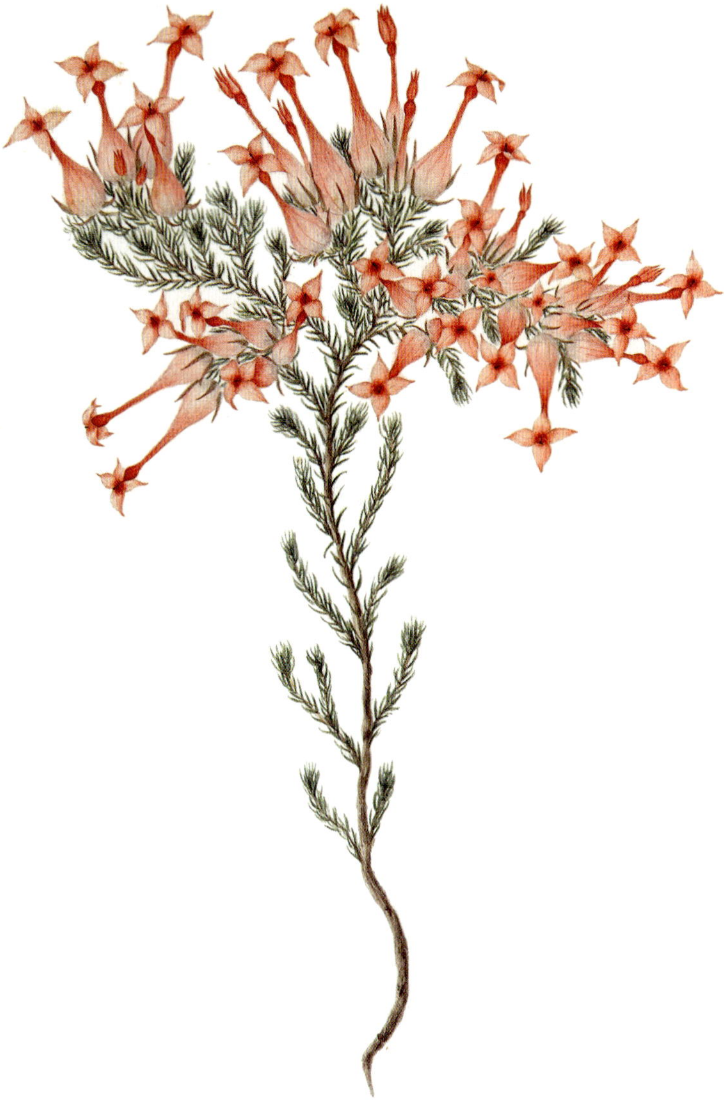

PLATE 27

Erica junonia, December 1891, watercolour, by Ada Marie Bodkin. The painting depicts the type specimen collected by Harry Bolus in the Koue Bokkeveld.

Having painted flowers during her childhood in England, Katharine enjoyed a modicum of experience, but in Natal with no local art teachers to help refine her techniques, she had to rely entirely on innate talent. By the time Katharine Saunders died in 1901 she had completed some seven hundred plant portraits, often embellished with annotations and dissections. Here too one

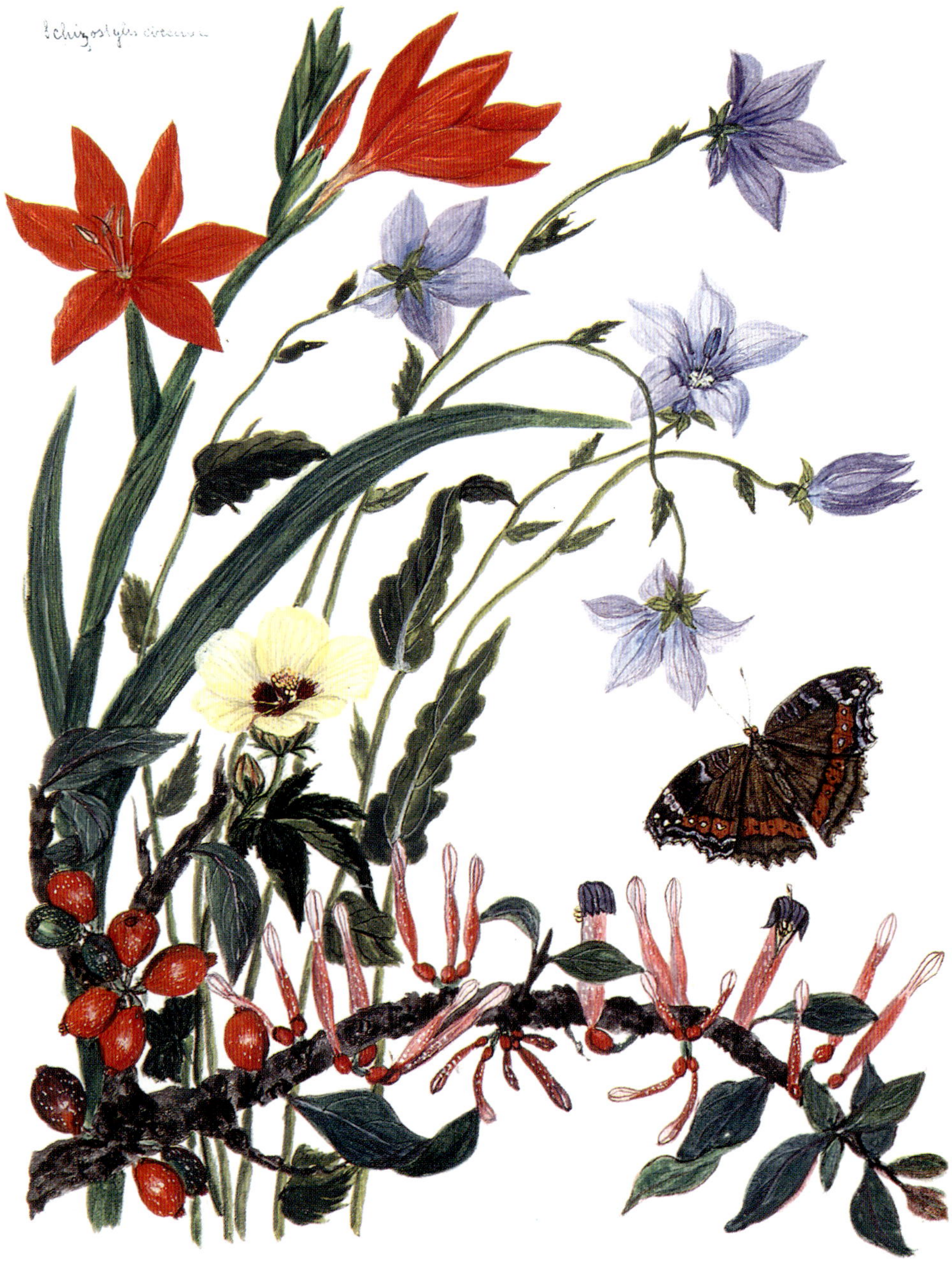

Plate 28 above and Plate 29 right

Two scenes of highveld grassland flowers, with spider, butterfly and locust, *c.* 1885, gouache, painted as a teenager by Edith Struben

sees the pervasive influence of her botanical friends and correspondents – Sir Joseph Hooker and W.T. Dyer at Kew, William Harvey in Dublin and Harry Bolus in Cape Town. To Katharine Saunders painting flowers was more than just creating a pretty picture; she too had learned to peel back the petals.

It was also in Natal that the first really ambitious project got under way to illustrate a large cross-section of plant species from a specific geographical region. In 1898 the first part of John Medley Wood's six-volume *Natal Plants* rolled off the presses of Bennett and Davis in Durban. Coming less than a year after Queen Victoria

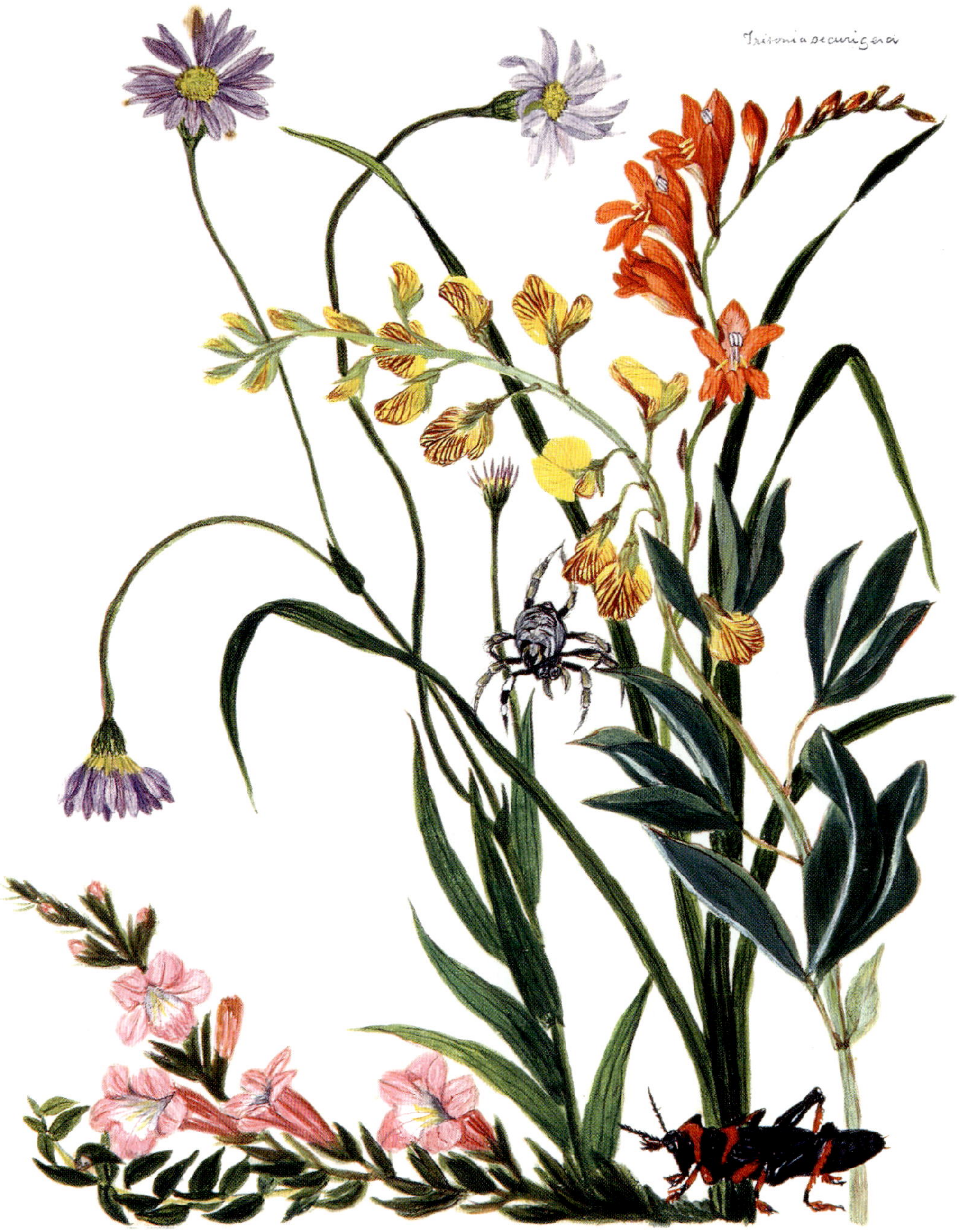

had celebrated her diamond jubilee, this was a proud achievement for colonial botany in Natal. Here were drawings by local artists, lithographed and printed by local craftsmen, published by local entrepreneurs – proof positive that even some of the remoter parts of the Empire were teetering towards technological independence. Frieda Lauth and Walter Haygarth provided the bulk of the initial artwork, with most of the later contributions coming from Millicent Franks. *Natal Plants* enjoys the distinction of being the first illustrated botanical work to have been wholly produced in South Africa. In all, five hundred and seventy-five uncoloured lithographs, faced by Medley Wood's accompanying text, had been published by 1912, when the series was eventually terminated.

Perhaps it was a feeling of complete security engendered by the *Pax Britannica* that gave the Victorian age such a profound air of confidence

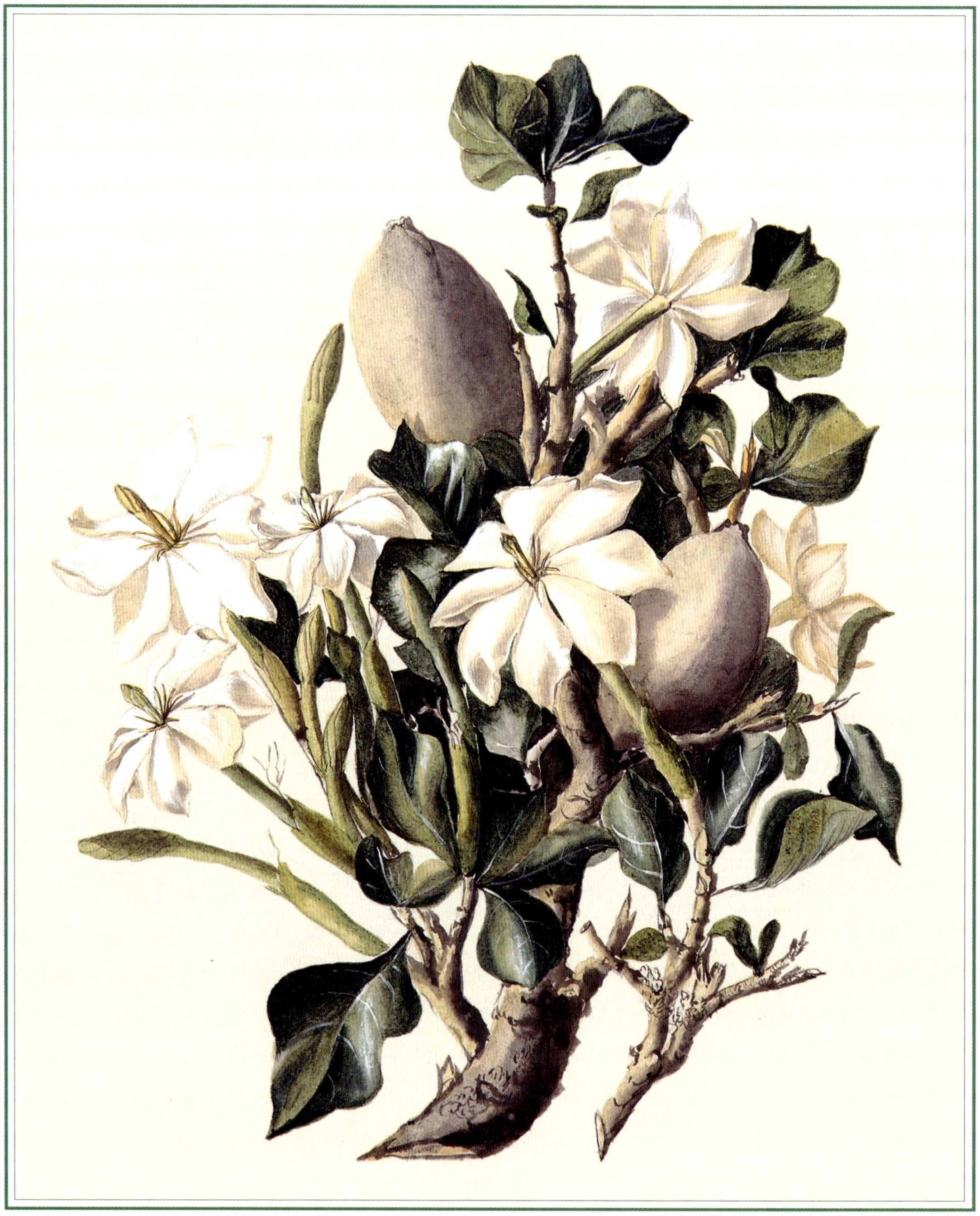

PLATE 30

Gardenia thunbergia, c. 1915, gouache and watercolour, by Edith Struben

in all its activities. But whatever it was, this was a period of extraordinary creativity even in so arcane a discipline as botanical art. In southern Africa, a rich indigenous flora was seen as a part of the great imperial treasure chest – an asset that had to be discovered, documented, preferably illustrated and, wherever possible, developed or commercialized. Thus it was that in towns, villages and frontier farms, a small committed group of botanically inclined artists rose to meet the challenge of the veld's spectacular largesse.

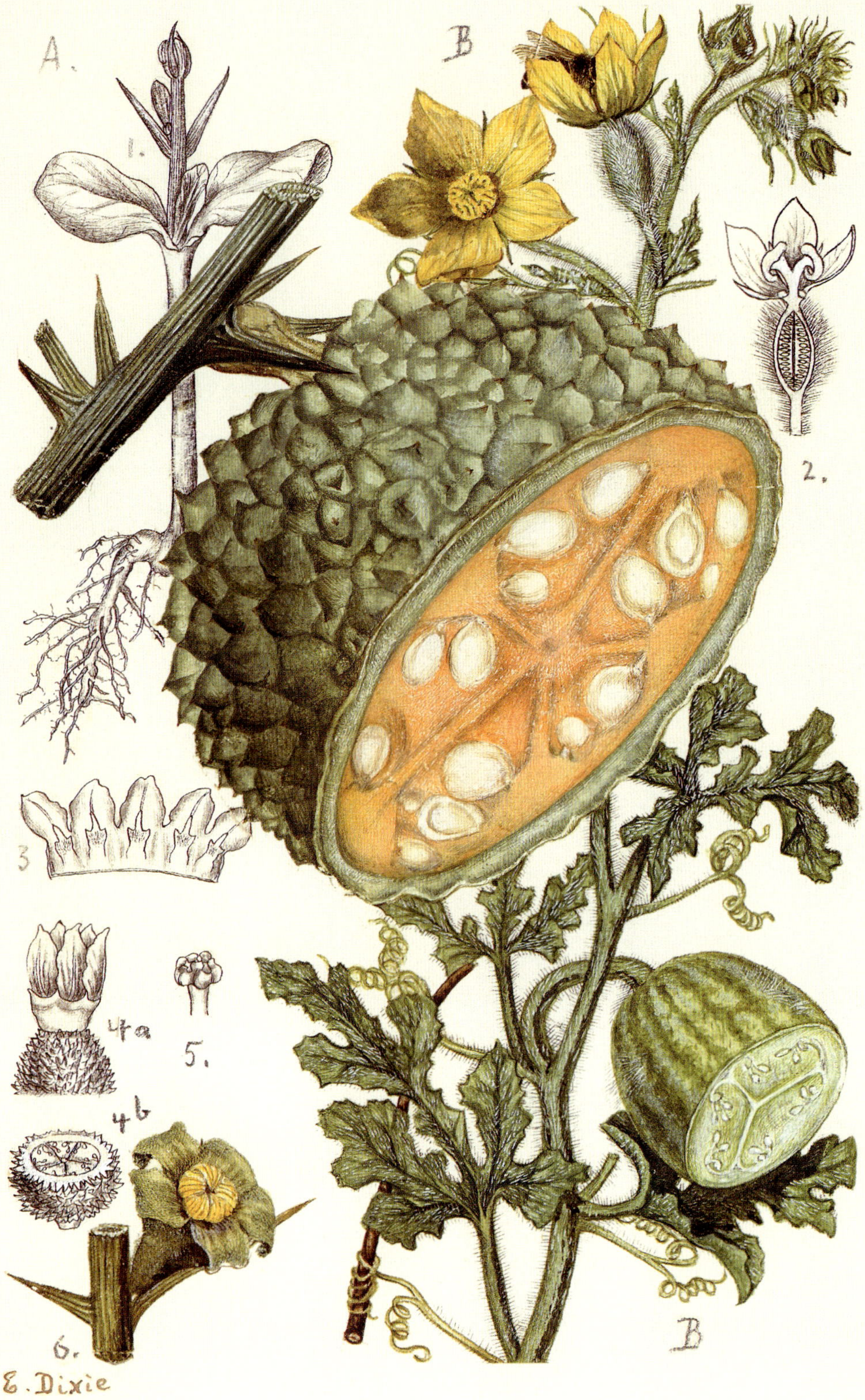

Plate 31

A plate of Cucurbitaceae featuring *Acanthosicyos horridus* (A) and *Citrullus lanatus* (B), *c.* 1925, gouache and watercolour, painted by Ethel Dixie for Rudolf Marloth's *The Flora of South Africa*, vol. 3 (Cape Town and London, 1932)

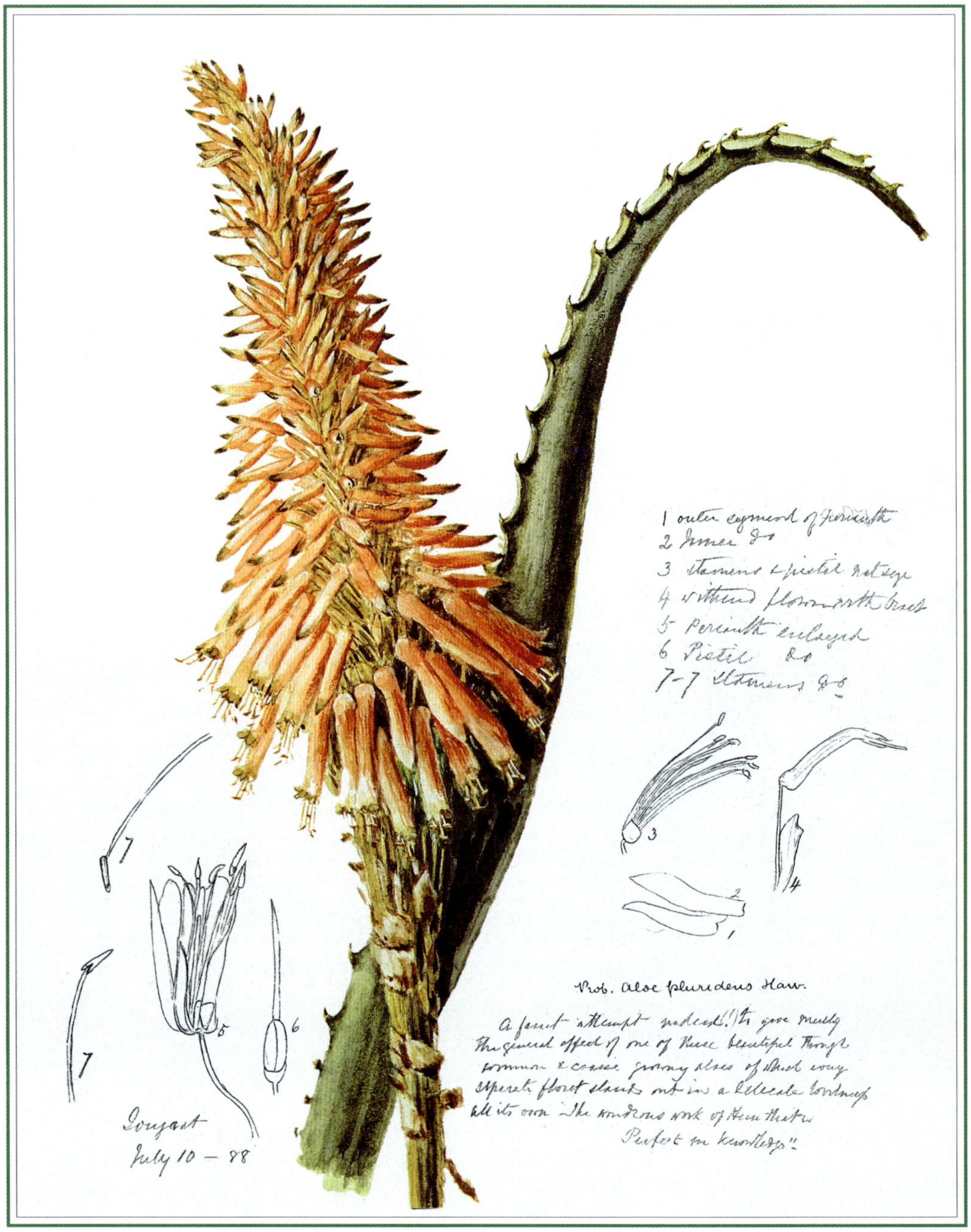

PLATE 32

Aloe arborescens (with dissections), 10 July 1888, watercolour and pencil, by Katharine Saunders; the artist has mistakenly identified the plant as '*Aloe pluridens*'

Marloth's *Flora of South Africa* and *Flowering Plants of Africa*

South Africa entered the twentieth century in the throes of the Anglo-Boer War but with the declaration of peace in 1902 the country again began to focus on creative activities. Gradually the economy gathered momentum as the region's great mineral wealth was further developed. Although this was a time when much of the country was intent upon grubbing up gold and diamonds, there was an increasing number of high-minded individuals bent upon investing

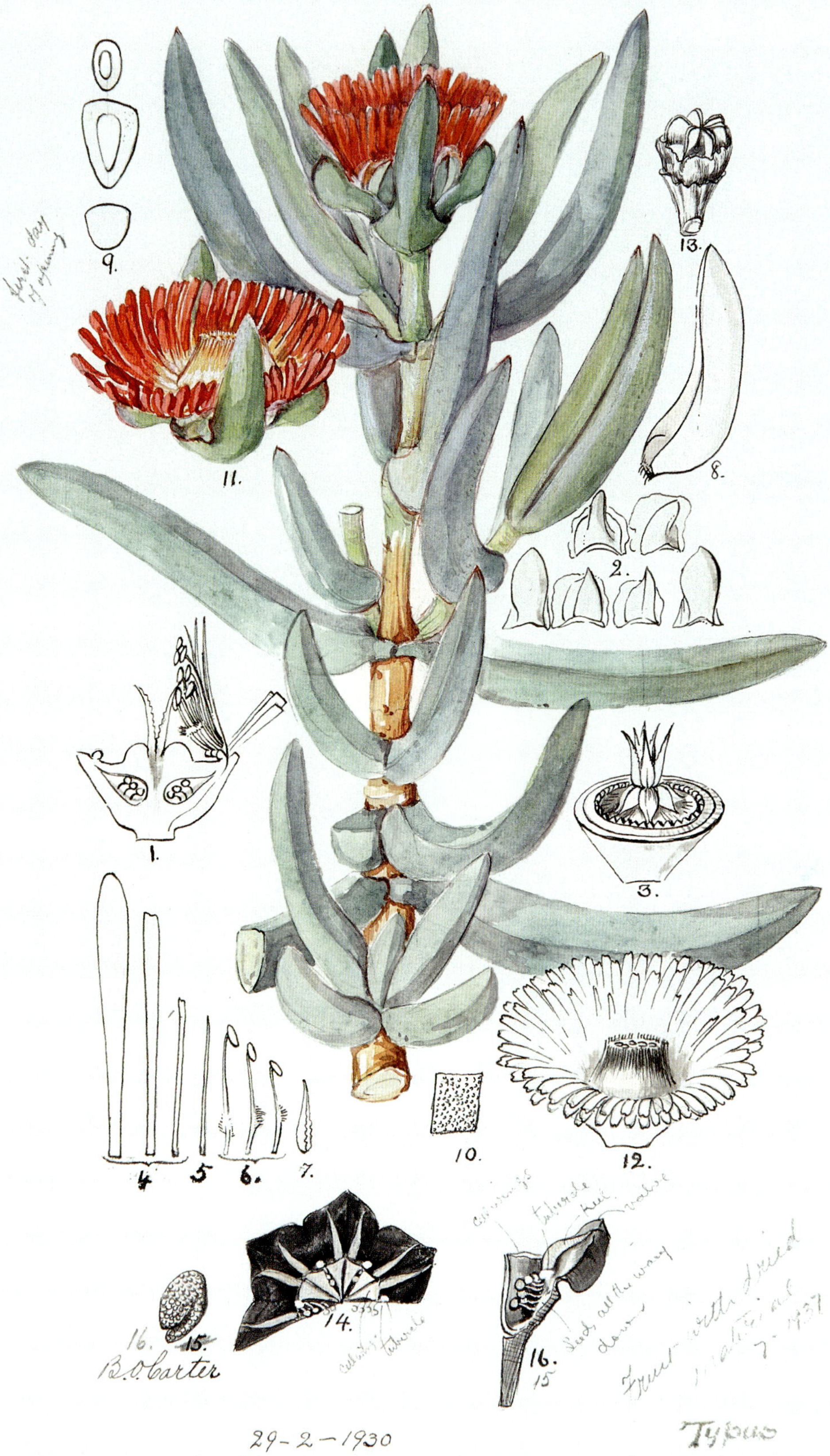

Plate 33

Astridia longifolia (showing detailed dissections), 29 February 1930, watercolour and black ink, by Beatrice Orchard Carter

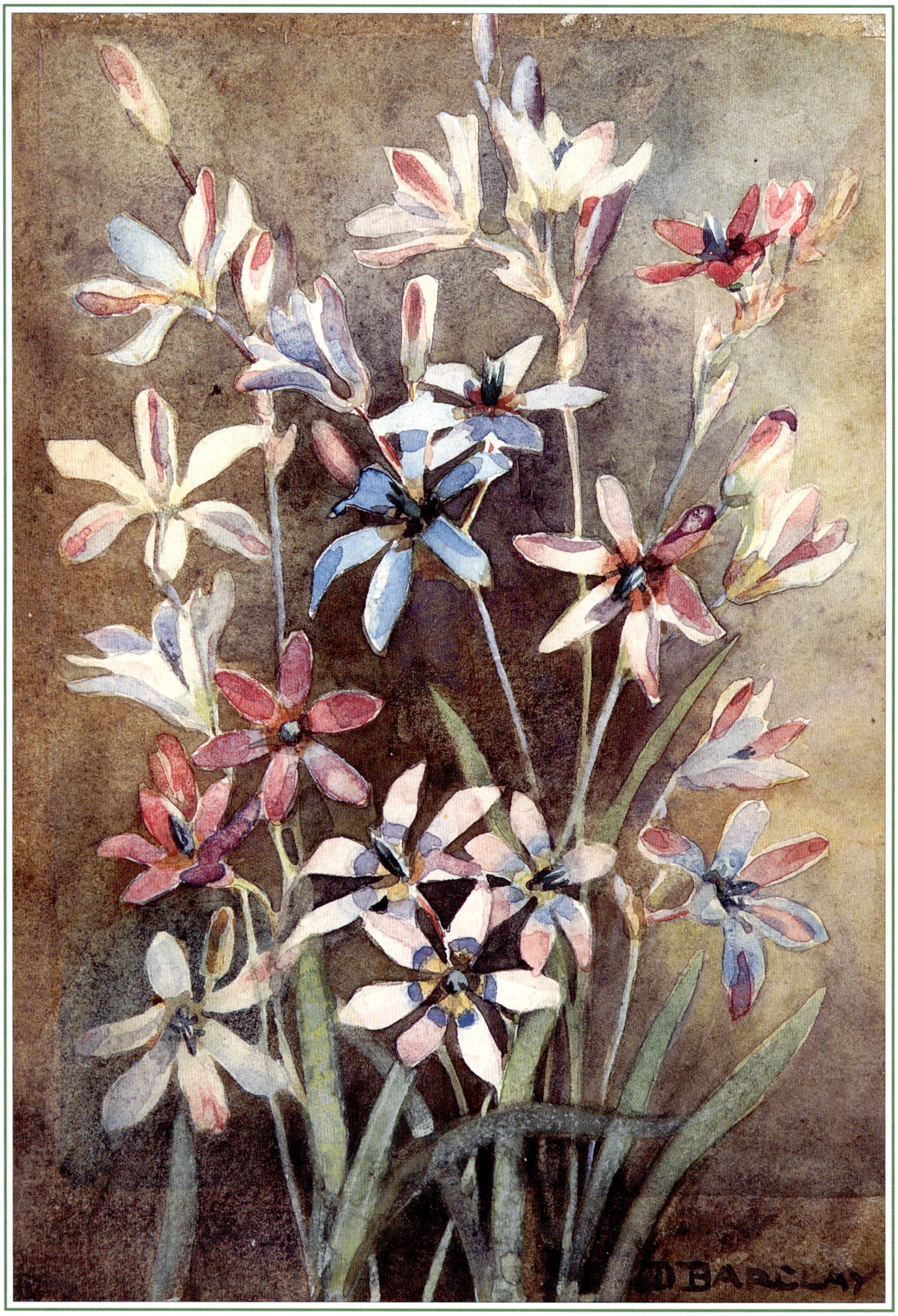

PLATE 34

Ixia monadelpha, c. 1925, watercolour, by Dorothy Barclay

some of this new wealth in cultural and scientific institutions or developing national parks, while others were actively drawing attention to one of southern Africa's less obvious but no less remarkable natural resources, the spectacular species-rich indigenous flora.

Foremost among the latter was the indomitable Florence Phillips (1863–1940) who conceived the idea of producing a richly illustrated guide to the flora of the four territories that were later to be united to form the Union of South Africa. Typically ahead of her time, the publication was to be botanically comprehensive, illustrated largely in colour but, above all, it was aimed at a broad, educated readership rather than scientists. As a child, Florence had been taken for long rambles across the veld by her naturalist father Albert Ortlepp. Now with her husband's considerable financial resources at her disposal, she was determined to underwrite a work worthy of the country's magnificent indigenous flora. Matters came to a head early in 1905. The ageing Harry Bolus was approached to prepare the text but he was too preoccupied with his orchid research and too wealthy to be dragooned by the domineering Lady Phillips. In the end her choice of Dr Rudolf Marloth (1855–1931), a Cape Town analytical chemist who taught chemistry and botany at Victoria College, Stellenbosch, proved to be a brilliant one.[25]

Marloth had an excellent record of botanical publications to his credit. Moreover, he was in the final stages of completing his massive seminal work, *Das Kapland*, which dealt with the ecology of the Cape Flora and was published in 1908. By May 1905 he had prepared a rough estimate of production costs of the proposed *Flora of South Africa*. Some three hundred plates were envisaged, while the total cost was estimated to be £3 000 of which £1 000 was allowed for 'Drawings, paintings and descriptions'.[26]

Here at last was a chance for local botanical artists to parade their talents in a book on southern African plants. Marloth lost no time in recruiting his chief contributor, Ethel May Dixie (1876–1973), a self-trained artist whose botanical paintings were already well known in Cape Town. At the time, Ethel Dixie was part of a circle of lady artists including Emily and Florence Thwaits (active 1880–1890) who were noted for their interest in botanical art. Miss Dixie was about thirty when work commenced on Marloth's *Flora* – which was ultimately issued in four volumes comprising six parts, between 1913 and 1932 – and would occupy her intermittently for nearly two decades.

Work on the huge project extended over many years with the First World War creating a major disruption, since most of the colour printing was undertaken in Germany. Although Marloth died in 1931, a year before the final part was issued, all the colour plates had by then been completed. Working in watercolour and gouache, Ethel Dixie contributed the major portion of the artwork, using freshly collected specimens as her subjects. The gouache component gave her work a rather solid quality that sometimes lacked the bright reflective sparkle so characteristic of many petaloid monocotyledons. Indeed, Marloth himself was evidently conscious of this problem and was at pains to make the following observations in the preface to Volume 4: 'Critics who may find that artists and lithographers have not done justice to some of our most gorgeous flowers will, perhaps, kindly consider that art cannot fully replace nature, and that the true brilliancy of a Disa or Nerine has never been reproduced on paper or canvas – and never will be. The reflecting surface of the paper is merely a dead passive material; but the living flower imparts an individual character to the light, first absorbing and then returning the rays into the air by means of innumerable lenses and mirrors, as explained under *Nerine sarniensis*.'[27]

Apart from Ethel Dixie, other artists were recruited, such as Esther Smith (active 1924) in Port Elizabeth, Florence Thwaits (then art mistress at the Huguenot College, Wellington, Cape), Millicent Franks (1886–1961) in Durban and Mr P. McManus (active 1920–1930) of Cape Town, all of whom made significant contributions. In a very real sense, Marloth's team consisted of the first truly professional botanical artists in southern Africa, their work having been specifically commissioned and paid for. Viewing the finished product, one can only

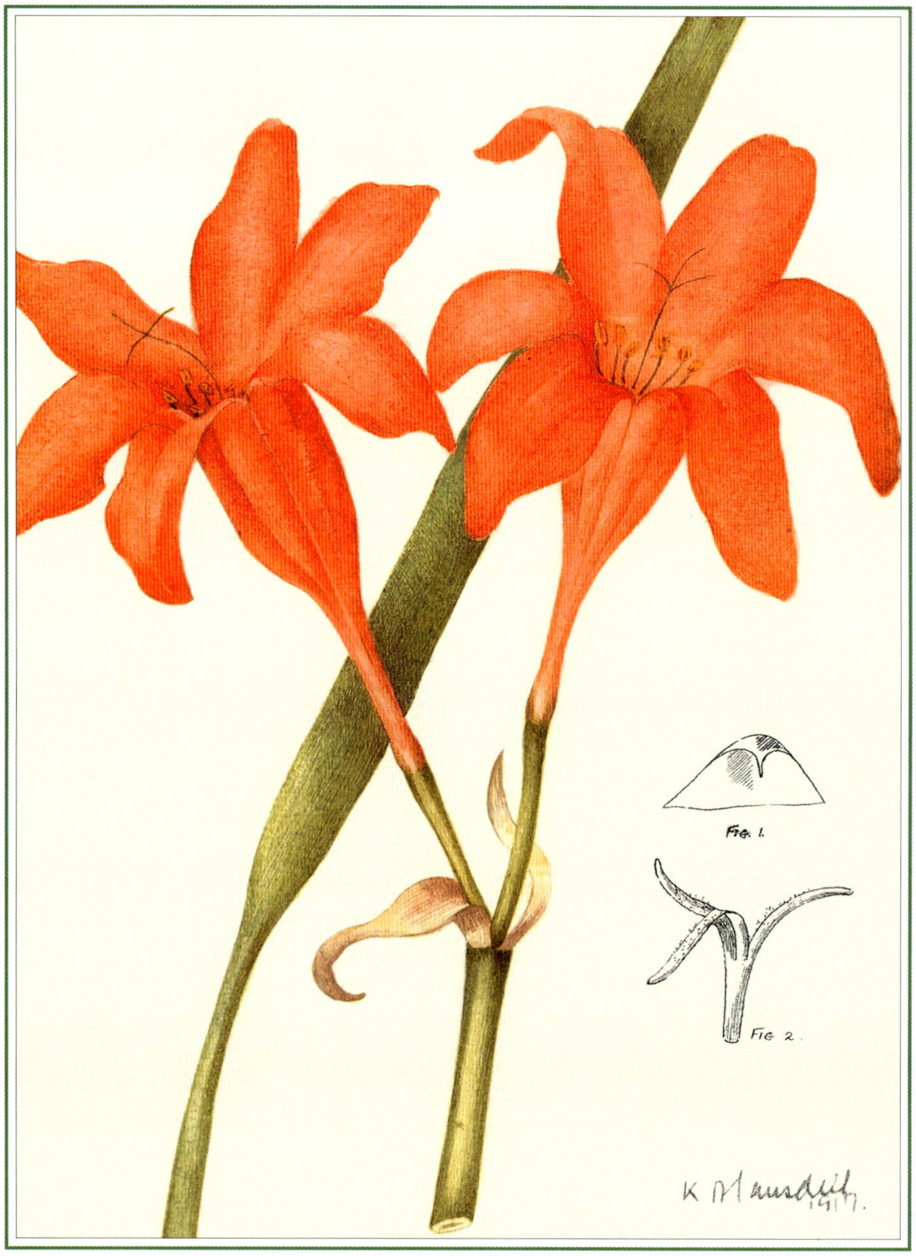

PLATE 35 above and PLATE 36 right

Cyrtanthus sanguineus: (left) Kathleen Lansdell's original watercolour, 1917, and (right) the lithographer's reworked version, published in *The Flowering Plants of South Africa,* 1921. Note the change in the positions of the leaf and left-hand flower.

marvel at the high quality of mechanical colour reproduction that was achieved by the German printers, Nenke & Ostermaier – then the world leaders in their field. Most of the colour proof plates were sent from Dresden to Cape Town, there to be laboriously corrected by Marloth himself before being returned to Germany. It was a slow time-consuming process that sorely tried the impatient Lady Phillips.

In the end Florence Phillips achieved her objective. Marloth's *Flora of South Africa* with its striking colour plates made a real impact on decision makers and thinking South Africans. It paved the way for an upsurge of popular interest

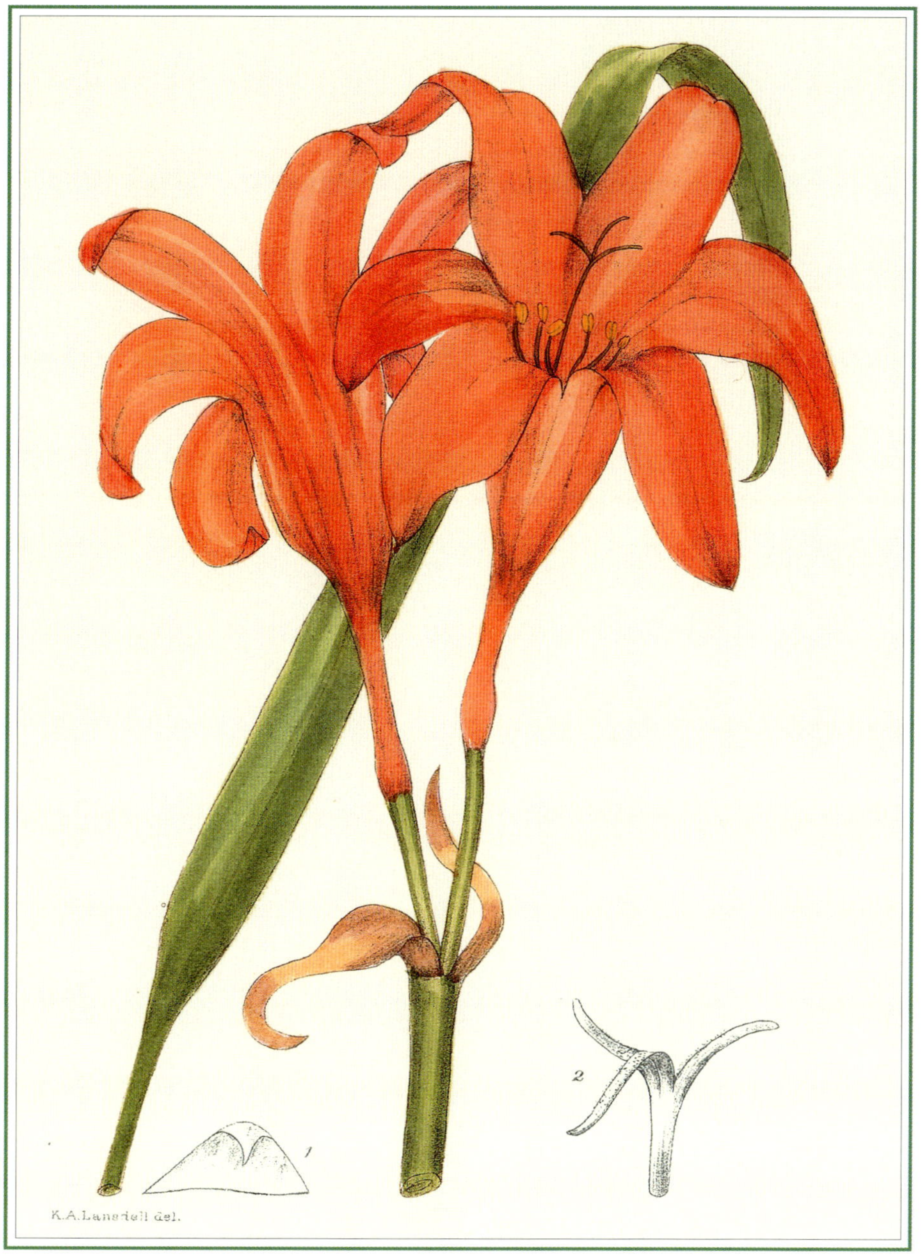

in the country's native plant life, becoming the forerunner of a steadily increasing flow of well-illustrated accessible botanical books that would appear in the post Second World War period.

After the death of Harry Bolus in 1911 the Bolus Herbarium, initially at Kirstenbosch and later at the University of Cape Town, steadily developed into a major centre for plant systematics under the curatorship of Louisa Bolus (1877–1970), niece of Harry Bolus. By 1915 the staffing had been expanded to include the London-born Mary Page (1867–1925) as botanical artist. Working to Mrs Bolus's exacting standards, Miss Page's exquisite, almost luminous, work is heavily embellished with dissections – perhaps excessively so (see Plate 115). Sadly, her prodigious output remains little known as most of her paintings are filed in the herbarium with the specimens she painted. Beatrice Carter (1889–1939), appointed in 1926 after Mary Page's death, continued in the same vigorously technical style emphasizing dissected leaves, bracts, floral parts and capsules, all so essential to Louisa Bolus's taxonomic requirements.

Apart from her scientific work, Mrs Bolus had other interests. Chief among these was the provision of educational material on southern African plants for children and adults. Again, artwork was required for this campaign. But here Louisa Bolus chose an artist with a completely different style, shrewdly anticipating that Mary Page's and Beatrice Carter's work was too technical, even intimidatingly scientific, to appeal to the general reader. So instead, Dorothy Barclay (1892–1940), a Cape Town-born artist was commissioned to prepare the colour plates that appeared in *South African Nature Notes* (from 1923), *South African Gardening and Country Life* (from 1923) as well as *A Book of South African Flowers* (1925) and *A Second Book of South African Flowers* (1936). Miss Barclay's free-flowing, soft, impressionistic images set against colour-wash backgrounds lacked firm lines or precise morphological details, yet they provided an impression of the species that was perfectly adequate for the level of readership at which these publications were aimed. In modern parlance, they were 'user-friendly'.

There is another interesting aspect to Dorothy Barclay's work. She was almost certainly the first southern African botanical artist to have her work mass-reproduced locally by mechanical colour-printing techniques. Letterpress colour printing in three primary colours from copper blocks was introduced in the mid 1920s by the Rustica Press, Wynberg, Cape, through their natural history and horticultural marketing arm, The Specialty Press.[28] The results may be seen in the four South African publications listed above, all featuring Dorothy Barclay's paintings. Here again we see the imaginative influence of Louisa Bolus in seeking inexpensive ways of mass-producing botanical illustrations in order to reach a wider audience.

Letterpress colour printing was not the most technically advanced printing process at that time, but it provided an acceptable alternative to the exorbitantly expensive procedure of colour printing being developed overseas and the even more costly task of hand-colouring lithographs.

While these activities were under way in the Cape, another visionary, Dr I.B. Pole Evans (1879–1968), chief of the department of agriculture's Division of Botany in Pretoria was planning a publication of even broader scope than Marloth's *Flora*. It would be a full-colour serial that would serve as a showcase for southern African botanical art, for decades to come. In June 2001, Volume 57 of Pole Evans's creation, *The Flowering Plants of Africa*, appeared in print. Eighty years ago, the first volume was issued, in 1921, under the slightly different title '*The Flowering Plants of South Africa*'.[29] At times its future has balanced on a knife edge, yet it has survived many uncertain times – the great depression, the Second World War, sanctions against South Africa in the 1980s, political transformation in the 1990s – but despite these vicissitudes, this important botanical journal continues to be published today.

Unashamedly modelled on the lines of the world-famous British journal, *Curtis's Botanical Magazine* which has been in continuous production since 1787, *The Flowering Plants of Africa* is unique due to its geographical focus on African, primarily southern African, plants. (*Curtis's Botanical Magazine* treats the flora of the entire world.) Forty species are featured in each volume (except from Volume 53 when the plate number was reduced to twenty), and every species is illustrated by a full-colour plate with an appropriate accompanying text. *The Flowering Plants of Africa* has played a notable role in documenting botanical biodiversity in the southern African region. Some 408 new species and 10 new genera have been described in its pages making it an important scientific reference. But it has also become a very significant medium for promoting the work of local botanical artists, with no less than 87 different artists being represented during its 78-year history.

This was not always so, as the early volumes tended to use the work of the Division of Botany's resident artists. Kathleen Annie Lansdell's (1888–1967) work dominated the first volume, followed later by Stella Gower (1894–1991) and still later by Cythna Letty (1895–1985) who contributed over 740 plates to the journal. Indeed, between 1945 and 1968 *The Flowering Plants of Africa* and Cythna Letty

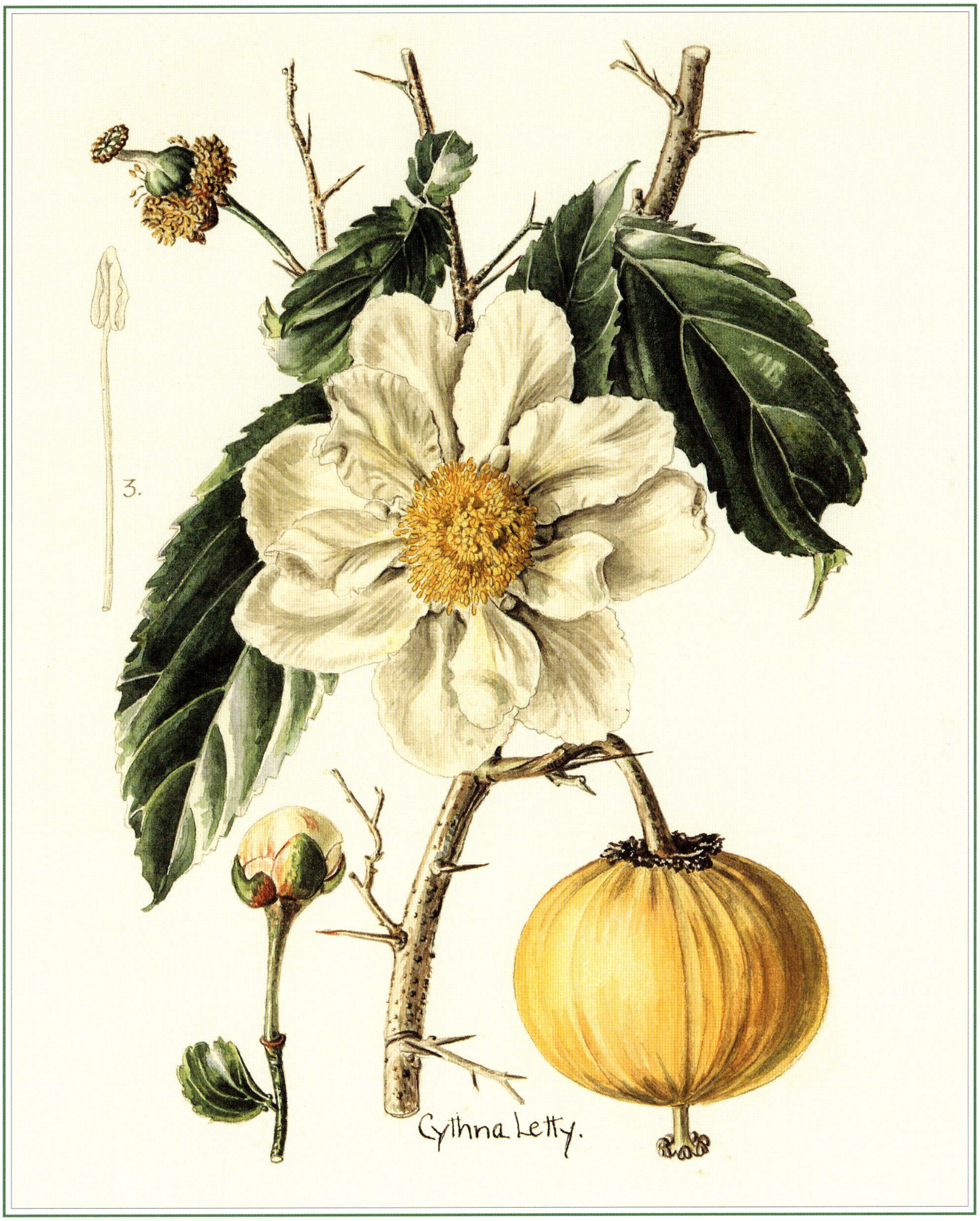

PLATE 37

Oncoba spinosa, January 1951, watercolour and pencil, by Cythna Letty, published in *The Flowering Plants of Africa,* 1951

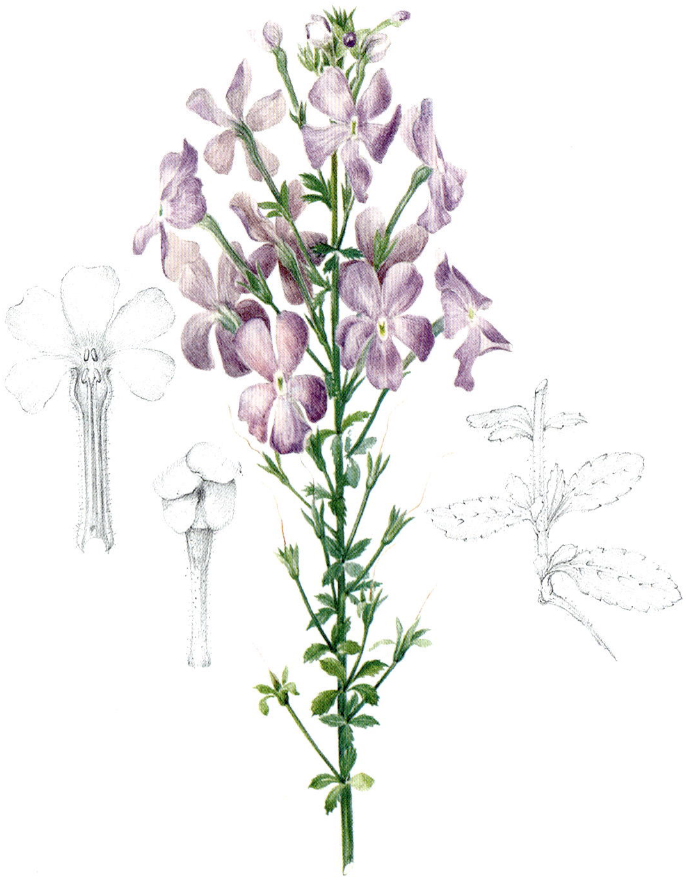

PLATE 38

Sutera grandiflora, c. 1920,
watercolour and pencil, by Stella Gower: reproduced in
The Flowering Plants of South Africa, 1924

were virtually synonymous. Such was the quality of her work that Letty became the first botanical artist in southern Africa who captured the public's imagination, and it must be acknowledged that her reputation was developed largely through her published work in *The Flowering Plants of Africa*. Her obvious empathy with her subjects and her brilliant technique in handling large glossy surfaces on fruits and leaves are the hallmarks of this remarkable artist's work.

Plans to produce a South African equivalent of *Curtis's Botanical Magazine* began in 1917 when I.B. Pole Evans wrote to the well-known London natural history publishers, Lovell Reeve, for quotations to undertake the lithographic work and subsequent hand-colouring of the plates. No government funding for the new journal was forthcoming so the determined Pole Evans turned to a number of well-to-do citizens interested in botany for their support. Four important allies were soon recruited. They were Mr C.G. Smith of Durban, Mrs Henry Burton (wife of the then general manager of the South African Railways), Sir Frank Reynolds and Mr Percy Ross Frames, all of whom contributed £100 and more towards the production of the first few volumes. Pole Evans's letter to Mrs Burton on Christmas eve, 1917, sets out his vision for the future journal:

'MRS BURTON, "VOORSPOED", RETREAT, CAPE COLONY; 24 DECEMBER 1917

Dear Mrs Burton,
You will probably remember the coloured illustrations of South African plants which Miss Lansdell has been doing for us. Last Saturday week when I was showing Mr C.G. Smith of Durban over our laboratory he was greatly taken with them and asked me why we did not get them published. When I told him there were no funds he immediately asked what they would cost and said that he would give £100 towards their publication. He also asked me to write and tell you that he had given a £100, and that he was sure if you wrote to Sir Frank Reynolds and Mr Ross Frames and told them what he had given he was sure that they would each do the same. I would like to publish these illustrations in a magazine devoted solely to coloured illustrations of South African plants. Such a magazine might be issued quarterly with about a dozen coloured plates, and I think a volume could be published annually for about 10/- a copy and each containing about 50 plates.

'At present I have no idea what such a publication would cost, but directly after Mr Smith's offer I wrote to a firm in London who publish a coloured botanical magazine for the Kew people asking them to quote me a price, etc.' [30]

About 4 years passed between conceiving the new journal in 1917 and the issue of the first volume in 1921. Production costs always loomed large. After revising their original costings, Reeves submitted a final quotation for 500 copies of Volume 1 of which the first 100 copies would be hand-coloured and made up at £320. For 250 coloured made-up copies, the price was to be £460.

No clear order seems to have survived as to whether 100 or 250 copies of Volume 1 were initially made up, although correspondence from The Specialty Press in Johannesburg who marketed the volume suggests that it was 250. Eventually, on 4 October 1919, the first batch of Kathleen Lansdell's paintings was sent to London. Production had at last commenced. At the outset a decision had to be made on paper size. The choice was between royal octavo (the format of *Curtis's Botanical Magazine*) and crown octavo. The latter was approximately 10 mm wider and, being more readily available, won the day.

N.E. Brown (1849–1934), senior botanist at the Kew Herbarium, was commissioned to check the proofs, but soon concerns were expressed regarding the accuracy of some of Miss Lansdell's plates and hence the correct identities of the species figured. Thus in February 1920 E.P. Phillips (1884–1967), senior botanist at the National Herbarium in Pretoria, suggested to Pole Evans that herbarium specimens of the subjects illustrated should be kept. This important principle has been consistently adhered to ever since.

Producing each volume – which was issued in parts – was a laborious triangular process which took place between Pretoria, London and Kew, with the original artwork and text being sent from Pretoria to Reeves where the text was set and the lithographers engraved the plates on stone, while N.E. Brown checked the proofs. Initially the lithographic artist engraved the outline image and shading of the original artwork onto a lithographic stone but later zinc plates were used. The resulting monochrome outlines were then painted in by skilled colourists from a master pattern copy of the original artwork. Sometimes the lithographers took liberties in their engraved interpretations of Miss Lansdell's originals. For example, in Plate 25 of *The Flowering Plants of South Africa* (1921), they have changed the left-hand flower of *Cyrtanthus sanguineus* to show a back view rather than an anterior view as in the original figure, giving what is perhaps a more balanced illustration. A further modification concerns the leaf, which disappears off the plate in the original but has been depicted with its characteristic downward droop in the improved lithograph. (See Plates 35 and 36.)

Finding suitable lithographers was also a cause for further delay, as Mr A.L. Soper, chairman of Reeves, reported to Pole Evans on 6 May 1920: 'But I regret to say that a very serious misfortune has arisen which must cause a further delay in the issue of the work. Mr Fitch, who has lithographed the plates of the Botanical Magazine for so many years and was also doing yours, has suddenly broken down and I almost fear that, although a month ago he appeared in good health and showed no sign of failing strength, he has drawn his last plate.

'The difficulty we are in is to find someone who combines his skill as an artist, as a lithographer, and his appreciation of the form and structure of plants. It would not be difficult to find lithographic artists, but they are only interested in plants in so far as they can be twisted into designs for wall papers and carpets, and are not qualified for scientific work. We have now three persons with some experience in botanical drawings who are learning lithography with a view to take up Mr Fitch's work and we are awaiting the result of their first trials.

'I fear therefore that I must ask you to exercise some little patience towards us in these unavoidable circumstances.' [31]

Thus in these early volumes of *The Flowering Plants*, the creative input of the lithographers and colourists had a significant impact on the appearance of the final published plate. Certainly, there were instances where the lithographers actually improved the quality of the final product by giving them greater definition. Problems with the quality of the original artwork also continued. Here it must be admitted that Kathleen Lansdell was not an especially competent botanical artist. Mr Soper expressed his reservations to Pole Evans:

'*17 May 1921* – But we are of course dependent upon the drawings and in January last I sent Dr Phillips (on his invitation) some observations on those recently received. They were distinctly

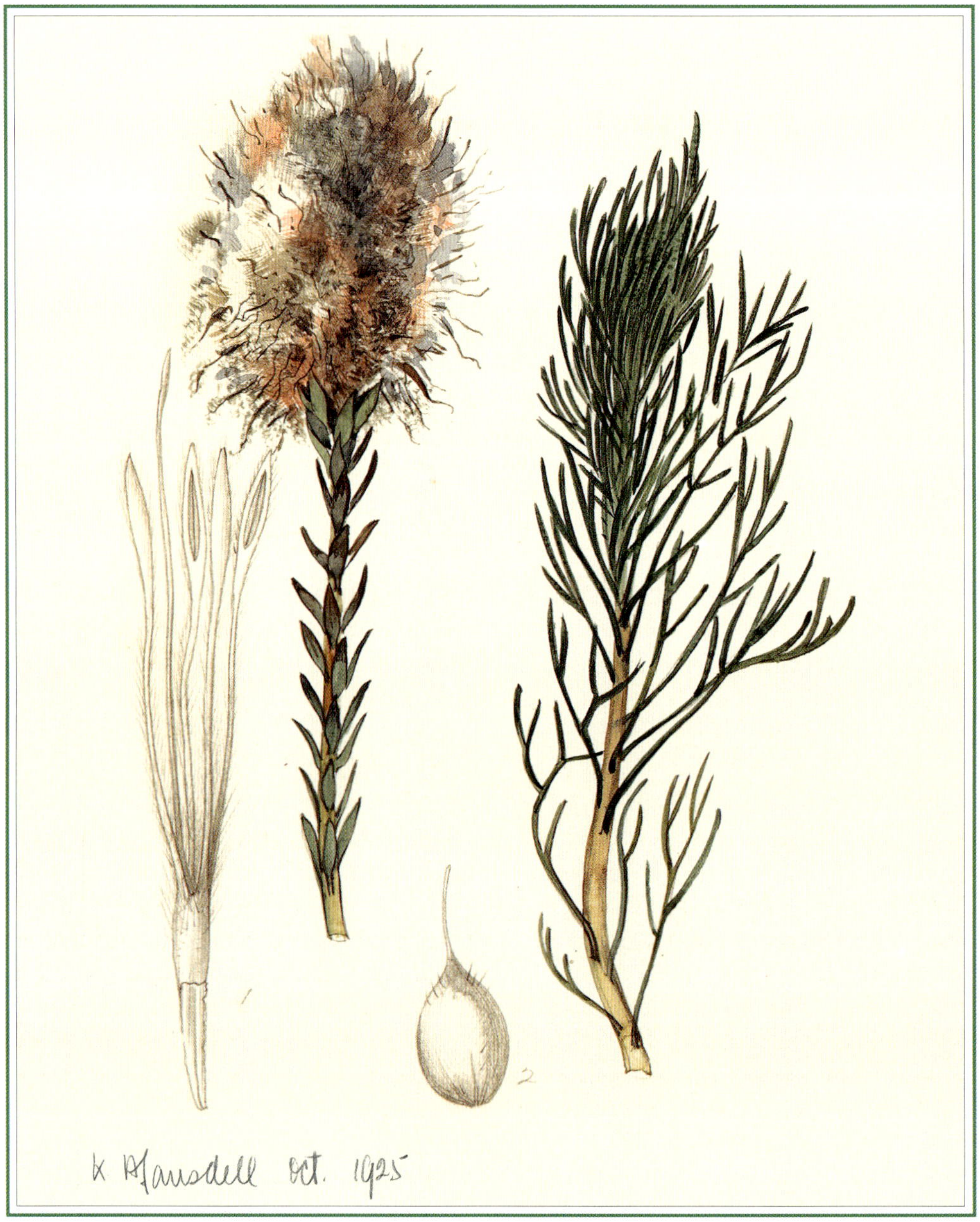

PLATE 39

Paranomus longicaulis, dated October 1925, watercolour: Kathleen Lansdell's painting of a partially decomposed specimen was published in *The Flowering Plants of South Africa*, 1926

better than the earlier ones; at the same time some of them appeared to me to be often lacking in definition and to suggest some want of appreciation of the value of botanical details. I am myself inclined to the opinion that every plate should contain a vertical section of the flower, and in that way exhibit its internal structure, arrangement and proportions as well as its external form: sometimes a cross section may be necessary; and a drawing of the fruit. With these it should be necessary, as a rule, to include illustrations of separate organs, unless for some special interest which cannot be brought out in the sections.' [32]

A response from E.P. Phillips to Pole Evans highlights this lack of interaction between artist and botanist in the initial years of the journal's history. Phillips observed: 'I agree with the criticism of Messrs Reeve & Co. but feel that what they complain of is due to the system under which the drawings are executed. Both Miss Lansdell and Miss Gower have often to put aside drawings partly finished when other official duties intervene. This in many cases makes it difficult for me to supervise the drawings while in progress. *22/6/21*.' [33]

To which note Pole Evans responded, 'Dr P., you can have Miss Gower exclusively at your disposal for this work.' [34]

Thus with further experience and Stella Gower employed almost fulltime as artist, matters slowly improved and *The Flowering Plants of Africa* eventually established itself in regular production. As the existing stock of coloured plates was depleted, the watercolourists on the staff of Reeves working from a pattern copy of the original, painted in the remaining lithographic impressions to make up the required volumes. This nineteenth-century method of colour reproduction continued until the middle of the Second World War when green and red overlays were added to the plates. Finally in 1944 the entire colour-printing process was undertaken mechanically.

But critical botanical supervision was apparently still not always forthcoming. Finding a representative selection of species from different parts of the region presented huge difficulties in the early days before the advent of airmail and plastic bags. Thus when Dr John Muir of Riversdale in the western Cape sent to E.P. Phillips fresh specimens of a great rarity, *Paranomus longicaulis*, by the time they reached the National Herbarium in Pretoria the densely floriferous inflorescences had decomposed into putrescent masses of fluffy mush. And that is exactly how Kathleen Lansdell painted her subject.

Why Phillips insisted on having these mouldering fragments painted and reproduced is something of a mystery but it gives an indication of the peculiar difficulties under which some artists have had to work and the often unreasonable demands made on them. This particular plate brought no credit to Kathleen Lansdell, but its inadequacies are clearly not her fault.

Demand for the new publication steadily increased to the extent that when stocks of Volume 1 were exhausted, a reprint was ordered. The reprinted Volume 1 is generally inferior to the original, especially in terms of the hand-colouring of the plates. It may readily be distinguished by examining the verso of the title-page which reads 'Made and printed by offset in Great Britain by William Clowes and Sons, London and Beccles'. This credit is absent from the original printing. A further distinction is that the attribution of Plates 37 and 38 is incorrect in the reprinted version. Stella Gower is cited as the artist in the original but in the reprint Plates 37 and 38 are attributed to K.A. Lansdell.

It was inevitable that the outmoded process of hand-colouring plates by Messrs Reeve would eventually come to an end as the skilled exponents of this craft retired. Thus from about 1950 hand-colouring of back numbers was carried out in Pretoria by technical assistants with artistic ability.[35]

The questions are often asked: will the exceptional improvements in colour photography condemn the botanical artist to redundancy? Do botanical books illustrated by artists serve a purpose in the modern world? It is true that some superlatively illustrated photographic books on the plant kingdom have appeared in recent years, but the emphasis of the photographer and botanical artist are usually different. As an aid to science, the image created by a competent botanical artist is an illusion in terms of depth, focus and emphasis. However minute or bulky the subject, every part is always depicted in perfect focus, with the artist usually discreetly accentuating diagnostic structures – something that is beyond the optical limitations of lenses or even the human eye. Thus the two media will almost certainly coexist as complementary rather than competitive entities. Economics will be the only limitation to the continued publication of works like *The Flowering Plants of Africa*.

Post-war years

After the Second World War, great advances in colour-printing technology resulted in a deluge of publications covering a broad spectrum of southern African plants, spearheaded by Mary Maytham Kidd's (1914–2001) classic *Wild Flowers of the Cape Peninsula*.[36] Half a century and three editions later it has sold nearly thirty thousand copies, a record for any botanical book in South Africa (see Plate 64). So great has been the reservoir of talent which emerged in the post-war years that it is impossible to discuss individual achievements in detail. Nevertheless, Cythna Letty (1895–1985), Hilda Mason (1909–1986), Auriol Batten (1918–), Irma von Below (1920–1984), Ellaphie Ward-Hilhorst (1920–1994), Elise Bodley van Wyk (1921–1997), Barbara Jeppe (1921–1999), Thalia Lincoln (1924–), Fay Anderson (1931–), Anita Fabian (1934–), Esmé Hennessy (1933–) and Gillian Condy (1952–) have made the most substantial contributions in recent years.

This list betrays a curious fact: local botanical artists in the twentieth century have almost entirely been women. There is no simple explanation for this interesting situation, but economics and social factors may be the most important. In eighteenth- and nineteenth-century Europe, either royal or aristocratic patronage supported the extraordinary talents of men like Ehret, Redouté and the Bauer brothers. Such reliable support has long since disappeared and in present circumstances, no man today could support a family on the meagre, uncertain earnings of botanical art. Then too, the very subject has probably been seen as too peripheral to mainstream art to represent a challenge in this often male-dominated sphere. A woman engaged in such a non-threatening activity as painting images of plants would therefore not have been seen to upset the *status quo* either domestically or artistically. Botanical art was a niche waiting to be filled.

How does one explain the emergence of such a large body of talent over three centuries, but more particularly in the last fifty years? Martyn Rix ascribed it to the presence of South Africa's species-rich flora, dominated by aesthetically dramatic plants that almost demand some sort of artistic response. He also saw the country's affluence, derived from its inordinate mineral wealth, as a source of support for artists. Both are indisputably valid reasons. But there is a third factor which has underpinned this specialized art form. It is a strong scientific-investigative ethos, derived ultimately from a deep-seated exploratory urge that has been at the very heart of Western civilization since Renaissance times. There are not many 'new world' countries where wealth, a spirit of scientific enquiry, and a resplendent native plant life all converge. South Africa is such a place, as is Australia. It is no coincidence that both have developed notable cultures of botanical art.

Doubtless it is due to advances in the quality and techniques of colour printing that the work of post-war artists has been brought to the attention of a wider audience. The inevitable corollary has been a growing demand among collectors for original examples of this specialized form of plant portraiture. Moreover, this interest has been intensified by our increasing preoccupation with environmental degradation and the concomitant loss of species into that black hole of eternal oblivion – extinction. Fine paintings by notable botanical artists of rare, beautiful or interesting plant species, especially those under threat of survival, are surely destined to become ever more sought after.

The great majority of artists mentioned here have in their work attempted to meet the requirements of science through accuracy and realism but without sacrificing their individual styles or innate need to interpret. Wilfrid Blunt encapsulated the essence of this seemingly paradoxical situation in his celebrated book, *The Art of Botanical Illustration:* 'Moreover, the botanical artist finds himself at once and always in a dilemma: is he the servant of Science, or of Art? There can, I think, be no doubt that he must learn to serve both masters. The greatest flower painters have been those who have found beauty in truth: who have understood plants scientifically, but who have yet seen and described them with the eye and hand of the artist.'[37]

Not only do botanical artists serve two masters, but they have nearly all unwittingly conveyed something of the mood and spirit of their age in almost all their creative works. The rather crude images of the seventeenth-century Dutch artists at the Cape tell us as much about the primitive state of science at the time as they do about the difficult conditions under which they were painted. Similarly, the works of artists like Harry Bolus and Mary Page reflect the level of scientific advancement in the late nineteenth and early twentieth centuries when better microscopes allowed the dissection and analysis of small floral structures. And who could deny that Auriol Batten and Thalia Lincoln with their subtle backgrounds showing plant portraits set against natural habitats or in associated plant communities, have unselfconsciously mirrored our late twentieth-century obsession with the environment and our need to see living organisms as part of an ecosystem? The future is unknowable but we may be sure that botanical artists who apply their pens and brushes in the twenty-first century will similarly capture the spirit of the new millennium.

1 M. Rix, *The Art of the Plant World* (New York, 1981), p. 218.
2 (Leiden, 1605).
3 G. Waterhouse, G.C. de Wet and R.H. Pheiffer, *Simon van der Stel's Journey to Namaqualand in 1685* (Cape Town, 1979).
4 Mary Gunn and L.E. Codd, *Botanical Exploration of Southern Africa* (Cape Town, 1981).
5 In South African Museum collection, Cape Town.
6 10 parts in 1 vol. (Amsterdam, 1738–1739).
7 *Dictionary of South African Biography*, 1, 1968, p. 168.
8 E.G.H. Oliver, 'Some observations on two early Cape florilegia', *Bothalia*, 13 (1&2) 1980, pp. 115–125.
9 S.J. van Oostroom, 'On some drawings of South African Proteaceae ascribed to Jan Hartog in the government herbarium at Leyden', in Mia C. Karsten, *The Old Company's Garden at the Cape and its Superintendents* (Cape Town, 1951), Appendix B.
10 Vernon S. Forbes and John Rourke, eds, *Paterson's Cape Travels, 1777 to 1779* (Johannesburg, 1980).
11 D.S. Evans, T.J. Deeming, B. Evans and S. Goldfarb, *Herschel at the Cape, 1834–1838* (Cape Town, 1969).
12 Brian Warner and John Rourke, *Flora Herscheliana* (Johannesburg, 1998).
13 See H.M. McKay, 'William John Burchell, botanist'; part 2: 'Burchell in South Africa', *Journal of South African Botany*, 7, 1941, pp. 61–76.
14 (London, 1849).
15 Ibid.
16 See Marianne North, *A Vision of Eden: the Life and Work of Marianne North* (Exeter, England, 1980).
17 *New Illustration of the Sexual System of Carolus von Linneaus; . . . the Temple of Flora, or Garden of Nature . . .* (London, [1799]–1807).
18 Gunn and Codd, *Botanical Exploration*.
19 Evans, Deeming, Evans and Goldfarb, *Herschel at the Cape*, p. 202.
20 2 vols (Cape Town, 1888, 1918).
21 3 vols in 4 (London, 1893–1913).
22 H.M.L. Bolus to John Rourke, personal communication, 1965.
23 *Dictionary of South African Biography*, 4, 1981, p. 630.
24 A. Bayer, *Flower Paintings of Katharine Saunders* (Maidstone, Natal, 1979).
25 P.G. Jordaan, 'Rudolf Marloth, South Africa's first trained botanist', *Lantern*, 13 (1) 1963, pp. 86–93.
26 Thelma Gutsche, *No Ordinary Woman; the Life and Times of Florence Phillips* (Cape Town, 1966).
27 Rudolf Marloth, *The Flora of South Africa* (Cape Town and London, 1915).
28 Hubert Elfers, ex Rustica Press, Cape Town, personal communication to John Rourke, 1999.
29 D.J.B. Killick and E. du Plessis, eds, *History of* The Flowering Plants of Africa *and an Index to Volumes 1 to 49* ([Pretoria, 1988]).
30 Correspondence relating to *The Flowering Plants of Africa:* Archives, National Botanical Institute, incorporating Archives of Division of Botany and Botanical Research Institute, Pretoria, File M163.
31 Ibid.
32 Ibid.
33 Ibid.
34 Ibid.
35 R.A. Dyer, 'A brief history of *Flowering Plants of Africa* and an index to volumes 1–36', appendix to *The Flowering Plants of Africa*, 36, 1964, pp. 1–4.
36 (Cape Town, 1950).
37 (London, 1950), p. 3.

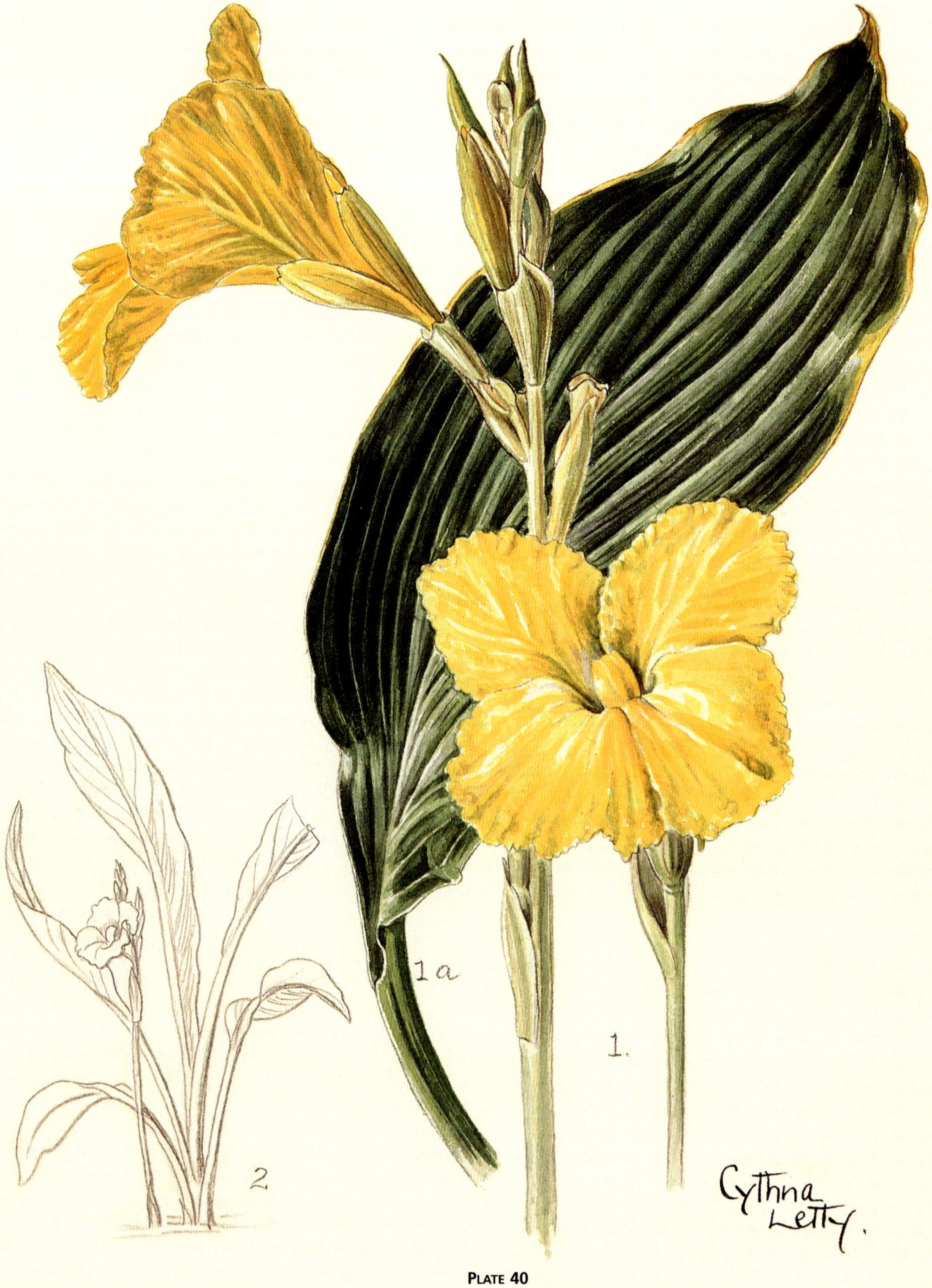

Plate 40

Cythna Letty, *Kaempferia decora* (wild ginger), [1953], watercolour and pencil: reproduced in *The Flowering Plants of Africa*, 1955

THE MIND AND THE EYE

DEE SNIJMAN

'The arts can be conveyed from generation to generation, and from nation to nation, without the use of words, but though biological understanding can, to a certain extent, be communicated through direct visual channels, it is kept alive and transmitted essentially by means of language.' [1]

There have been times in the history of humankind when changes in the global environment have led to great shifts in thought. One of the major issues gaining momentum today is a concern for the future survival of Earth's living things – the myriad forms of life that have shaped the atmosphere, climate, soil, and water and which stabilize our planet's intricate global ecosystem.

Fundamental to this questioning is the realization that we need to know more about the world's biological diversity. So far systematic biologists – sometimes called 'taxonomists' – have described some 250 000 plant species from around the world.[2] Of these, approximately 20 000 flowering plant species are native to southern Africa[3] and at least 80% are unique or endemic to the region.[4] With this local concentration of plant species southern Africa is richer than many other regions of biological importance – even those from the lush tropical areas of Africa, India, Indonesia, Malaysia, Australia, and South America. Yet, despite concerted efforts to name and document Earth's botanical diversity, taxonomists continue to discover fascinating new plants, least of all in southern Africa, where at least 290 plant species were described in the short period between 1994 and 1998.[5]

Such a dazzling wealth of plant species may suggest that southern Africa's flora is boundless. Yet recent surveys reveal clear signs of large-scale disruption and destruction of species. In particular, the most recent *Red Data List of Southern African Plants* indicates that 58 named plants have recently become extinct and as many as 3 435 plant species are threatened with extinction.[6] Sadly, many of these – the peacock moraeas, gladioli, leucadendrons, ericas and haworthias – are highly ornamental plants and their disappearance would amount to an enormous loss, not only in aesthetic terms but also for their monetary and functional worth.

In many societies, the direct economic value of plants often accounts for the greatest interest in plants. Well-known examples are the freesias, gladioli, ixias and nerines which launched the Dutch bulb industry, the buchus that yield essential oils for scents and flavourings, the wild legumes used as rooibos and honeybush tea, and the many fynbos species that provide fresh and dried flowers for South Africa's export cut-flower industry. Nevertheless, the most vital benefit of our plant-life is its ability to maintain the integrity of the region's ecosystem and, in a severely water-stressed region like southern Africa, this includes its ability to sustain a clean water supply – a frequently overlooked service that is easily disrupted by alien plants.

Under the current rapidly changing environmental conditions, many rare plants face imminent extinction merely through their inability to migrate. Yet, ironically, this sedentary habit simultaneously benefits those who wish to learn more about plants and interact rationally with them, for it allows us to explore plants close up and enables artists to capture their transient floral beauty on paper – as in the images of this volume. Indeed, in terms of their amazing range of form, colour, and texture, the plants featured here may fascinate and hold one's attention, yet their isolated images merely hint at what is known about the richness and complexity of southern Africa's flora.

To offer the viewer an alternative way of interpreting plants and their images, the focus here is twofold, both on the plants themselves and on the botanical preoccupations that have influenced their representation. By providing an outline of the developments in plant description, portraiture, and classification in southern Africa from their early beginnings, the picture that should be revealed is one of relentless exploration, discovery and documentation – sometimes fitfully, sometimes feverishly – but always infused with delight at the wonders of southern Africa's flora.

The language of plant description

Descriptions of species in the written form are the foundation of all biology, for they delimit the smallest group of individuals that are distinguishable from all others. In tracing the discipline's development and its influence on botanical art, it is apparent that plant descriptions initially relied on the acute observation of only a few plants. But as exploration of the natural world produced increasing numbers of new plants, the process involved greater selection and ordering of information to allow the relationship of each new species to be determined. Ultimately, descriptions lead to an understanding of the form, structure and behaviour of living things and these set the stage for enquiry into the reproduction and development of organisms, their relation to the non-living world, and to one another.

The first written observations of plants date back almost two millennia to the manuscripts of Theophrastus (370–*c.*285 BC). Inevitably, the discipline went into decline during the struggle for survival in the Dark Ages but it has continued since, for an unbroken span of four and a half centuries. In general, visual plant descriptions precede written accounts. In southern Africa, however, the earliest known plant images of a *Brunsvigia* and an aloe-like plant – attributed to San rock painters – are believed to be less than three hundred years old,[7] which suggests that they may have been made after the first written records of the region's plants appeared (see Plate 8).

The southern Cape species, *Protea neriifolia*, (see Plate 42) provides the starting point for southern African plant descriptions, when a flowerhead, collected by Dutch sailors, reached the Netherlands in 1597. Here it became the focus of attention of Carolus Clusius (1526–1609), professor of botany at Leiden, who published his observations of it in his *Exoticorum Libri Decem*,[8] an important work that refers to several other plants from Africa.

Of the flowerhead he wrote: 'not a fruit, but the head of a certain very graceful genus *Carduus* . . . it is imbricated by many scales in the way of constantly covering one another . . . Amongst them had appeared something like oblong membranous ligules, plain and rather narrow, of a greyish colour. The upper parts are broader and outside white, but inside of brown colour, and adorned along the margins with very soft hairs of a very elegant purple seam, which after some time degenerates into black. In the middle of the head was a dense heap of very hairy stamens of yellowish colour, between which were hidden narrow pointed spines of 2 inch length. On their lower part was a woolly seed attached which, not unlike the seed of *Nerium*, was dark in consequence of the reddish colour but immature.'[9]

Since Clusius's publication, proteas have come to characterize the Cape's unique flora. They have consequently acquired considerable importance and much of what is known about these plants is due to successive generations of taxonomists' efforts to record and document their diversity in natural history collections. In the 1600s, however, botanical exploration was still in its infancy and reference collections were not yet part of the scientific heritage. Although Clusius had the important distinction of discovering more than two hundred new plants, these were mainly from Spain, Portugal, Austria, and Hungary,[10] and this limited knowledge of the world's flora led him to determine the protea as a thistle (*Carduus*) – a similar-looking but unrelated plant that was known to him in Europe. In addition, the sexual function of the inner parts of the flower had not yet been discovered (by

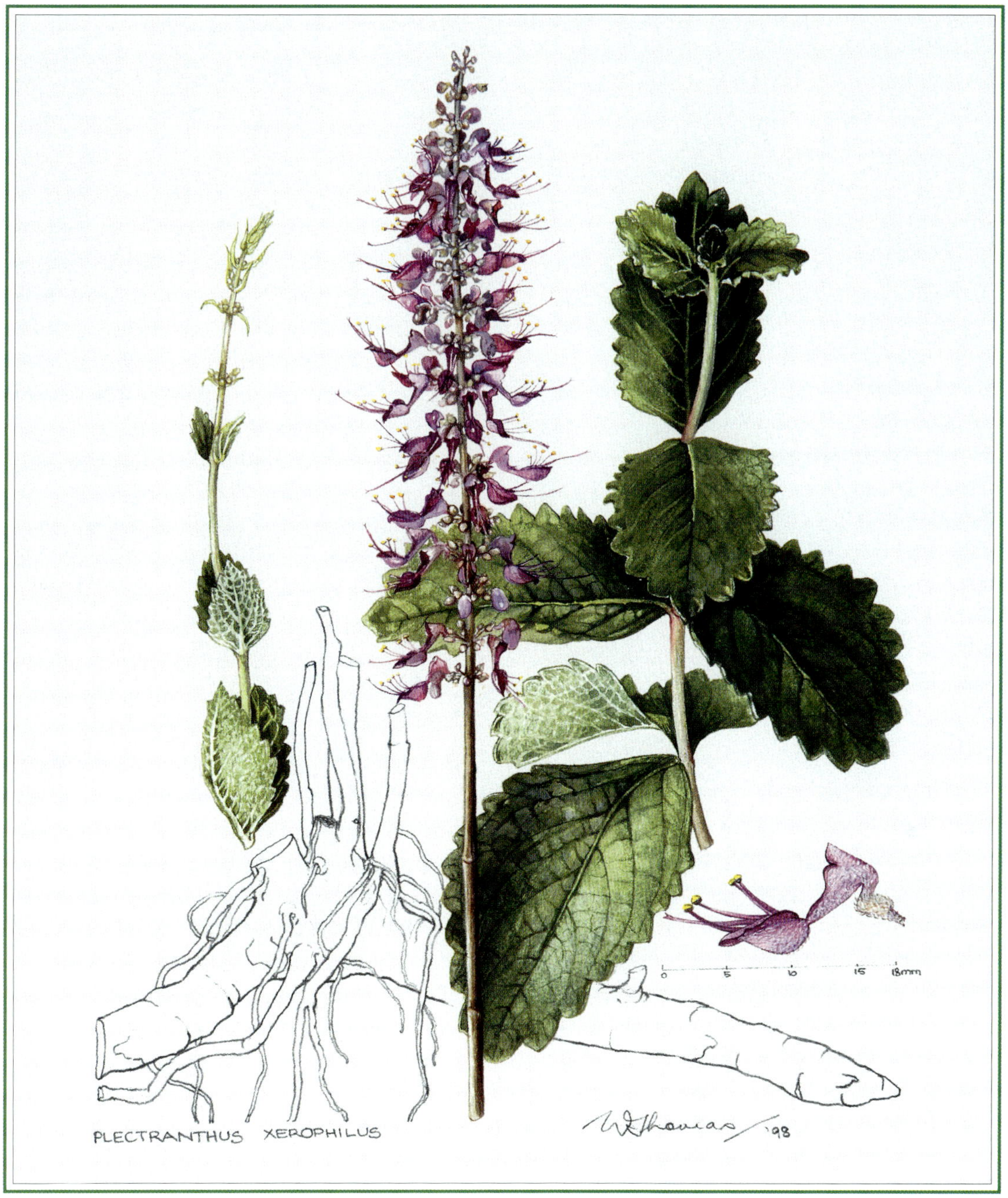

PLATE 41

Vicki Thomas, *Plectranthus xerophilus,*
a bushveld species belonging to the mint family, 1996, watercolour and pencil

Rudolf Jakob Camerarius [1665–1721]), which further limited his interpretation and generally hindered the development of plant description up to the end of the seventeenth century.[11]

Accompanying Clusius's colourful and detailed description is an illustration of the flowerhead, depicted on a blank background (Plate 42). When published, its primary value was most probably one of curiosity. Since then, however, the illustration has been chosen as a type – an important reference point or tag – for the name *'Protea neriifolia'* and, as such, it has acquired special value in itself.[12] Clusius's discipline and enlightened understanding in documenting the flowerhead is made all the more remarkable by the fact that over two centuries were to pass before the plant was actually recognized as the species *Protea neriifolia* by one of Britain's most versatile botanists, Robert Brown (1773–1858).

By the 1700s intellectual curiosity replaced commercial interests as the primary motivation for many European explorers, and several botanists embarked on expeditions to collect and record plants from around the world. At the same time there came a growing realization that natural history collections held great importance for the progress of science. One of these early explorers was Robert Brown, who travelled to Australia on HMS *Investigator* to study the natural history of the southern hemisphere. He called at the Cape on his return voyage to Britain in 1801[13] but, ironically, *Protea neriifolia* was not one of the species he saw. When writing on proteas, his observations of the species were thus limited to pressed plants and Clusius's original description.

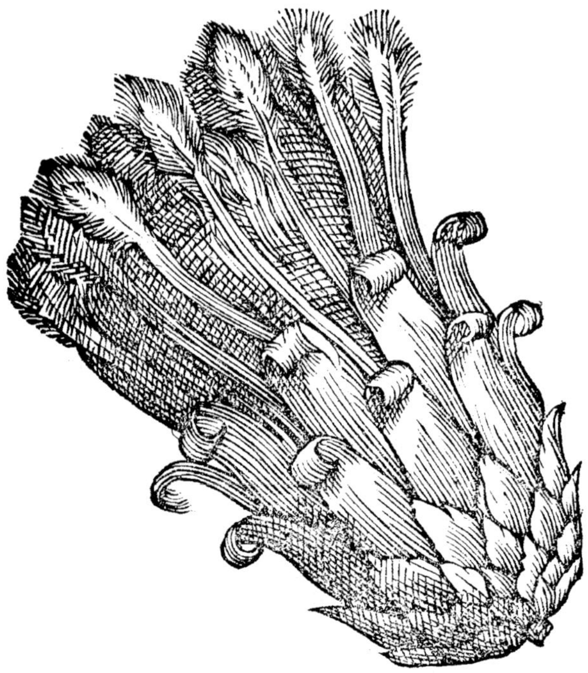

Plate 42

The dried *Protea neriifolia* flowerhead described by Clusius in *Exoticorum Libri Decem* (Leiden, 1605)

Like Clusius, Brown followed the same order of describing plants from their base to their apex and from the outermost to the innermost parts of the flower – a sequence still followed today. He briefly documented his observations of *Protea neriifolia* as 'leaves smooth and linear-lingulate, margins opaque, on the outside with subsimple hairs at the base near the branch, inner bracts of involucre with apex not very broad, outer surface silvery-sericious with tufts of glossy black hair on the margin, feathered awns rise on the inside above the blades of the calyx, style hairy. Apparently distinct from the closely related *P. lepidocarpon*'.[14]

Clearly Brown managed to designate structures, such as 'involucre', 'awn' and 'calyx', where Clusius had fumbled for words to describe them. But of greater significance was his new conceptual approach. Whereas Clusius described the features of a single individual, Brown described those shared by many individuals, and these he grouped together in one species. This important change is reflected in his purposeful suppression of verbs – a technique still used in formal descriptions today – that serves to diagnose the species, not the individual plant, as the basic biological unit of study. For the first time, Brown's concise description allowed *Protea neriifolia* to be compared consistently with other *Protea* species, yet his cool, technical approach, compounded by the absence of an illustration, largely removed the new information from plant enthusiasts.

Carl Linnaeus (1707–1778), professor of natural history in Uppsala, Sweden, was the primary force behind this revolutionary course in plant description. During the eighteenth century, he came to realize the need for plant descriptions to fulfil the requirements of direct comparison, one with another, and he duly proceeded to standardize the language of botany. His methods were promptly accepted and proved to be so efficient that in his lifetime he described as many as 9 000 species of plants and animals,[15] of which 1 360 plant species were southern African in origin.[16] Despite the fact that Linnaeus claimed early authorship of a major component of the Cape's flora, his most successful student, Carl Peter Thunberg (1743–1828), who visited

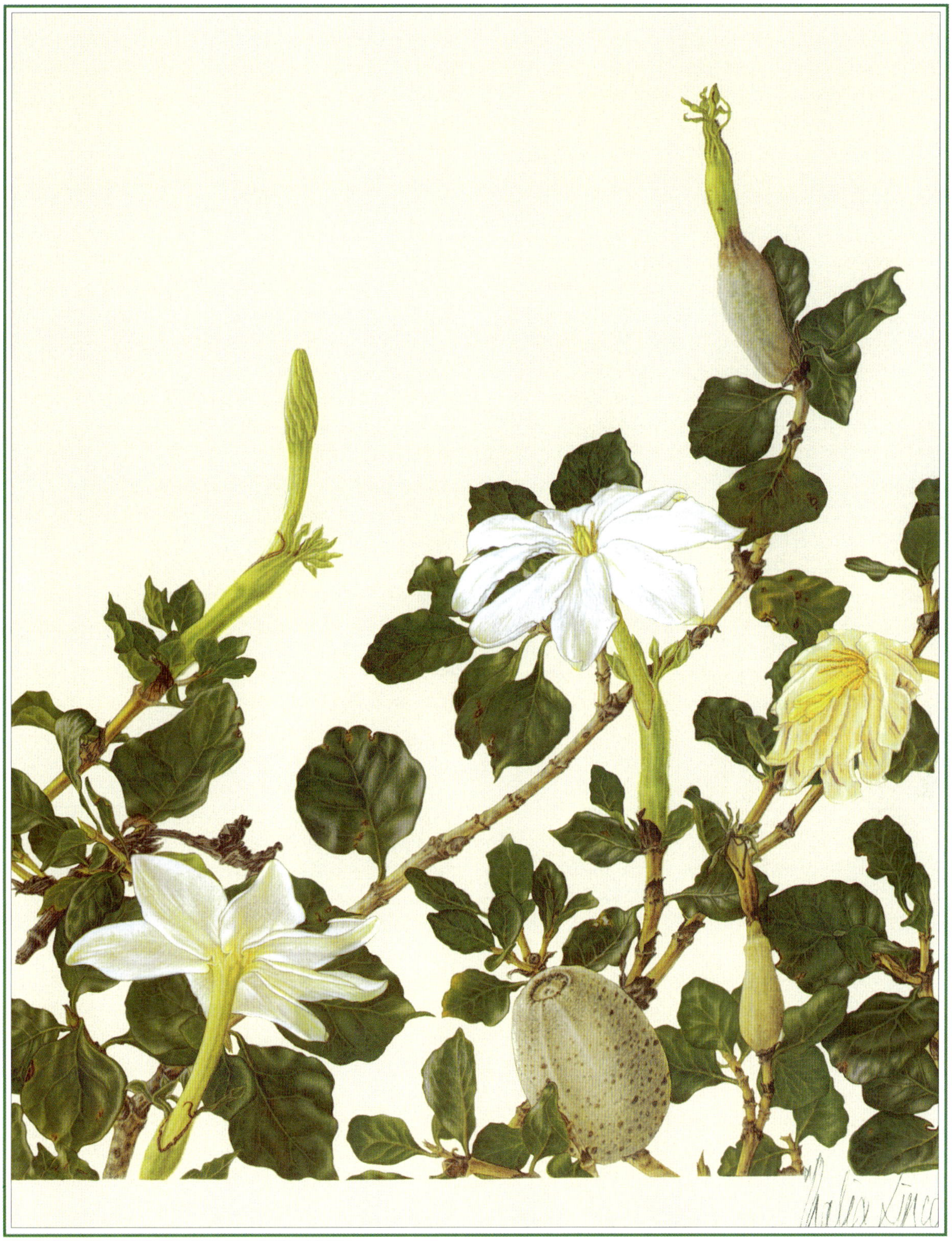

PLATE 43

Thalia Lincoln, *Gardenia thunbergia,* c. 1994, coloured pencils on Bristol board: reproduced in Thalia Lincoln, *Flowers of Southern Africa* (Johannesburg, 1995). See also Edith Struben's representation of this species (Plate 30)

the Cape briefly between 1772 and 1775, managed to describe a further 719 new Cape species, using the Linnaean method.[17] Such was his contribution that he is commemorated in the genus *Thunbergia*, which includes *Thunbergia alata* (the widely grown black-eyed Susan), and in many species, such as *Gardenia thunbergia* (see Plates 30, 43).

From the eighteenth century on, plant descriptions progressively changed from vague but often picturesque accounts to technically rich and accurate treatments but, as many critics have pointed out, they became rather lifeless commentary.[18] This western tradition of plant description spread everywhere, until taxonomists around the globe worked from the same set of terms, speaking a common botanical language. It was through these developments at the end of the eighteenth century that botanical Latin took root and came to be set apart from both classical Latin and the commonly spoken language of the time.[19]

As botany became more technical, the curiosity in plants was kept alive for plant enthusiasts mainly by artists. Prior to the publication of books devoted entirely to southern Africa's indigenous flora, plants of the region were mostly made known through the lavishly illustrated botanical periodical, *Curtis's Botanical Magazine*, which was first published in 1787 and still continues to be published by the Royal Botanical Gardens at Kew in England. From the outset, the small dimensions of the periodical curtailed William Curtis's (1746–1799) choice of subjects but he overcame the problem by occasionally using a larger, folded plate for the artworks. Because of their immediate impact, several southern African plants were among the first species to be shown in this manner. The first, the bird-of-paradise flower, *Strelitzia reginae* (Plate 121), was introduced by Francis Masson (1741–1805), Kew's first collector at the Cape.[20] Other Cape plants that followed soon after were *Amaryllis belladonna*,[21] and *Protea cynaroides*[22] the plant which was to become South Africa's floral emblem.

In response to a growing local appreciation of southern Africa's rich diversity of plant life, the periodical *The Flowering Plants of Africa*, which assumed the same style as *Curtis's Botanical Magazine*, was launched locally in 1921. Over time it mirrored the trends in plant description that developed in the botanical literature elsewhere. Initially, many of the descriptions were copied from existing botanical works, but eventually they acquired originality as new species began to be described in the series. Unlike *Curtis's Botanical Magazine, The Flowering Plants of Africa* has so far only used one folded plate, for the exceptionally tall-stemmed Madagascan species *Aloe helenae* painted by Gillian Condy (see Plate 116).[23] But apart from this example, most artists complied with the periodical's editorial policy and favoured small plants – herbs and succulents – rather than plants with individual forms not suitable for small compositions (16 cm x 21 cm).

Ironically, many of southern Africa's unique plants – those belonging to endemic families – are trees or large woody shrubs.[24] Yet despite their great botanical interest, they have seldom been illustrated. Some exceptions in *The Flowering Plants of Africa* are watercolours of *Brunia* by Stella Gower and Edith Burges,[25] *Glischrocolla* by Fay Anderson[26] and *Erythrophysa* by Betty Connell (Plate 44).[27] Apart from the size of a plant, the choice of species for illustration was further influenced by the research interests of southern African botanists and the ease of transporting living plants to artists in a sufficiently fresh state to be painted. Thus many of Namaqualand's ephemeral annuals – the beetle daisies, diascias and nemesias, which are so admirably suited to illustration – are yet to be featured in *The Flowering Plants of Africa*.

The twentieth century also saw growth in the numbers of locally trained botanists who combined forces with resident artists to produce detailed monographs on families and genera that hold particular importance in the documentation of southern Africa's flora. Such monographs, sometimes years in preparation and based on a firsthand knowledge of plants in the wild, are prime sources of information for all biologists. One of the first was *Ericas in Southern Africa*, illustrated mainly by Irma von Below (Plate 45) and Fay Anderson, with a few small watercolours by the second author, Ted Oliver.[28]

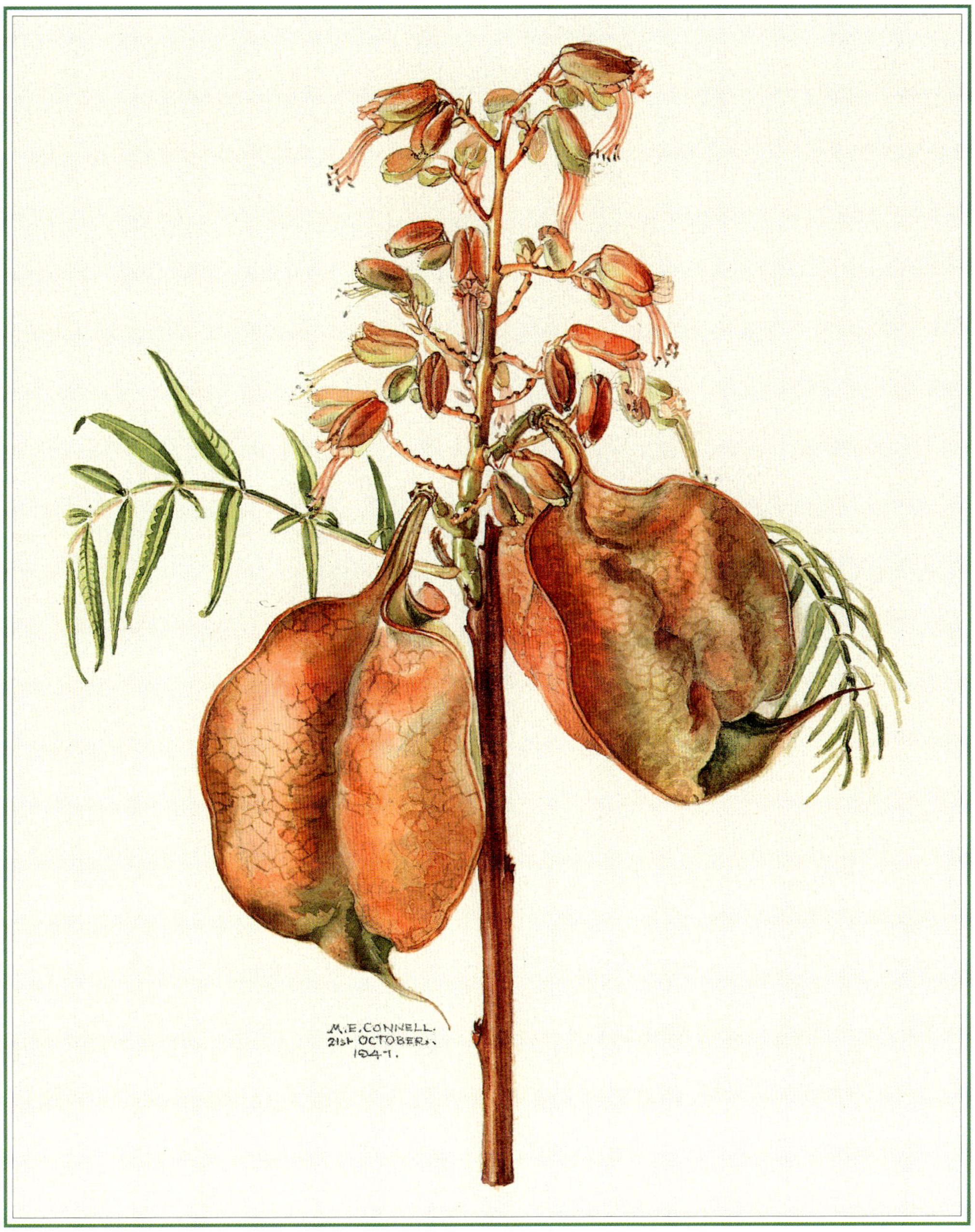

Plate 44

M.E. (Betty) Connell, *Erythrophysa transvaalensis,* watercolour: reproduced in *The Flowering Plants of South Africa,* 1942

Southern Africa is renowned for its richness of ericas. The Cape region alone has more than seven hundred species, sometimes so densely concentrated that as many as five species have been counted in a grid measuring 50 cm x 50 cm[29] – an area just six times that of a single plate in *The Flowering Plants of Africa*.

The publication of *The Genera of the Mesembryanthemaceae*,[30] illustrated by Mary Page (1867–1925), Beatrice Carter (1889–1939) and others, appropriately concentrated on the second largest group of plants in southern Africa – the mesembryanthemum or vygie family. At least 660 mesemb species are known in

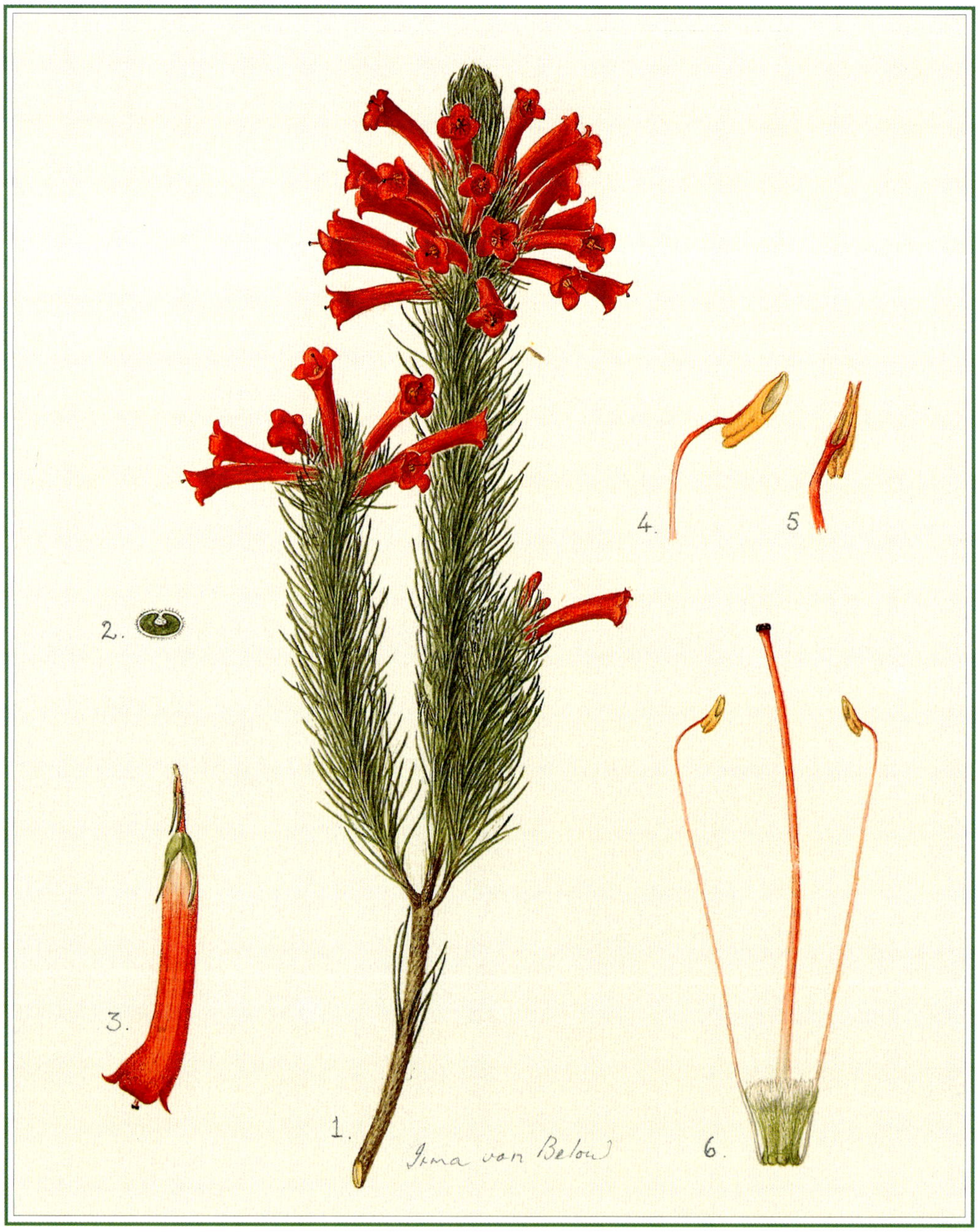

Plate 45

Irma von Below, *Erica vestita,* watercolour:
reproduced in H.A. Baker and E.G.H. Oliver, *Ericas in Southern Africa* (Cape Town, 1967)

the Cape[31] and so many are represented in Namaqualand and the Little Karoo that this entire area is known as the 'succulent karoo biome'. (As a biological subdivision of the earth's surface, a biome is characterized by its particular type of dominant vegetation, its associated animals and ambient climate.) Unlike the ericas, however, the mesembs have relatively uniform flowers – flat and wide open with numerous, radiating, shiny petals – but extremely variable leaves and fruits, which are exquisitely adapted to cope with seasonal aridity. Superficially, many of the fruiting capsules may appear similar but they are intricately and diversely designed to retain their seeds until conditions for germination are optimal – a complex dispersal strategy which has only recently been fully understood. Despite this, Mary Page and Beatrice Carter often included the capsules in their illustrations, not only as an aid to identification but also to reflect their pleasing structural symmetry.

Because of the extra care required in pressing succulent plants, they are usually poorly preserved in herbaria (exceptions are W.F. Purcell's collections in the Compton Herbarium, Kirstenbosch: see Plate 59), and dried specimens need to be complemented by colour photographs or drawings. The publication of Beatrice Carter's and Mary Page's illustrations (see Plates 33, 46 and 115) greatly helped plant enthusiasts and professional botanists gain access to long sought-after information on some of these fascinating succulents, including the stone plants, bokbaai vygies and sour figs that are such well-loved and widely cultivated elements of southern Africa's flora.

Among these twentieth-century monographs is John Rourke's *The Proteas of Southern Africa*, illustrated by Fay Anderson, Lura Ripley and others,[32] a work which shows how a knowledge of plants in the wild gave new momentum to plant descriptions in southern Africa. The account of *Protea neriifolia*, for instance, includes a formal English botanical description (since today Latin is used only when species are described for the first time), followed by a lengthy commentary that fully covers the history, the distribution and the habitat – the distinct environment in which a plant lives – and the cultivation of the species. The text accommodates the precise, technical requirements of botanists but also satisfies the curiosity of plant enthusiasts. We gain new insight into the living conditions of proteas – frequently ravaged by fire on mountains, in coastal plains or in peaty swamps – and their relationships to insects, birds and mice. Fay Anderson's paintings place equal

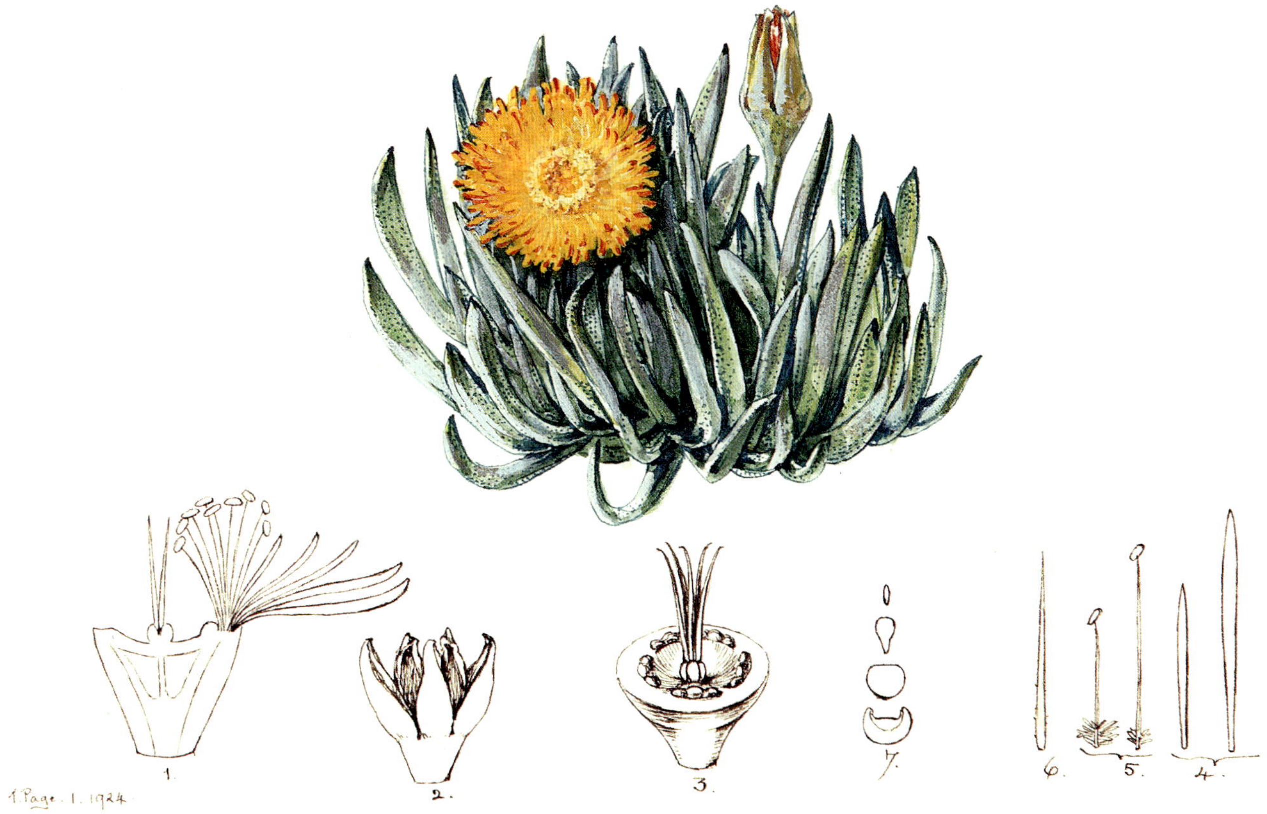

Plate 46

Mary Page, *Hereroa concava,* watercolour and ink

emphasis on the state of proteas in the wild. The leathery leaves of *Protea neriifolia*, for instance, are slightly damaged by herbivorous insects, and this attention to detail gives authenticity to the image, without detracting from the foliage's true form.

Other copiously illustrated monographs produced in the latter half of the twentieth century focus largely on southern Africa's most ornamental plants, particularly its diverse monocotyledonous flora. These range from agriculturally important plants like grasses[33] to highly prized horticultural subjects like *Agapanthus*,[34] red hot pokers,[35] crinums,[36] gasterias,[37] moraeas[38] and watsonias.[39] A group illustrated by Fay Anderson is the woody Iridaceae,[40] particularly interesting in that these plants are the only shrubs in the entire family.

Ironically, although the dicotyledons form the largest part of southern Africa's flora, relatively few groups have been lavishly treated. The few publications so far available are on erythrinas,[41] mimetes[42] and pelargoniums.[43] The artists of these monographs include Thalia Lincoln, Ellaphie Ward–Hilhorst, Fay Anderson, Esmé Hennessy, Gertrude Laurence, Winsome Barker, G. Joyce Lewis and Esther Smith, whose individual styles are as varied as the subjects of study. Some illustrations, like the watercolours of *Babiana* by bulb specialist G. Joyce Lewis[44] (Plate 84), are detached studies that meet little more than the specific needs of scientific description. But many others combine artistic skill with scientific accuracy, and reflect the artists' obvious delight in the plants that have been portrayed.

To focus on plant structure – particularly enlarged floral detail – most artists retained a white background in their work. But with a knowledge that species of *Mimetes* may easily become extinct through too-frequent burning, Thalia Lincoln and John Rourke chose to break with this tradition and were unstinting in their efforts to depict all the known mimetes in their natural landscapes (Plate 107). To achieve this, they hiked long distances to see the remote montane communities where mimetes had managed to survive. John Rourke's aim was to assess the variation within the populations, and possibly to observe their visitors – Cape sugarbirds and Orange-breasted sunbirds – while Thalia Lincoln's purpose was to capture the spirit of the plants that she selected to draw.[45]

The work of Auriol Batten, whose combined knowledge of botany and art is probably unsurpassed by her contemporaries, reveals not only an intimate knowledge of plants in the wild but also a concern for close collaboration with botanists. Species of *Gladiolus* are notoriously difficult to name – and paint – often because their flowers are so transient. But through close contact with iris specialists Peter Goldblatt and John Manning, who studied the diagnostic features of gladiolus,[46] Auriol Batten achieved images of depth and clarity that capture the ephemeral beauty of their flowers and provide sufficient detail to assist the reader to interpret the formal written descriptions (Plate 88). She devoted equal care to *Dierama: the Hairbells of Africa*.[47] Her field observations in this study enabled her to include pencil sketches of the species' growth habit and habitat together with the detailed colour paintings of the plants themselves, so adding a further dimension to Olive Hilliard and Bill Burtt's text (see Plate 48).

Interestingly, in a review of the past two hundred years of botanical paintings at the Royal Botanic Gardens, Kew, the particular absence of any style that was developed there to show plants in their natural habitats was noted – an omission which is believed to stem from the perception that such works constitute a branch of landscape painting that lacks the rigour and precision of 'real' botanical illustration.[48] Unlike their European counterparts, Thalia Lincoln, Auriol Batten and their southern African botanical colleagues were uniquely placed to work close to nature. They seized the opportunity of presenting the whole view of plants – without compromising on their accuracy – and of combining their texts and illustrations to convey something of the mutual interdependence of southern Africa's plants. As a result, southern Africa today is richly endowed with several botanical monographs all of which allow one to perceive plants anew, not only as bearers of beautiful flowers but also as part of the intricate fabric that constitutes the region's biomes.

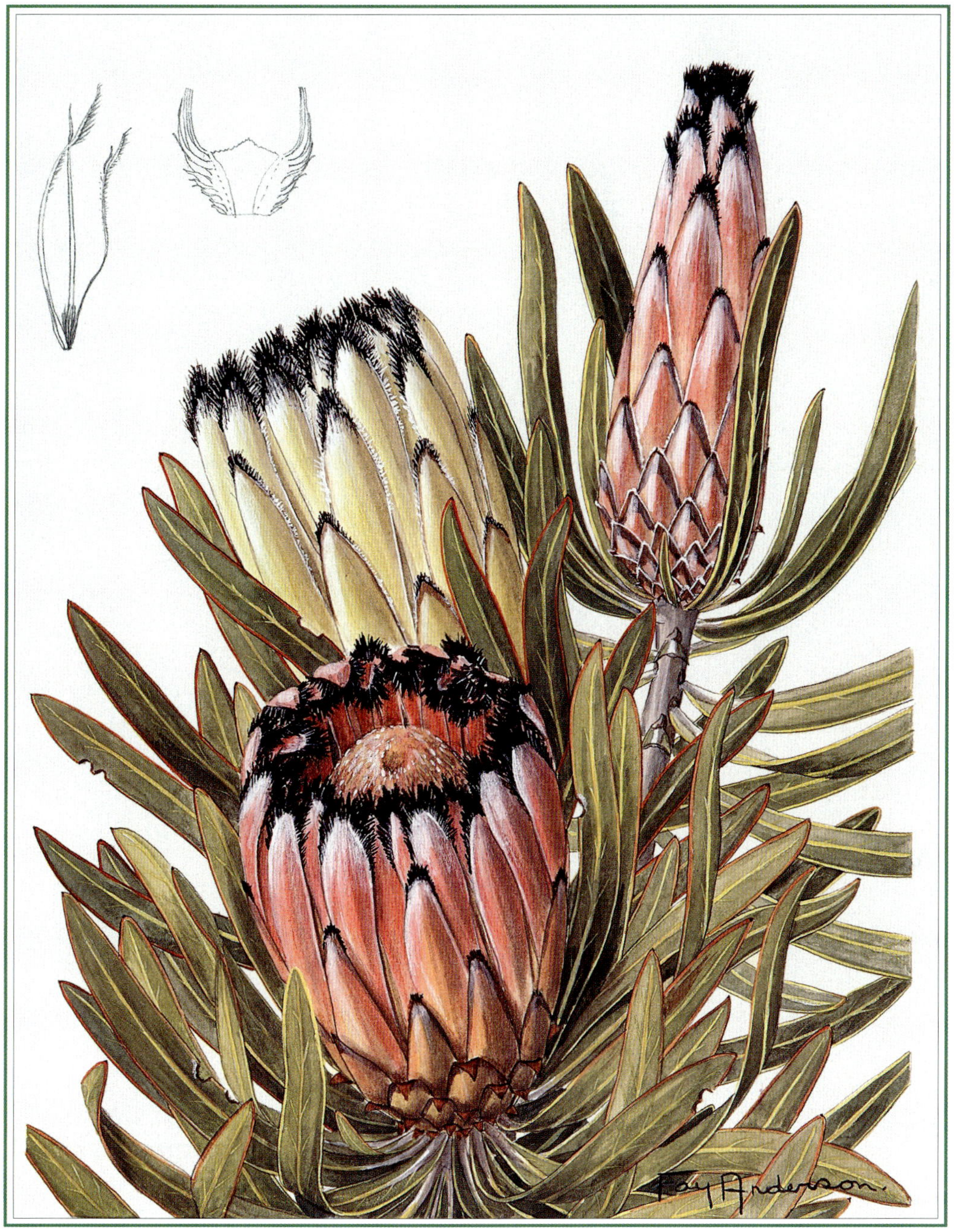

PLATE 47

Fay Anderson, *Protea neriifolia*, watercolour and pencil: reproduced in J.P. Rourke, *The Proteas of Southern Africa* (Cape Town, 1980)

The quest for order

Plant description leads inexorably to plant classification, the method of ordering vast amounts of information into a usable system. Contemporary systems of plant classification are the result of the labours and insights of numerous botanists over several centuries. During the past two centuries in particular, the classification of plants has undergone the greatest change. Several trends, however, are evident: the gradual development of a hierarchy of plant groups; the gradual recognition of major plant families; the

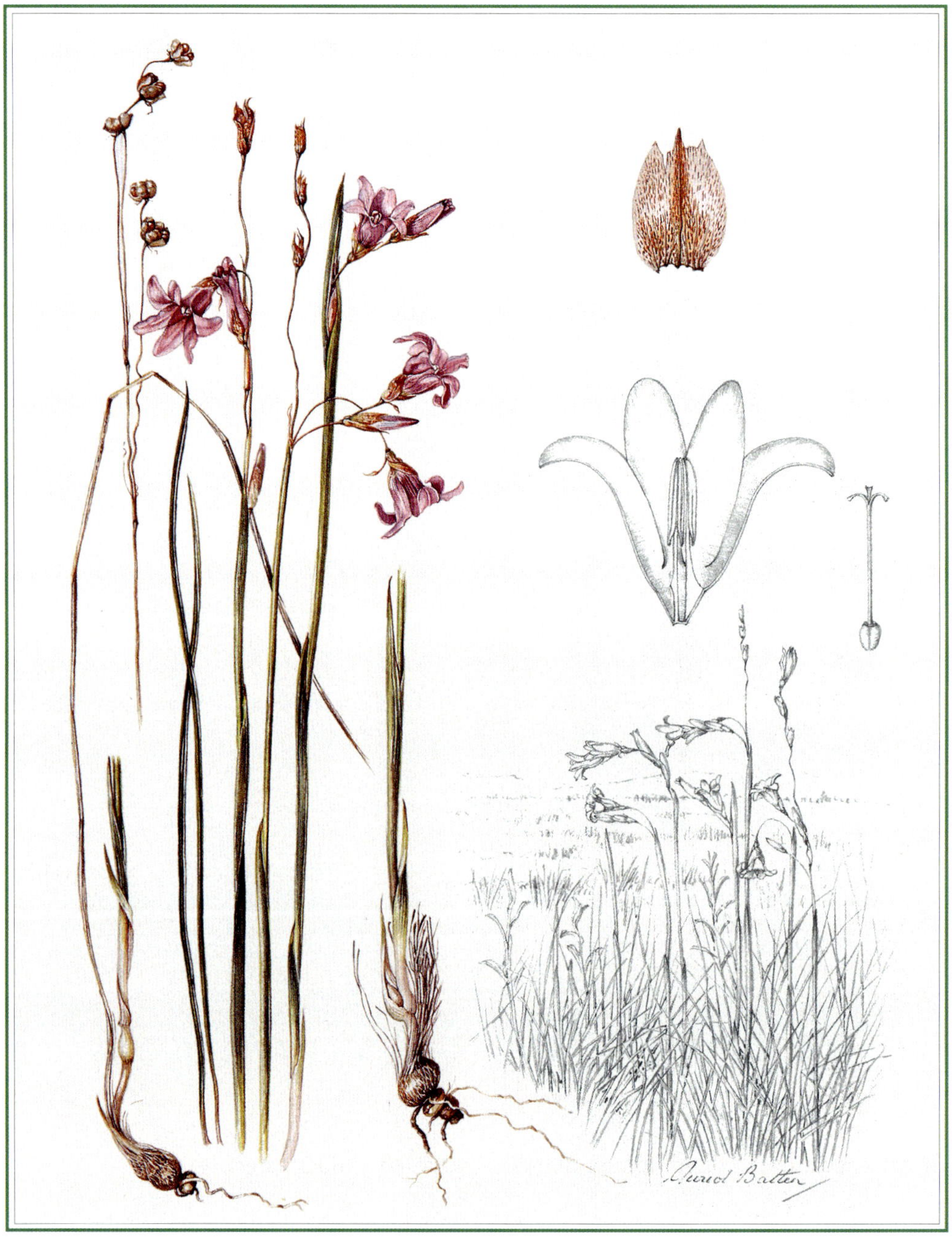

PLATE 48

Auriol Batten, *Dierama trichorhizum,* 1987, watercolour and pencil: reproduced in O.M. Hilliard and B.L. Burtt, *Dierama: the Hairbells of Africa* (Johannesburg, 1991)

use of increasing numbers of characteristics to classify plants; and fundamental changes in the philosophy regarding the purpose of classification.

By the mid eighteenth century, Linnaeus had became the first botanist to provide an internationally acceptable taxonomic framework in which plants (and animals) could be arranged into classes, orders, genera and species. In so doing, he grouped flowering plants into 23 classes (based on the number, union and relative length of the stamens or male organs), each class being subdivided into orders based on the pistils (or female organs) in the flower – the traditional sequence of studying the flower from the outside to the inside. The orders in turn were divided into genera and species. The stimulus to focus on floral parts, and to regard them as more important than other characteristics for taxonomic purposes, arose from other advances made during the seventeenth and eighteenth centuries – the development of glass lenses to magnify the fine detail of specimens, and the discovery of the sexual function of the inner parts of the flower. Whereas Linnaeus is believed to have used these numerical characters for simplicity and precision, subsequent botanists, like Antoine Laurent de Jussieu (1748–1836), justified their own choice of stamens and pistils as characters for classification on the grounds that they perform the highest function of nature – to reproduce.[49] Linnaeus's approach to classification – a rigid method based on a few, discrete characteristics – has since been discredited, but his conceptual framework for hierarchical classification is still in use today.

As classification systems came to be based on detailed floral structure, the practice of illustrating plants as a whole – including the roots that were of special importance to herbalists in the sixteenth century – diminished. Copious details of floral parts, essential to the plant's identification, began to be added to paintings and drawings. This influence is still seen in many illustrations today, particularly in representations of bulbous plants, where the underground organs are either omitted or not given the same attention as the reproductive parts. Ellaphie Ward-Hilhorst's paintings of bulbous plants, especially her *Haemanthus* portraits,[50] are exceptional examples of how all parts of the plant can be perfectly balanced. In her watercolour of the north-western Cape species, *Haemanthus dasyphyllus*, for instance, not only are the bulb, leaves, flowers, and fruits depicted, but also the uniquely wrinkled contractile roots that maintain the bulb at an optimal depth below ground are recorded (Plate 61).

During France's political revolution of 1789, a new philosophy arose in that country, and botanists came to question the purpose of systematic classification. Michel Adanson (1727–1806), one of the most original botanical thinkers of that century (and after whom the baobab *Adansonia digitata* is named), created a scheme based on the equal use of as many features of the plant as possible. Adanson's assumption that no feature is more important than any other resulted in the first major classification to be intentionally natural in its approach, so species that looked alike were grouped together on the basis of a suite of characteristics. The word 'natural' has held its own special meaning in different phases of human culture, but during Adanson's time, the word was used in relation to patterns of thought. A natural classification – one that would allow the most inductive statements to be made about the things classified – was thus free of hypotheses to explain phenomena that are not open to direct observation.[51]

A further development in France, initiated by Pierre Magnol[52] (1638–1715), was the recognition of plant families, a category in classification that encompasses genus and species. The use of

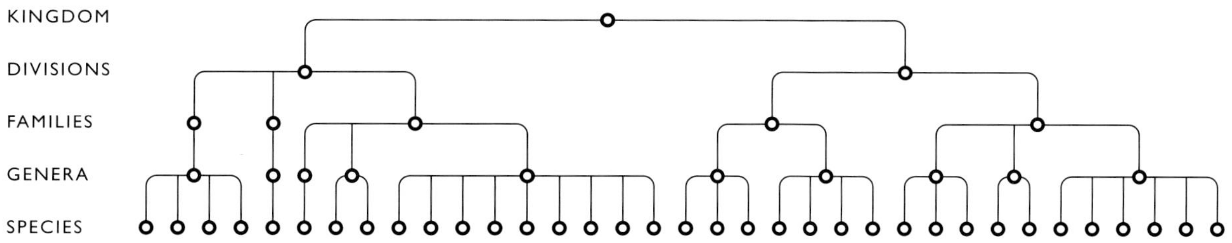

PLATE 49

Hierarchical structure of the Linnaean system of plant and animal classification in which species are grouped into successively more inclusive categories (genera and families) to denote relationships

family names, however, originated somewhat later, from Jussieu's *Genera Plantarum*.[53] Families were intended to reduce the important units in classification to a number and size that could easily be memorized.[54]

The idea of a natural system for botanical classification had become thoroughly entrenched in the minds of botanists by the middle of the nineteenth century and the last and the best elaborated of the pre-evolutionary natural systems was conceived by George Bentham (1800–1884) and Joseph Hooker (1817–1911).[55] Based primarily on floral structure – particularly the arrangement of the flowers on the stem, floral symmetry, the degree to which parts are fused together, and the position of the ovary – the system for the dicotyledons, for instance, starts with plants with many, free, floral parts, as in *Anemone* (the ranunculus family) and ends with the simple-flowered *Salix* (willows), which lack petals and have few stamens.

In the mid nineteenth century Charles Darwin (1809–1882) had proposed that all species have descended from a common ancestor [56] – one of the most significant advances in science of the past 150 years. Consequently, new classifications were developed which could be explained along evolutionary lines. The scheme of Heinrich Engler (1844–1930) and Karl Prantl (1849–1893),[57] for instance, is the reverse of Bentham and Hooker's. They assumed that simple structures were primitive. To reflect this, their system starts with the simple-flowered willow family and ends with the daisy family, whose flowers are typically arranged in dense, flattened heads and have petals fused together – characteristics which were perceived to be the most complex of all the flowering plants (see Plate 51).

Changes in classification are reflected in several local Floras, which are essentially descriptions of all the wild plants of a country, compiled in such a way that readers may easily identify any plant within that region. The *Flora Capensis*, the first comprehensive reference on southern Africa's flora to be written in English, adopted Bentham and Hooker's system and became the standard work on the region's plants for about fifty years from the end of the nineteenth century.[58] One of the most interesting local Floras based largely on the Englerian system is Rudolf Marloth's *The Flora of South Africa*.[59] Its 4 volumes contain 180 colour plates, the originals from which they were made having been painted from living plants by Ethel May Dixie, Esther Smith, Florence Thwaits, Millicent Franks and Peter McManus. It also includes photographs taken by Marloth himself.

With the pioneer publications by Warming [60] and Schimper,[61] the new philosophy of ecology – the study of plants in relation to their environment – found expression in Marloth's *Flora*. Several of the colour plates show some of the unique relationships that Marloth had noted to exist between certain groups of plants, birds and insects (see Plates 52, 68 and 96). In contrast, the photographs emphasize the concept of plant communities – recurrent assemblages of plants which form an obvious feature of a landscape. Although not commonly used in combination, the colour paintings and photographs in Marloth's *Flora* complement each other well – the paintings focus on structural detail of individual species whereas the photographs reflect plant associations. Interestingly, several of the photographs provide us with a glimpse of some early southern African landscapes, as yet untouched by agriculture, afforestation and urban development.

Classification continues to preoccupy the thoughts of botanists, and schemes are still being sought that will allow the evolutionary relationships of plants to be better revealed. In the last part of the twentieth century, cladistics, developed by the German entomologist Willi Hennig,[62] became the primary method to derive phylogenetic classification – that which reflects the evolutionary history of a group of plants. With the extensive technological advances of the present, great amounts of information have been accumulated on plants, ranging from anatomy to genetics and chemistry. Cladistic methods use as much of this information as possible to construct an evolutionary tree from the features of the plants to be studied. These features are ordered into ancestral and advanced conditions and the species that share advanced attributes are grouped together.

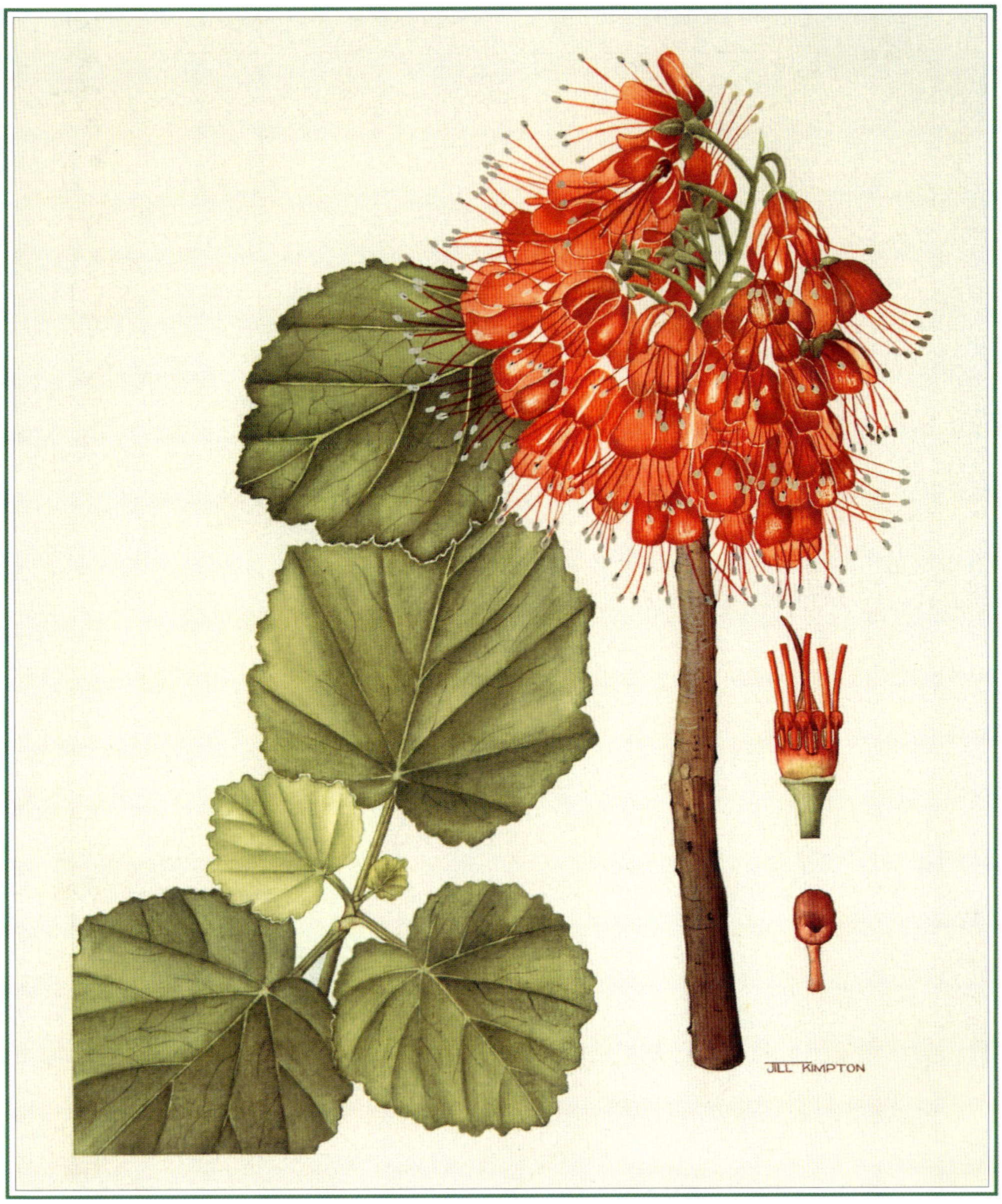

PLATE 50

Jill Kimpton, *Greyia sutherlandii,* watercolour: reproduced in *The Flowering Plants of Africa,* 1977

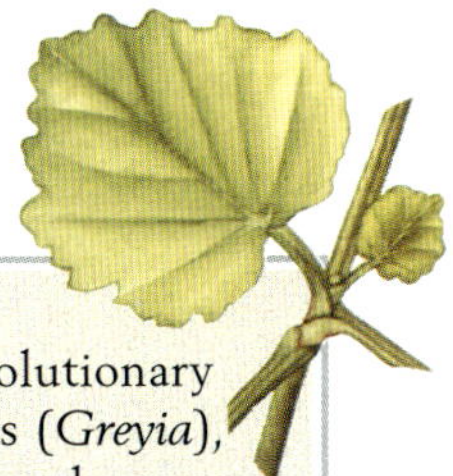

Since the size of plant families today is not determined by convenience but rather by evolutionary affinity, some like Greyiaceae, which is endemic to southern Africa, includes only one genus (*Greyia*), whereas the daisy family, the largest in the world (and southern Africa), contains over a thousand genera. The earliest plant families to be recognized as such were those which had long been associated – some for centuries previously – as symbols in medieval art and literature. The more powerful the symbol, the larger was the family – as in Rosaceae and Liliaceae. Also predominant were the Brassicaceae (which includes cabbage, mustard and many other crop plants) and Apiaceae (the carrot and parsley family). In contrast, families that were of low cultural importance, such as Cyperaceae (the sedge family), were given little recognition. The family is still the focal point in plant classification and several hundred are now recognized.

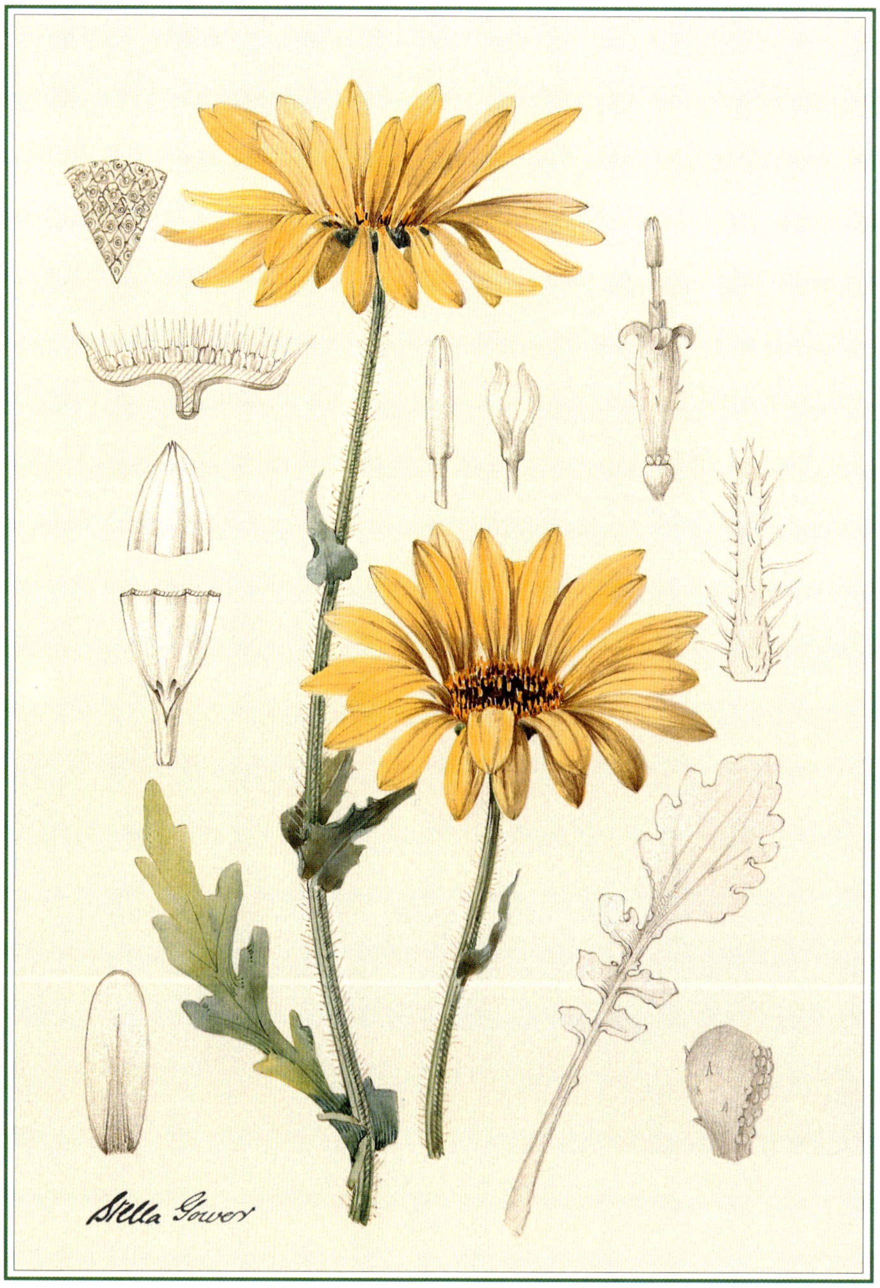

PLATE 51

Stella Gower, *Arctotis fastuosa* (= *Venidium macrocephalum*), a well-known Namaqualand daisy, watercolour: reproduced in *The Flowering Plants of South Africa,* 1923

At this stage of analysis, the basal branches of the evolutionary tree remain unresolved but they are known to include the magnolias and nymphaeas (water lilies) since these plants have many more ancient and primitive features than other flowering plants (see Plate 53). Some primitive features in magnolias are the wood (which is without specialized vessels to conduct water), the flowers (which have spirally arranged free parts on an elongated axis), and the incompletely closed fruit walls. Among the dicotyledonous plants, well-known families like the Ericaceae (erica family), the Lamiaceae (mint family) whose species have been so admirably painted

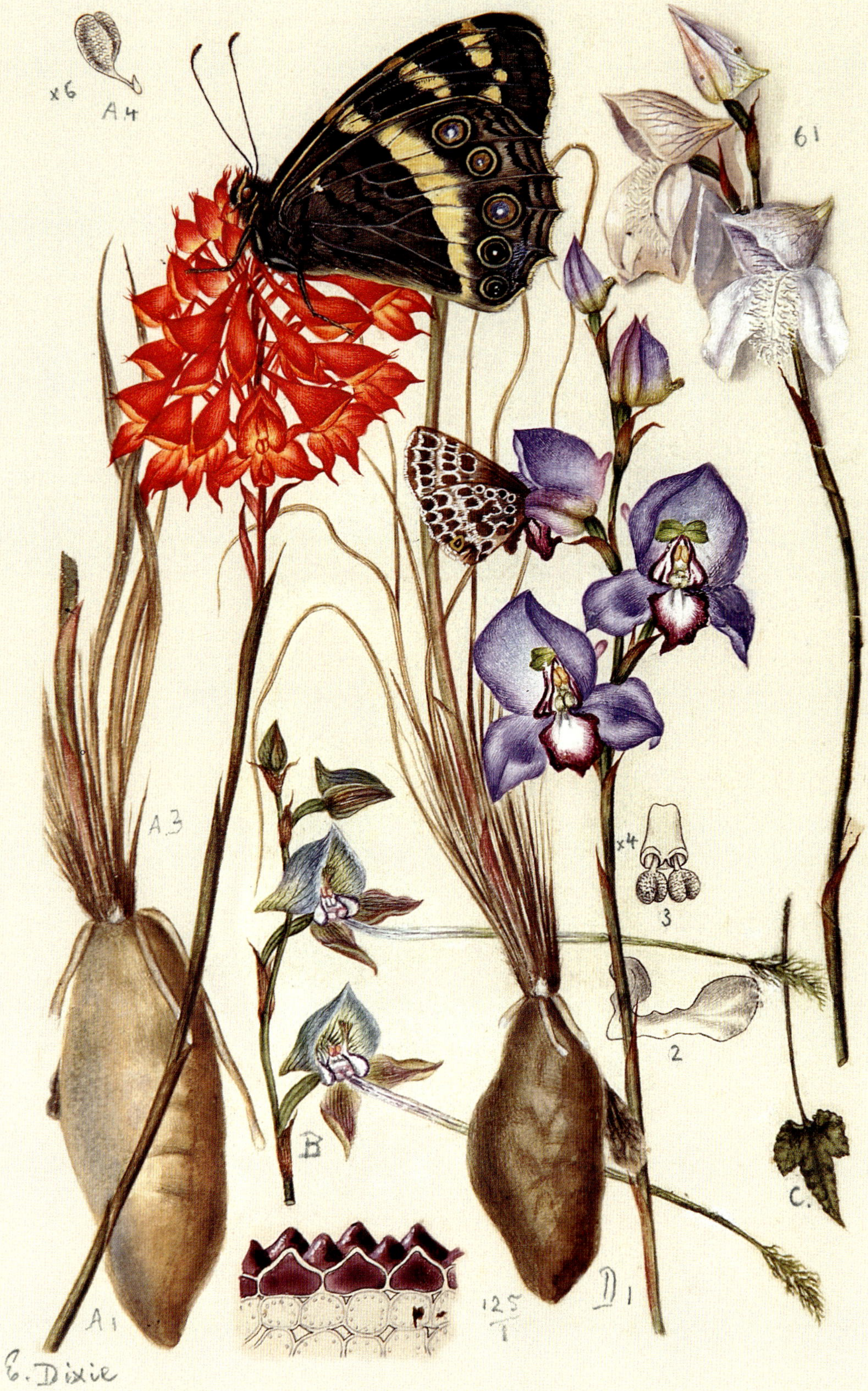

PLATE 52

Ethel Dixie, Cape orchids and butterflies A: *Disa ferruginea* with *Aeropetes tulbaghia* (Table Mountain Beauty); B: *Disa multifida* (= *D. charpentieriana*); C: *Disa spathulata;* D: *Disa graminifolia* with *Lepidochrysops trimeni* (star blue); E: *Disa barbata;* watercolour and pencil: reproduced in Rudolf Marloth, *The Flora of South Africa*, vol. 4 (Cape Town and London, 1915)

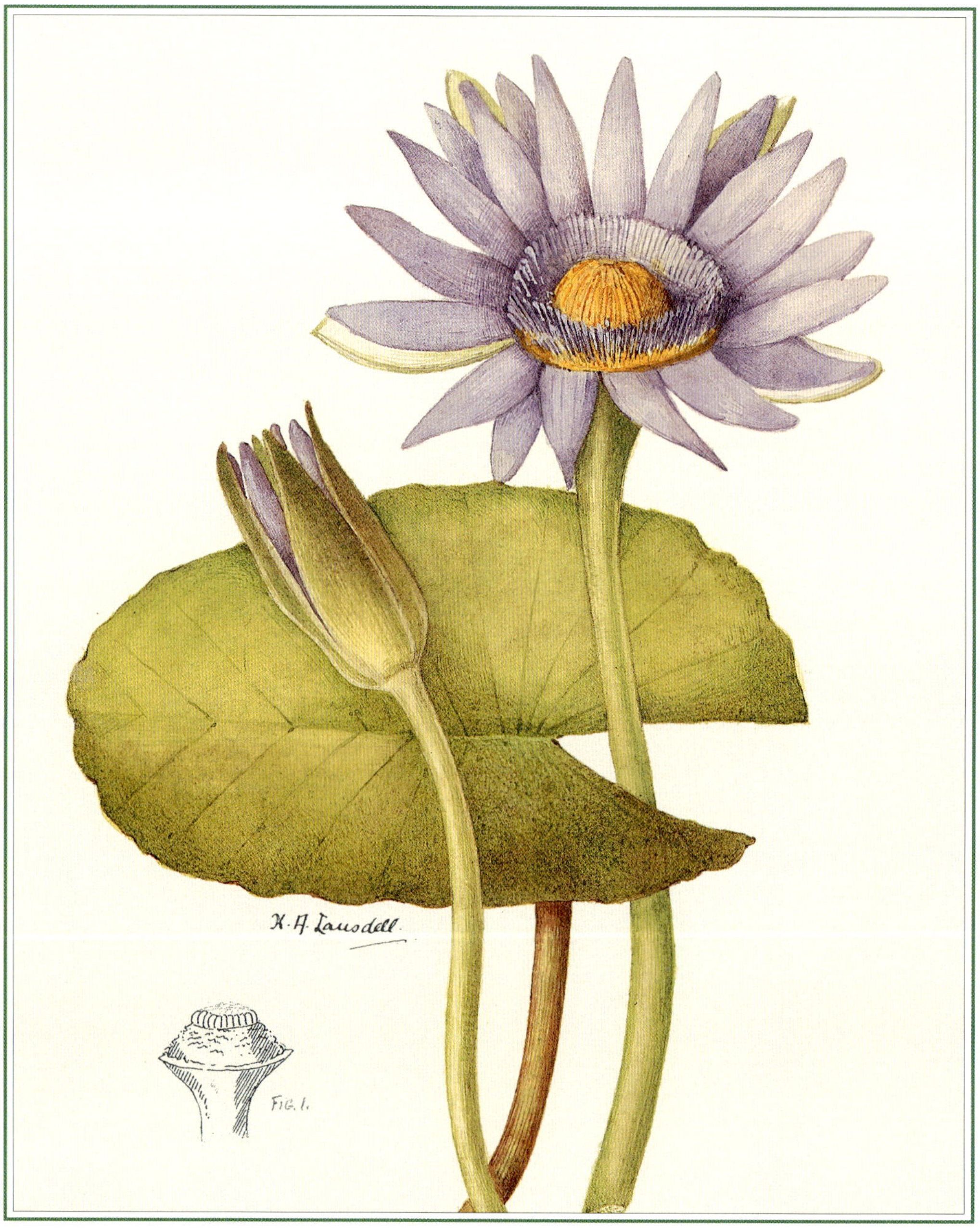

PLATE 53

Kathleen Lansdell, *Nymphaea nouchali* (= *N. stellata*), watercolour and ink: reproduced in *The Flowering Plants of South Africa*, 1921

by Vicki Thomas (see Plate 41), the Apiaceae (carrot family) and the Asteraceae (daisy family) remain most highly placed on the evolutionary tree. In the monocotyledons, some of the plants that occupy the terminal branches are the Commelinaceae (spiderwort family), the Zingiberaceae (ginger family) (see Plate 40) and the simple-flowered Restionaceae (see Plate 54) and Poaceae – the reed and grass families – (see Plate 55) which are highly specialized for wind pollination. With few exceptions all these families are well represented in *The Flowering Plants of Africa*.

It can readily be appreciated that the grouping of plants according to their evolutionary

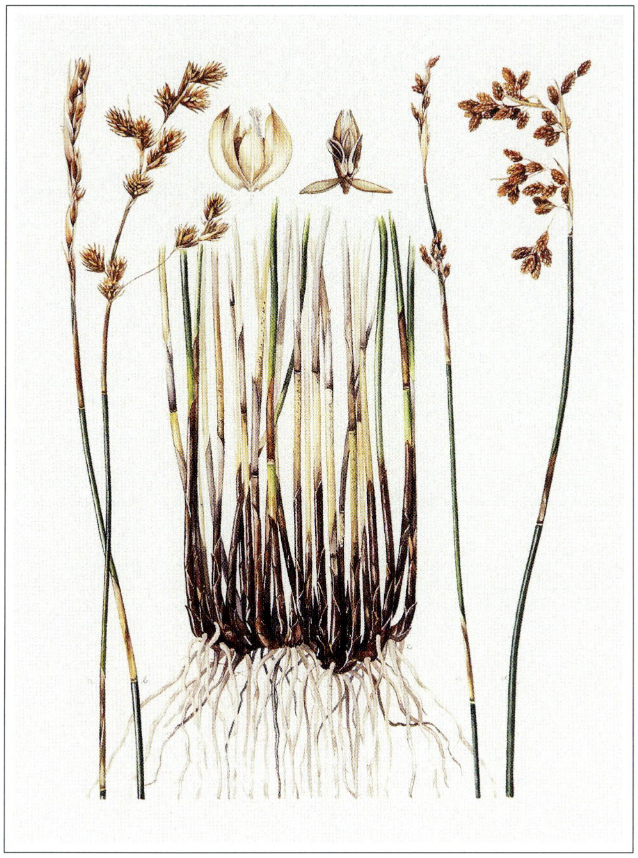

PLATE 54

Vicki Thomas, *Thamnochortus erectus*, 1994, watercolour; restios such as this have rarely been painted, despite their predominance in the Cape fynbos

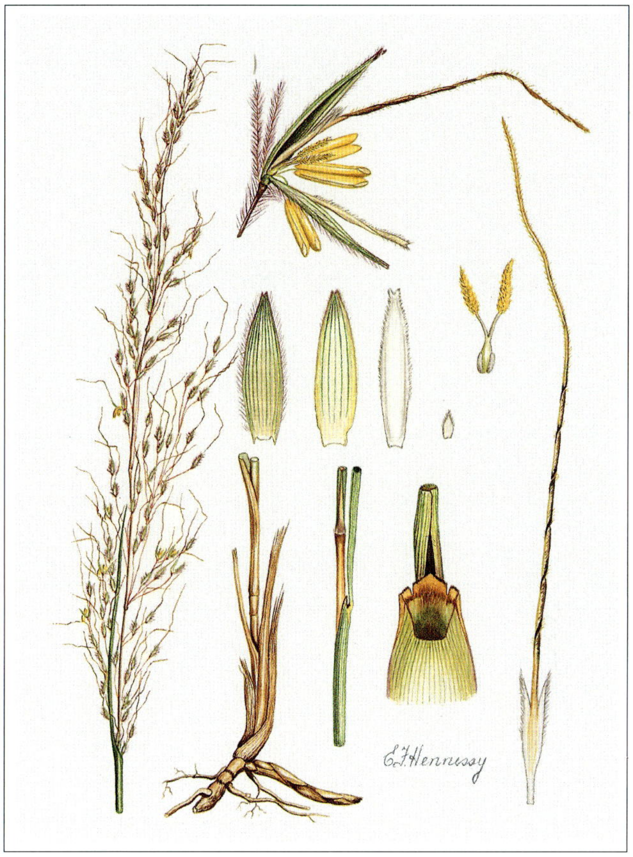

PLATE 55

Esmé Hennessy, *Sorghastrum stipoides* (with floral dissections), watercolour and pencil: reproduced in *The Flowering Plants of Africa*, 1989

ancestry is far more predictive of characteristics than would be a grouping that did not reflect accurately the relationships of those species. This simple concept has profound implications for humankind's efforts to apply knowledge about species. If that knowledge is to be organized in databases to maximize our ability to predict the biological properties of species, and thus use those properties more efficiently and economically, then we will want to store and retrieve that knowledge within the conceptual framework of an evolutionary classification. This underscores the fundamental importance of discovering, to the maximum extent possible, the evolutionary relationships of all species – an objective which motivates most botanists today.

Throughout the ages, the study of the relationships between major groups of plants is matched by many fascinating graphic analogies in which natural relationships are expressed as nets, webs, maps and branching trees. Some have even made use of ribbons to join plant groups which showed affinity one to another, while bows in the ribbons marked the limits of the family.[63] Modern artists working on scientific projects involving many plant families, still face the discipline of arranging images according to the affinities of plants to each other and not in relation to humankind. But it is often within these confines that plant diversity can be most clearly revealed. The arrangement of images to introduce the structural ground-plan of a plant family, with subsequent images revealing the increasingly specialized diversity of form, colour and texture in each, speaks of evolutionary history and invites endless speculation on their

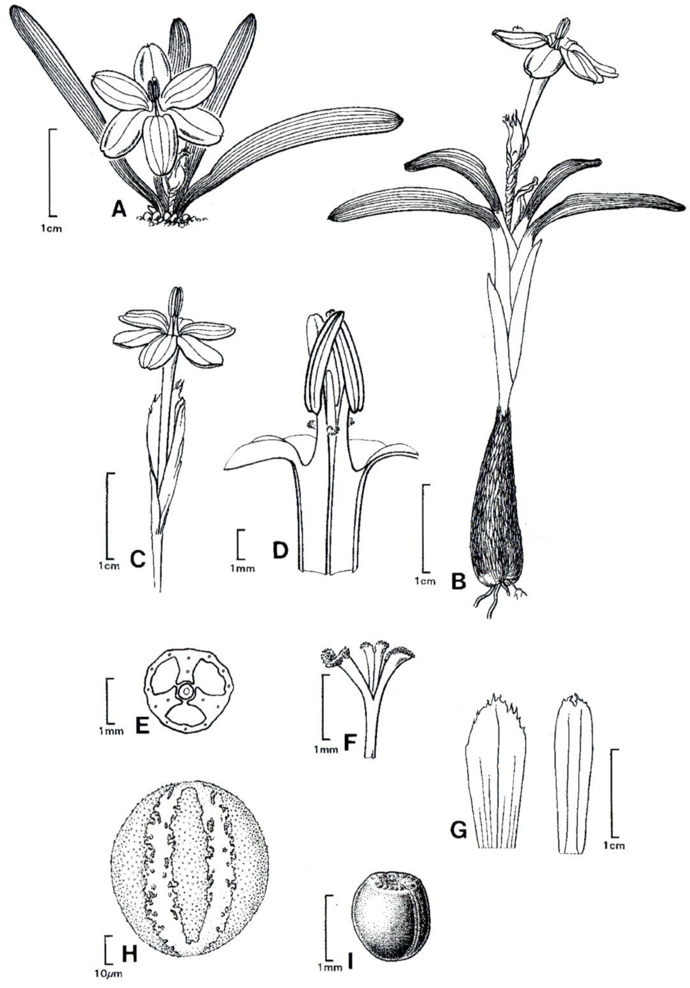

PLATE 56

John Manning, *Ixia acaulis*, 1992, pen and ink: reproduced in Peter Goldblatt and John C. Manning, '*Ixia acaulis*, a new acaulescent species of Iridaceae: Ixioideae from the Knersvlakte, Namaqualand, South Africa', *Novon*, 1993

origins. The success of portraying plant diversity in this way depends, however, upon true collaboration between botanist, artist, and publisher.

Plant collections

Natural history collections are a significant part of the scientific and cultural heritage of all nations. They document past botanical research and at the same time provide the essential basis for botanical research in the future. The attraction of making collections has always characterized humankind: in the sixteenth and seventeenth centuries, for example, the European nobility amassed private collections known as 'cabinets of curiosity' which were kept primarily for their aesthetic and material value.[64] Eventually this gave way to the use of formal natural history collections for scientific study. Luca Ghini (*c.*1490–1556), a professor of botany in Bologna, Italy, is thought to have been the first person to dry plants under pressure and mount them on paper to serve as a lasting record.[65]

The word 'herbarium' originates from the collections of dried herbs that enabled the early European herbalists to identify plants used for medicines, perfumes and flavourings. Initially, the dried plants were sewn into book-like herbaria but, as the exchange of specimens or the depositing of specimens in other herbaria became common practice, this early closed system gave way to an open-ended one of separately mounted sheets stored in folders – a method which is still current. As in the past, the service ethic remains relevant today and continues to strengthen the importance of herbaria to society.

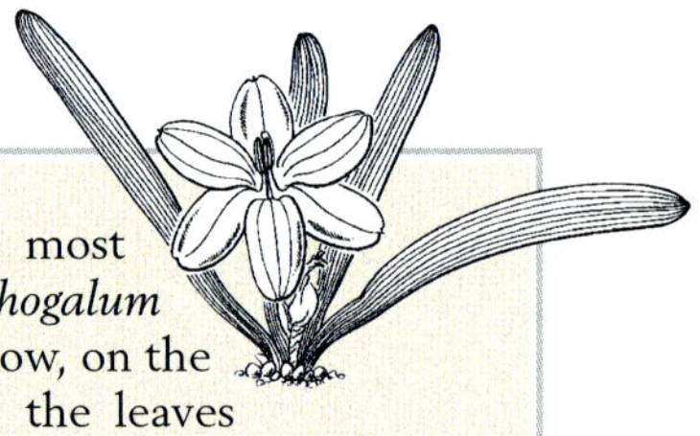

There is growing consensus among botanists about which classifications are most acceptable. Up until 1985 such well-known plants as *Aloe, Bulbine, Ornithogalum* (chincherinchee) and *Gloriosa* were all considered to belong to the lily family. Now, on the basis of the underground storage organs (which vary from tubers to bulbs), the leaves (which are large or rudimentary), the seed, the embryo, the phytochemistry, and the genetic structure of these plants, they are considered to belong to one of several families. *Gloriosa superba*, for instance, now belongs to the family Colchicaceae.

Many fine line drawings accompany the recent monocotyledonous studies, for example this rendering by John Manning of a dwarf *Ixia* (Plate 56). Black and white illustrations may lack the charm of watercolours but the eye of the scientist finds great satisfaction in their clarity and richness of botanical detail.

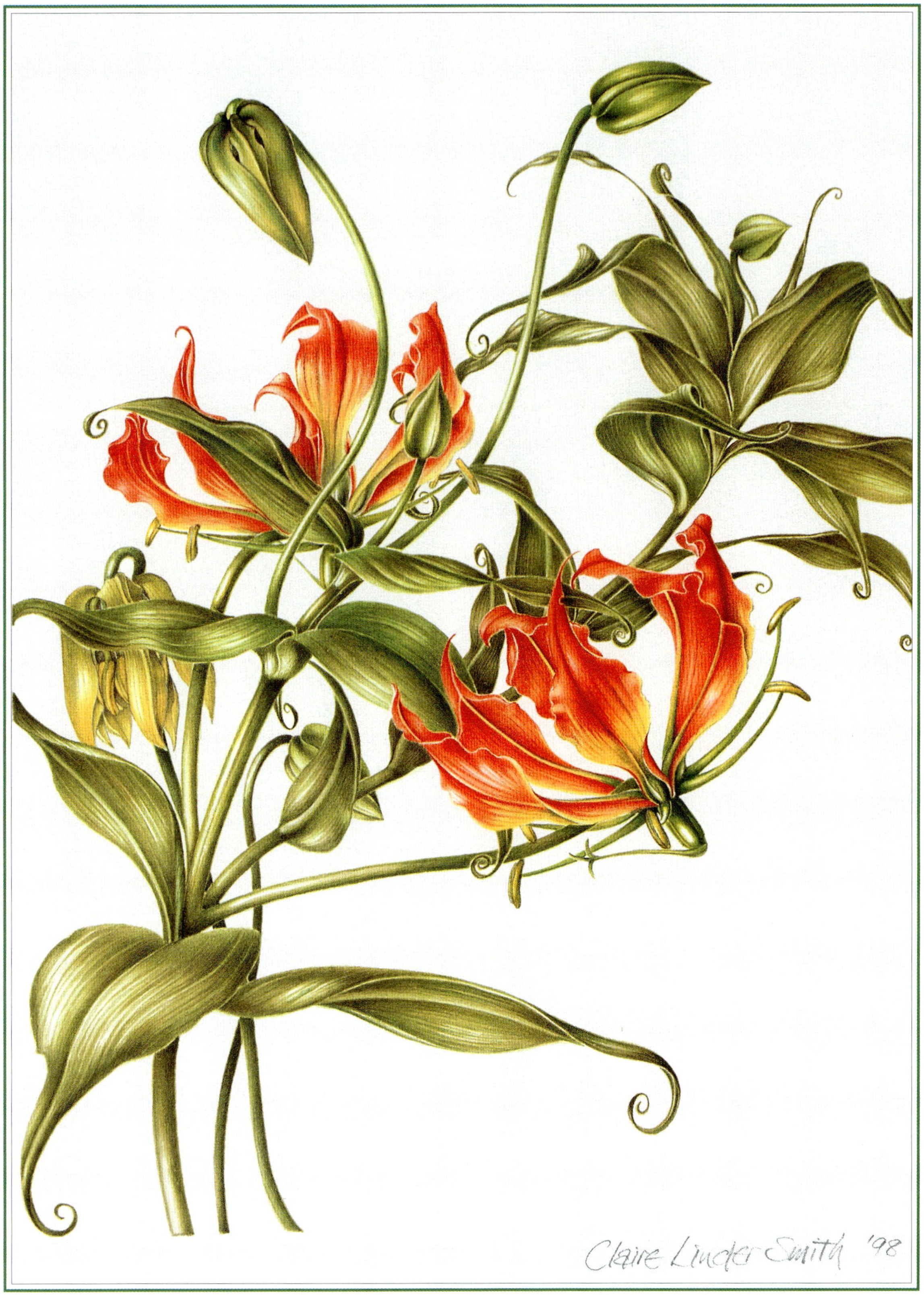

PLATE 57

Claire Linder Smith, *Gloriosa superba,* 1998, watercolour: permanent display, The Blue Train, 1999

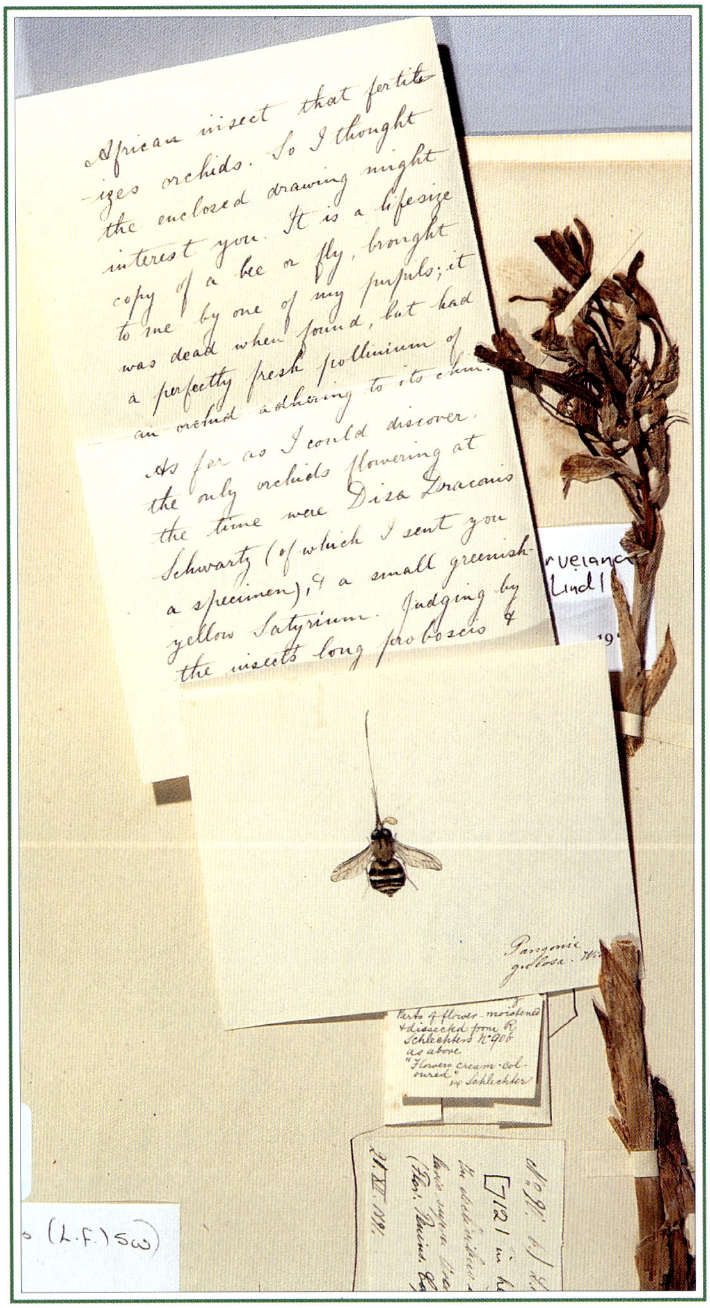

PLATE 58

A herbarium sheet bearing the painting of a horsefly *(Philoliche gulosa),* by Celestine du Plessis, 1911, watercolour, 9 cm x 6 cm

The best herbarium specimens are those which have been most rapidly killed and dried – a practice which has often fuelled the argument that taxonomists mistakenly believe that a plant can be taken from its natural habitat without any significant loss of its essential properties.[66] This denial of the living whole is further exemplified in Linnaeus's rejection of the flower colour, smell and taste as criteria unfit to distinguish species.[67]

Many collections up until the beginning of the twentieth century reflect the Linnaean approach, so their information content is limited almost entirely to plant structure. But a renewed understanding of the plant in its entirety has since ensured that specimen labels today carry abundant information, particularly notes on habitat, ecology, and characteristics of the living plants – such as colour and variegation – which are eventually lost after the plant has been dried. Some plants, however, never lose their smell, even long after they have been dried. Wooden herbarium cabinets that once stored buchus (*Agathosma*) and wild garlic (*Tulbaghia*) remain pleasantly but pungently aromatic many years later.

Herbaria contain not only plant specimens but also drawings and correspondence from the past, which are often of great interest to researchers of the future. During research at the Bolus Herbarium, University of Cape Town, pollination biologist Steve Johnson found a tiny painting of a horsefly accompanied by a letter dated 1911 addressed to orchid specialist, Harry Bolus. The correspondent, Celestine du Plessis, a teacher at the time, wrote: 'I thought the enclosed drawing might interest you. It is a life-size copy of a bee or fly . . . it was dead when found, but had a perfectly fresh pollinium [pollen mass] of an orchid adhering to its chin. As far as I could discern, the only orchids flowering at the time were *Disa draconis* and a small greenish yellow *Satyrium*.' This exciting observation that the beautiful *Disa draconis*, a Cape orchid, is pollinated by horseflies temporarily went unnoticed, but its value was fully appreciated once pollination studies on orchids began eighty years later.[68]

Although plant collecting is vital in the quest

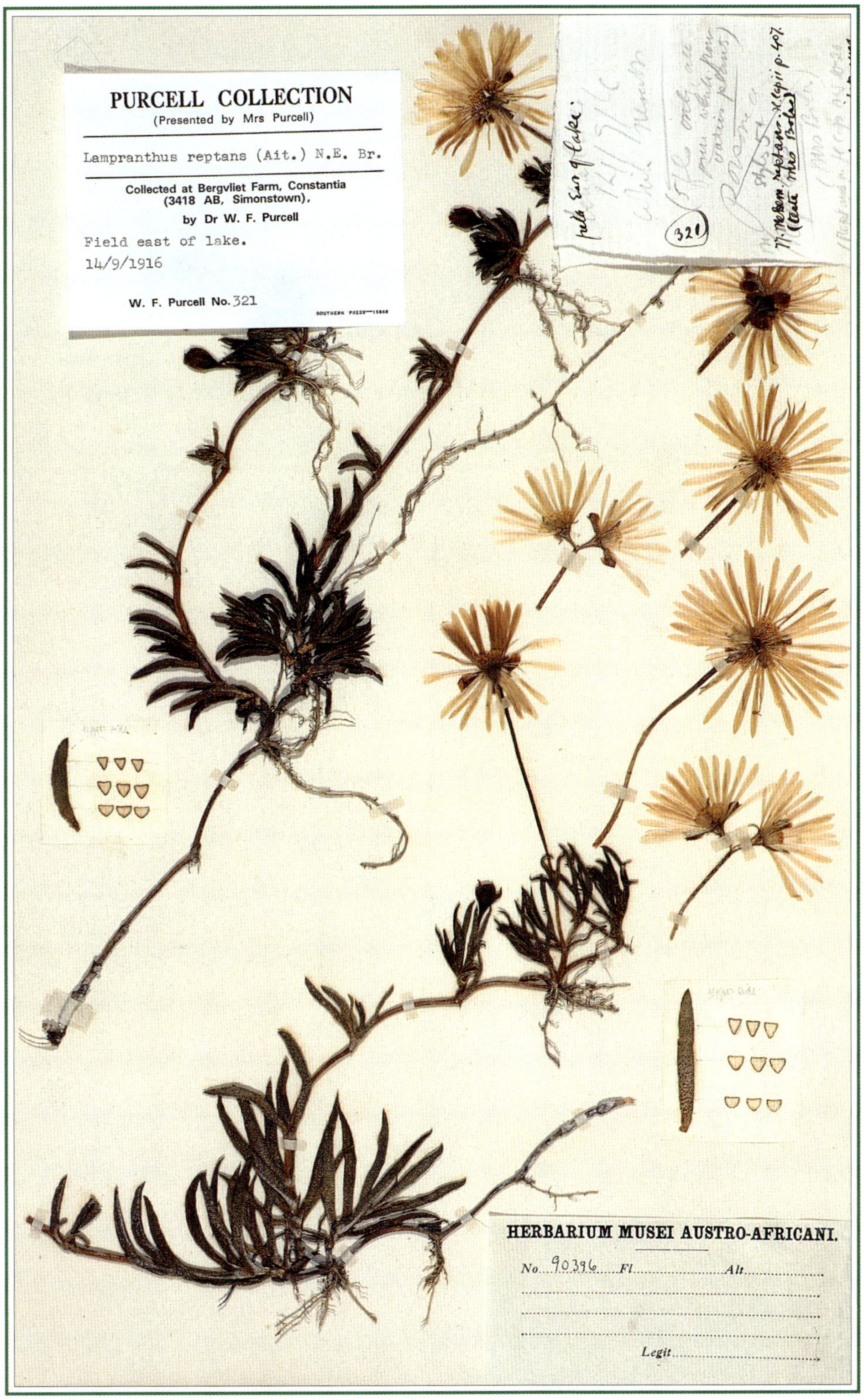

PLATE 59

W.F. Purcell, Pressed collection of *Lampranthus reptans*.
Specimens mounted on herbarium sheets are always laid out to allow maximum observation of the plant material. Good-quality specimens include root systems, vegetative structures, and carefully laid out flowers and fruits in sufficient quantity to display size ranges within the plants.

for knowledge, the practice may sometimes act as a prelude to – and an excuse for – exploitation. The comments of William Hooker (1785–1865), a past director of the Royal Botanic Gardens at Kew, reveal that he anticipated some of today's ethical considerations in removing wild plants from their natural habitats. He noted in 1836 that British gardens and nurseries could pride themselves at one time on having many choice specimens of Cape proteas but that they had been lost through neglect.[69] Two centuries ago it was permissible to collect plants freely in nature but today legislation protects all wild plants in southern Africa and permits are required to collect only limited numbers of plants in their natural habitats. Although protection of species' habitats is most important, the legal protection of attractive and desirable plants like succulents, bulbs and orchids, which are particularly vulnerable to collectors, is also given priority.

Ideally, plant portraits – whether as paintings or photographs – ought to serve in a small way as substitutes for ornamental plants collected from the wild. Yet, they are also known to whet the public appetite for new and more exotic plants. Succulent specialist Steven Hammer was once asked to help count a confiscated shipment of the rare species *Haworthia wittebergensis*[70] which had been intercepted at the Laingsburg post office – boxes there had been marked 'apples' and the astute postal official knew that it was not the season for the fruit. The count revealed seven hundred plants, all of which had been ripped from their rocky habitats – a criminally gross scale of field collecting undertaken to satisfy the acquisitiveness of succulent collectors.[71] Modern botanists, particularly those studying rare species, often try to protect the area they study through collaboration with nature conservation authorities and private landowners, but when publishing their studies they are faced with the dilemma of whether or not to reveal full field data. On the one hand the data may improve the scientific understanding of the plants, on the other they may pose a potential threat to the remaining wild populations.

Interestingly, the composition chosen for mounted plant specimens is often repeated in illustrations, particularly those of shrubs and trees. The elegant forms of palms lend themselves naturally to whole plant portraits but few trees have otherwise been depicted in their entirety. Instead, artists – like their botanical counterparts – often resort to representing trees as fragments, in much the same way that the branches, twigs, flowers and fruits would be spread out on a herbarium sheet, thereby compromising the character of the living plant. A notable exception is Thalia Lincoln, who consistently creates a sense of wholeness by focusing on the flowering and fruiting tips of shrubs and trees, allowing them sufficient room to 'grow' into the space above[72] – as in her drawing of *Gardenia thunbergia* (see Plate 43).

When viewed in this context, Ellaphie Ward-Hilhorst's composite image of *Haemanthus dasyphyllus* (Plate 61), in which the flowers, fruits and leaves are separated from one another, may superficially appear to mirror a herbarium sheet. But a knowledge of the plant's growth habit confirms the accuracy of this interpretation. *Haemanthus dasyphyllus* occurs in the arid north-western Cape, where its success depends on the storage capacity of its large bulb.[73] Within a week of good autumn rain, the plants flower and fruit in quick succession, which is sufficiently early to allow the seeds to germinate and produce leaves to build up the young bulb's reserves during the favourable winter-rainfall season. In Ward-Hilhorst's rendering of the plants, the discrete reproductive and vegetative phases are thus aptly depicted. She shows infinite patience not only in recording the plant's minute structural details – the papery bulb coverings, the barring on the undersurface of the leaves, the waxy floral bracts, and the translucent berries – but also in waiting for the appropriate season to paint them.

Of names and species

The species is the basic unit by which we classify animals and plants and, ultimately, plant portraiture is a representation of the species. Without a name, however, little can be said with accuracy about any species. The

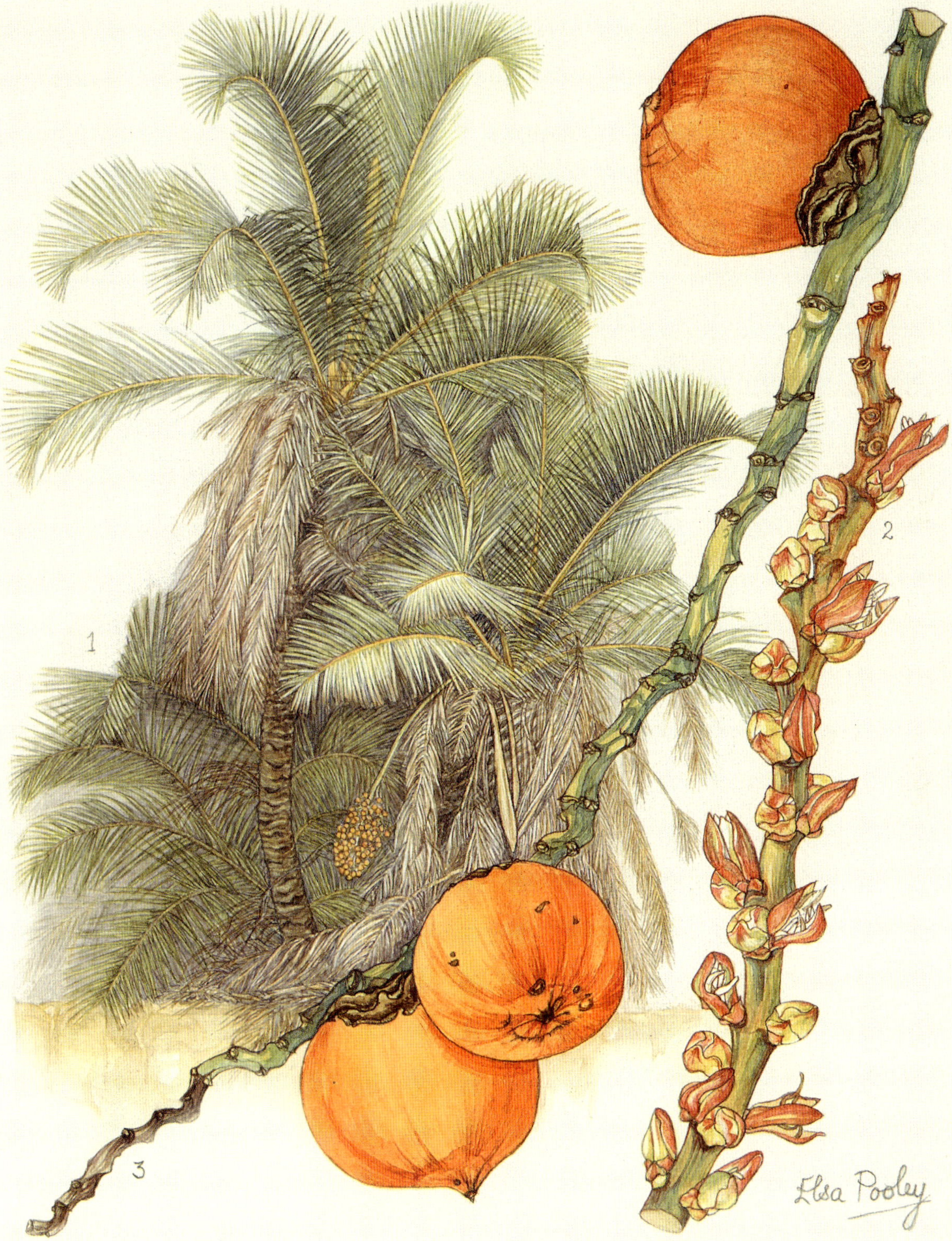

Plate 60

Elsa Pooley, *Jubaeopsis caffra* (Mkambati palm), watercolour: reproduced in *The Flowering Plants of Africa,* 1991

key factor in biology, therefore, is the consistent use of a scientific name – in the form of a two-word name – for each species. The silver tree has the name '*Leucadendron argenteum*', which conveys the concept that the tree belongs to the genus '*Leucadendron*' and has the species epithet '*argenteum*' (meaning 'silver'). Most specific epithets indicate something characteristic about a species, such as colour of the flowers or fruits ('*Haemanthus albiflos*' indicates 'white-flowered'); habit of growth ('*Bowiea volubilis*' points to twining); shape of the petal or leaf ('*Nerine platypetala*' describes 'broad-petalled'); discoverer of the species (*Gladiolus comptonii* is named after R.H. Compton, major collector of southern African flora, and founder of the Compton Herbarium at Kirstenbosch); or place of discovery ('*Celtis africana*' reveals Africa as source). Latin binomial names originate from the nomenclature contained in Linnaeus's *Species Plantarum*,[74] a system which has remained intact for almost two and a half centuries.

If a name is merely a symbol of communication, then it is valid to enquire why we should use this relatively complicated system of names. Why not use more accessible common names which often endear the plant to those who speak the language of the vernacular name? There are several succinct reasons why this is so. Common names are not universal; in most parts of the world, particularly in southern Africa, relatively few plant species have common names; and often two or more plants are known by the same common name. Scientific names are the definitive means by which people around the world can talk about any plant in any language. Thus the correct Latin name for a plant is the primary reference base for the storage, retrieval and use of all botanical knowledge.

Drawing the limits

Conflicts regarding the application of names for species are unfortunately common, and much will depend upon how scientists view species. The rationale for recognizing species is the question at the heart of botany. Concepts of species have differed through time, and even today a number of different concepts are currently in use. The two that symbolize a deep argument in the scientific community are the biological species concept, and the phylogenetic species concept.

The biological species concept, its forerunner [75] initiated by the eighteenth-century zoologist, Georges Buffon (1707–1788), states that species are groups of individuals that more or less freely interbreed, and that are reproductively isolated from other groups. The biological species concept has been successful because it is in accord with the most plausible mechanism for speciation – the inability of isolated populations to interbreed. Adaptation to local conditions, or random drift, that occurs when isolated populations form, results in heritable genetic differences between populations. These differences accumulate over several generations and should the populations meet again, they no longer interbreed, and hence they become separate evolutionary entities or species.

The greatest practical problem with this species concept is that it cannot be universally applied. Southern Africa has numerous examples of disjunct or non-contiguous populations of plants which apparently belong to the same species. Where few differences distinguish these populations, some botanists may regard them as all belonging to the same species, while other authorities may circumscribe each of them as a separate species. Another problem relates to the treatment of hybrids. When isolated populations are once again brought into contact, it should be possible to apply the criteria of the biological species concept to test whether speciation has occurred. Often some degree of hybridization takes place. If hybrids are not disadvantaged, the populations will coalesce and form a single species. If the hybrids are disadvantaged, the species boundaries are reinforced. This process takes many generations to play out, however, and during this time the taxonomic status of the populations is hard to resolve.

The other main contender to the current species debate is the phylogenetic species concept. According to this concept, a species is the smallest aggregation of populations diagnosable by a unique combination of features in comparable individuals. It is free from assumptions about how – not whether – evolution occurred.

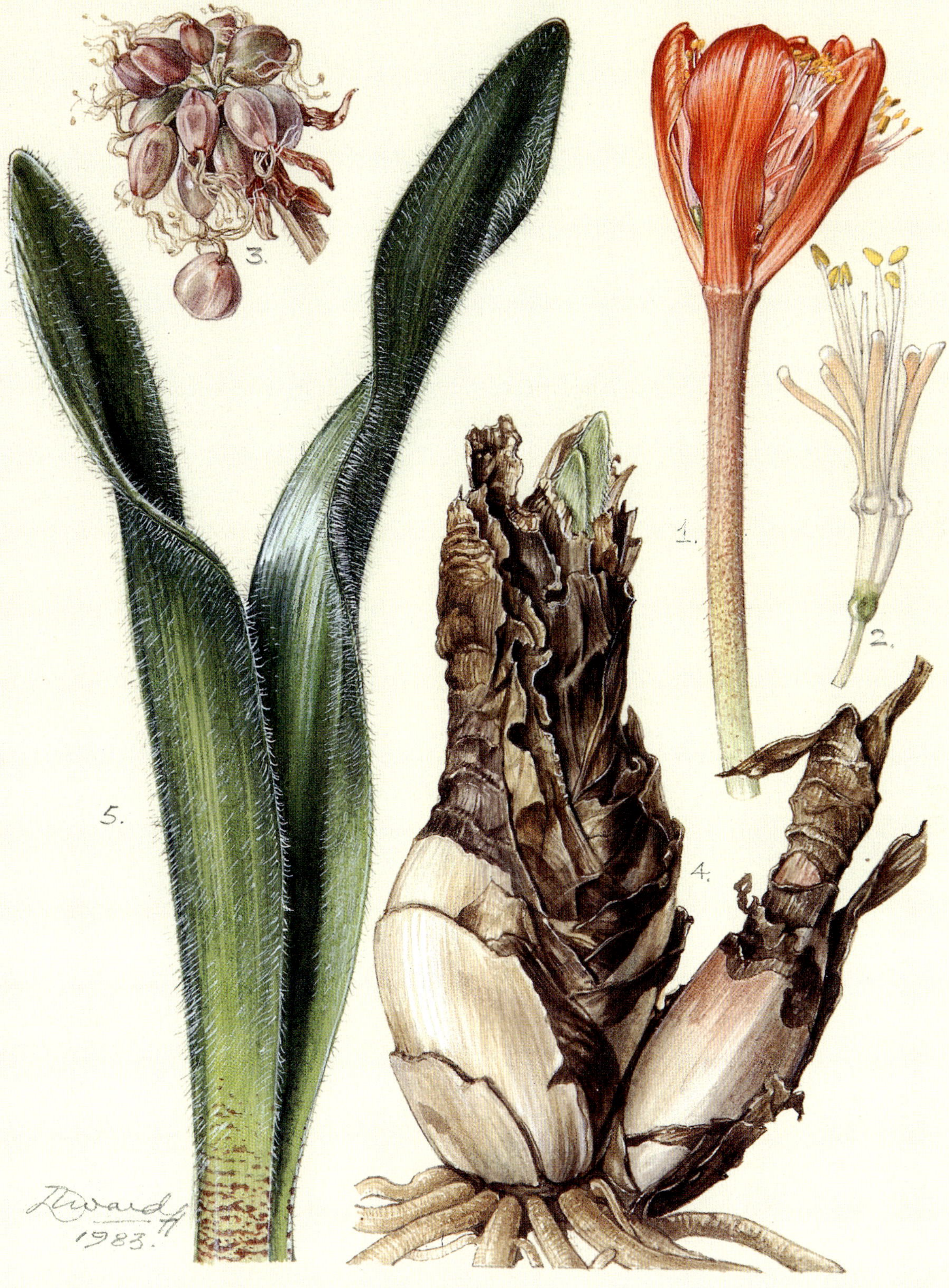

PLATE 61

Ellaphie Ward-Hilhorst, *Haemanthus dasyphyllus*, 1983, watercolour: reproduced in D.A. Snijman, 'A revision of the genus *Haemanthus* L. (Amaryllidaceae)', supplement to *Journal of South African Botany*, 1984

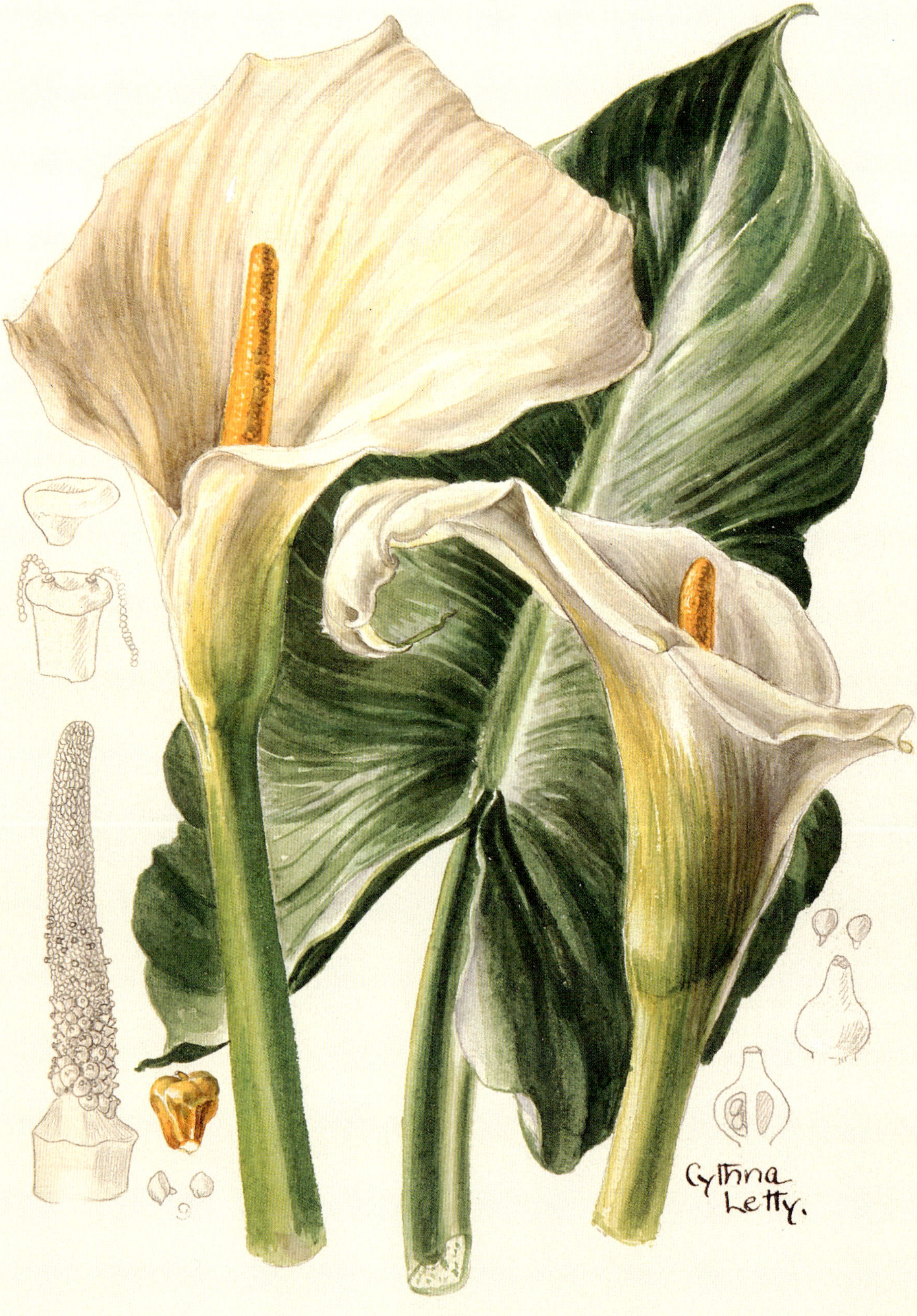

PLATE 62

Cythna Letty, *Zantedeschia aethiopica,* watercolour and pencil: reproduced in C. Letty, 'The genus *Zantedeschia*', *Bothalia*, 1973

The application of this definition, however, is difficult. The vast majority of botanists use qualitative and quantitative differences in plant structure to distinguish species, but not all structural features are equally important and those which change in direct response to varying conditions of the environment, such as shade or full sunlight, are not considered reliable – emphasis is placed on those features which are hereditary. In the case of quantitative characters, 'gaps' in the values can be used to distinguish species. The situation becomes more complex with the simultaneous consideration of more than one character. In addition, the use of quantitative characters to distinguish phylogenetic species requires rigorous sampling and analysis and all too often the amount of plant material available to a botanist is fragmentary and numerical analysis is simply not practicable.

Because there is not yet universal acceptance of a single species concept, its application remains the focus of debate in the systematics community and botanists routinely apply a number of different species concepts, sometimes even in one work. What remains constant, however, is that the name of a species is permanently attached to a type specimen. Unlike a conceptual type, however, a type specimen need not represent the most typical element of a species, as in the case of Clusius's illustration of *Protea neriifolia* (Plate 42).[76]

Botanist and botanical artist

In their joint endeavours there is no doubt that the relationship between botanists and artists is most frequently one of mutual benefit. Ideally, the constant critical examination of a plant and its image leads the mind and the eye to perceive the plant's true properties and allows both the text and the artwork to be refined. Several artists have worked particularly successfully with botanists, and their artwork has been highly acclaimed. For instance, Kew's great botanical artist, Walter H. Fitch (1817–1892), worked with William Hooker, and in the Cape, Mary Page, who specialized in illustrating mesembryanthemums, worked under the demanding eye of Louisa Bolus (1877–1970), curator of the Bolus Herbarium from 1903 to 1955.

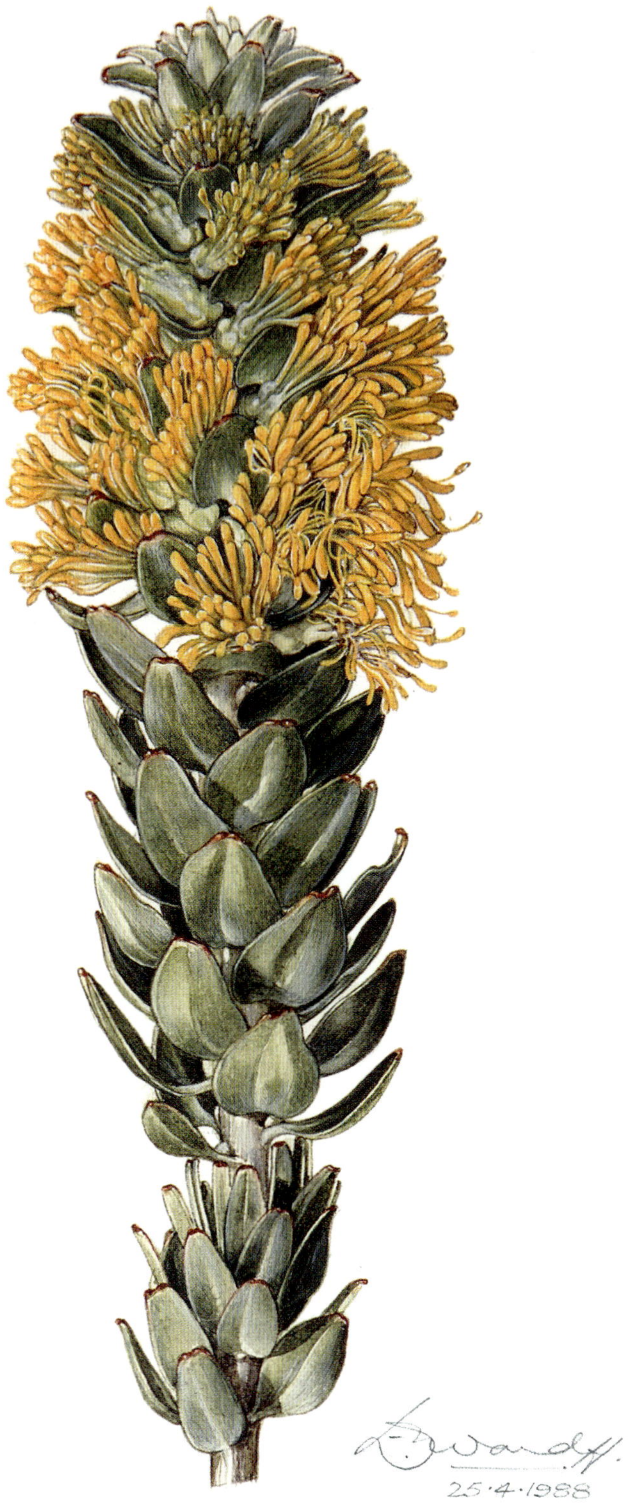

PLATE 63

Ellaphie Ward-Hilhorst, *Mimetes chrysanthus*, 1988, watercolour: reproduced in J.P. Rourke, '*Mimetes chrysanthus*', *Veld & Flora,* 1988

Yet, despite many other examples of art's considerable contribution to botany, the modern scientist's acknowledgement of the collaborating artist may frequently appear to be understated – especially when compared with the effusive texts in sixteenth-century herbals. Unfortunately, this apparent shortcoming usually arises from the restrained style (which is often interpreted as dull and colourless by the critics of science) that is demanded by scientific journals – but not from the authors' indifference.

Cythna Letty, one of the most prolific artists of *The Flowering Plants of Africa*, actually gained considerable botanical expertise through lack of guidance from her botanical colleagues. When she first took up her position as botanical artist at the National Herbarium in 1927, she was obliged to teach herself how to use a binocular microscope, to prepare dissections of flowers, and to identify the plants that she was required to illustrate.[77] In this way she acquired sufficient taxonomic knowledge to study a plant group of her own choice and this culminated in the illustrated publication, *A Revision of Zantedeschia*,[78] of which she was the sole author and artist.

A notable example of how art has contributed to publicizing scientific discovery is seen in the example of *Mimetes chrysanthus*, a remarkable new species in the protea family. In 1987, just six years after John Rourke had completed his monograph on the Cape genus *Mimetes* with Thalia Lincoln, several populations of an unknown mimetes were found on the mountains bordering the Little Karoo. Rourke formally described the new species one year later,[79] but the impact of the discovery was made primarily through Ellaphie Ward-Hilhorst's watercolours of the golden mimetes (Plate 63) which were widely reproduced on greetings cards, in *Veld & Flora*,[80] and in *The Kew Magazine*.[81]

The visual interpretation of local floras for field guides is another endeavour where artists have contributed considerably to popular botanical literature. Several artists are the sole authors of field guides – even when the plant identifications have been contributed by eminent botanists – as in Mary Maytham Kidd's wildflower guide to the Cape Peninsula[82] and Hilda Mason's *Western Cape Sandveld Flowers*.[83] All field guides use composite plates and where many unrelated species from different families have to be arranged together, the quintessential 'look' of plants, often with very diverse growth forms, has to be captured. This challenge differs considerably from that of systematically depicting many species from a single genus, where the focus is often on subtle differentiating features and painstaking detail.

Many illustrations serve as a record of species that were in cultivation in the past and in some cases these images hold great scientific value. Where no permanent herbarium specimen was made for a species that was described and depicted for the first time, the illustration is then chosen as the type specimen to which the name of the species is inexorably attached. Thus the only way to fix the identity of the species named *Veltheimia viridiflora* by Jacquin was to refer to the illustration of that plant in Nikolaus von Jacquin's monumental publication depicting plants cultivated at the imperial gardens at Schönbrunn in the eighteenth century (Plate 66). Despite the plate's antiquity it captures sufficient detail to allow the species it depicts to be equated unequivocally with the earlier-described *Veltheimia bracteata*. Although the joy of researching species through the early literature is a far cry from the excitement of discovering a new species, the pursuit holds great recompense for the bibliophile.

It is widely acknowledged that the greatest difference between the scientist's and the artist's approach to plant portraiture is the emphasis that each places on accuracy, not only of plant structure but also of plant identity – a difference which is reinforced by the acquisition policies of museums of art and scientific institutions. For instance, The Hunt Botanical Library in Pittsburgh, Pennsylvania, an international repository for original examples of contemporary botanical art and illustration, will accept a portrait of a plant that is 'sufficiently botanical as to be identifiable as to its genus'.[84] The National Botanical Institute, in contrast, accepts illustrations for *The Flowering Plants of Africa* (and hence for its art collection) only if the plant used in the illustration is preserved as a voucher specimen and

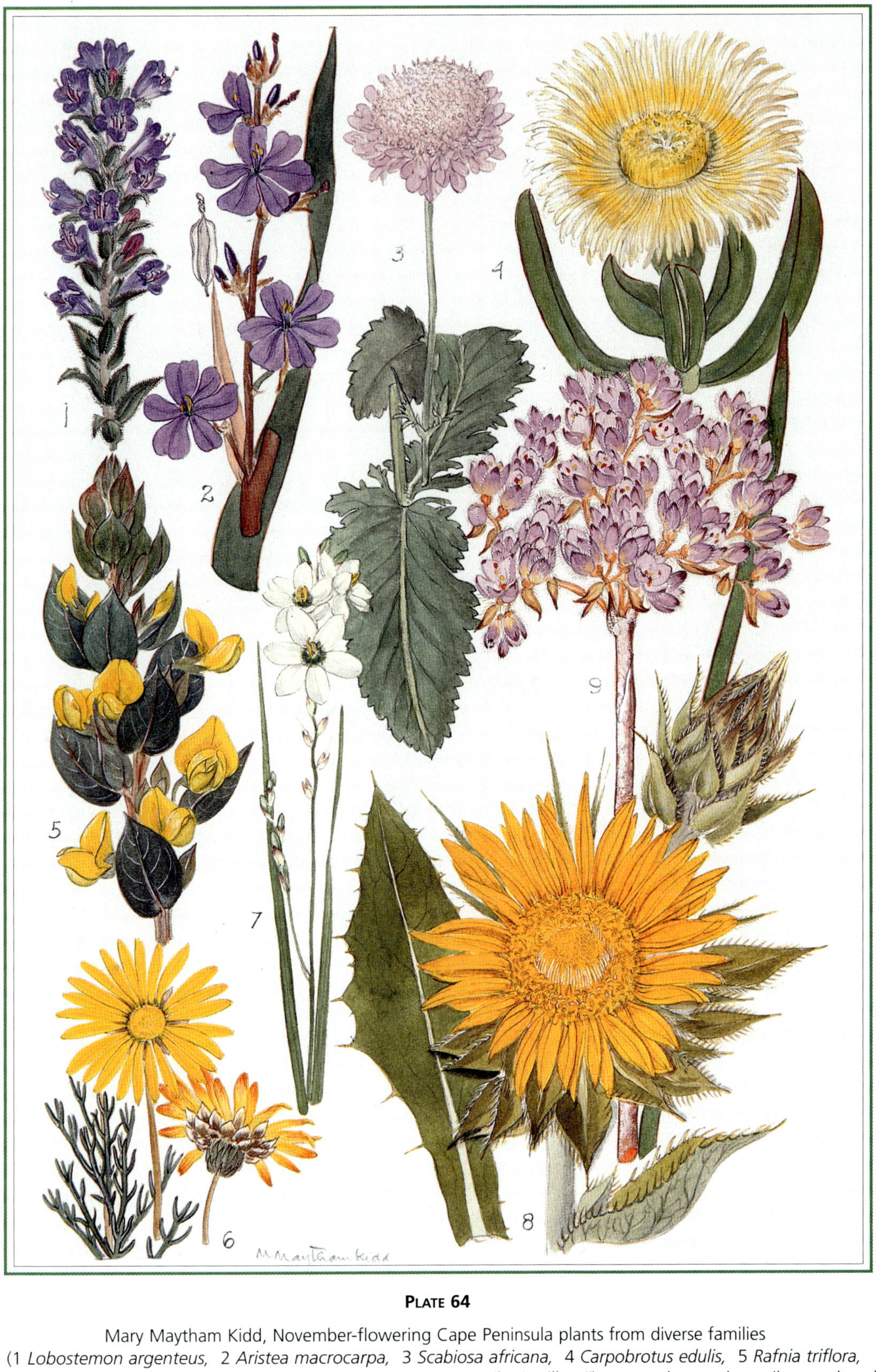

PLATE 64

Mary Maytham Kidd, November-flowering Cape Peninsula plants from diverse families (1 *Lobostemon argenteus*, 2 *Aristea macrocarpa*, 3 *Scabiosa africana*, 4 *Carpobrotus edulis*, 5 *Rafnia triflora*, 6 *Ursinia crithmoides*, 7 *Ixia polystachya*, 8 *Berkheya armata*, 9 *Dilatris pillansii*), watercolour and pencil: reproduced in M. Maytham Kidd, *Cape Peninsula: South African Wild Flower Guide 3* (Cape Town, 1983), 3 ed.

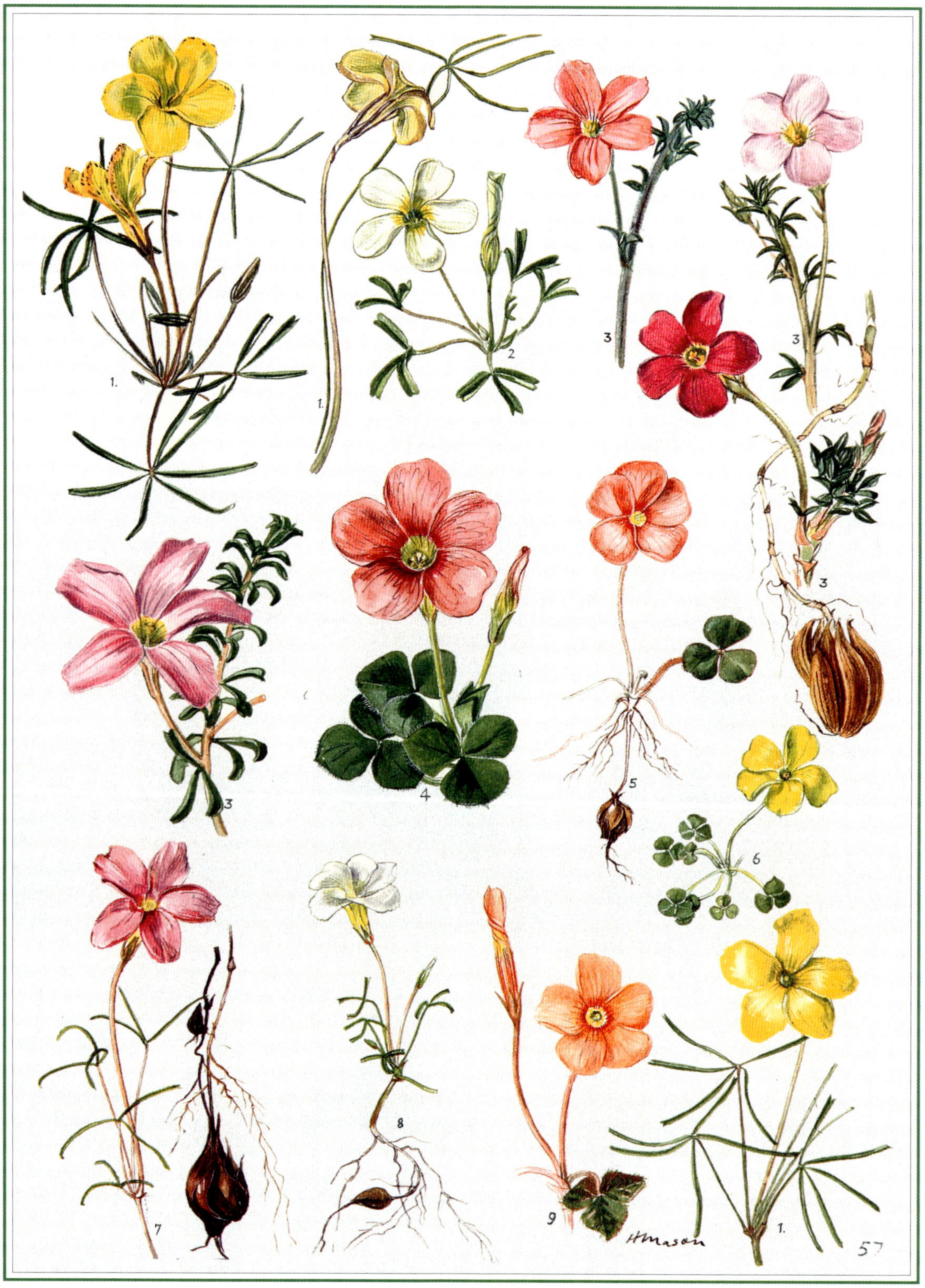

Plate 65

Hilda Mason, *Oxalis* species (1 *O. burtoniae*, 2 *O. tenella*, 3 *O. hirta*, 4 *O. purpurea*, 5 *O. eckloniana* var. *sonderi*, 6 *O. luteola*, 7 *O. polyphylla*, 8 *O. pusilla*, 9 *O. eckloniana* var. *hopefieldiana*), watercolour: reproduced in H. Mason, *Western Cape Sandveld Flowers* (Cape Town, 1972)

housed in a recognized herbarium, so that the botanist's or artist's interpretation of the species' identity can be further researched if necessary.

As explained by John Manning and Peter Goldblatt, in their chapter 'A greater glory: flower form and function', much can be inferred about the ecology of a plant from a depiction of its structure, though only a limited number of plant illustrations will be sufficiently detailed to provide an accurate identification of the subject. The absence of a voucher specimen is less of a handicap in regions like western Europe, where many representatives of the flora are sufficiently well known to be easily recognizable, but it is particularly problematic in southern Africa where – because of its rich flora – many species are still poorly understood. In addition, few artists emphasize the plant's name within the picture frame, so immediate access to all that is known about the species is often limited. Thalia Lincoln is exceptional in purposely weaving the scientific name into the main body of her compositions, thereby emphasizing the important connection between a plant and its name. Moreover, she is one of the few botanical artists of southern Africa who strives to saturate the plant's name with visual appeal (see Plate 124).

A plant portrait which possibly exemplifies the different concerns of artist and botanist is that of *Nerine sarniensis* (Plate 67). This oil painting is a visual delight of massed blooms but the number of stamens in many flowers is less than the number typical of all species in the amaryllis family to which *Nerine* belongs. In essence, the image should be viewed primarily as an aesthetic statement of the shape, colour, and texture of nerines but its lack of structural accuracy cannot obscure its subject's true identity – the widely known Guernsey lily. Had the artist chosen to portray a lesser-known plant, however, its true identity may have remained a mystery.

In effect, this and the many other beautiful images presented here, show how the artist brings expertise on space, colour and texture, whereas the systematist – the analyst of structural variation – contributes a knowledge of the features that delimit the species. This latter knowledge is a taxonomic imperative that recognizes species as the fundamental units of biological diversity and the primary currency for setting conservation priorities.

Epilogue

As evidence of the irreversible loss of so many species increases and with an awareness of the critical importance of maintaining the integrity of the biosphere, there is growing concern about the small number of systematists currently available to study the large numbers of species about which so little is known. A major outcome of the United Nations Conference on Environment and Development, which was held in Rio de Janeiro in June 1992, was the formulation and acceptance of the Convention on Biological Diversity. Many signatories to the Convention recognize that the shortfall of trained taxonomists is a major impediment to the present need of society to understand its relationship to Nature. In response, global funding agencies have been prompted to facilitate renewed systematic research to generate more complete taxonomic information – an initiative which has benefited the southern African subcontinent since 1998.

Natural history philosophers have suggested that the steady depreciation of systematics relative to other subdisciplines in botany has stemmed from its intrinsic nature, which, as its long history has shown, has had to adapt its conceptual framework repeatedly to accommodate biology's advances. In particular, the flow of information about species, through new technologies and previously unknown plants, has necessitated reordering descriptions, re-evaluating relationships and revising schemes of classification. Inimical to this process is that plant names seldom remain static. Thus taxonomy may appear to be beset with arcane terminology and ritualized descriptions on the one hand, yet ironically to be unstable and lacking rigour on the other.

Ultimately, humankind seeks a system that will allow the recognition of naturally occurring groups of organisms. In folk taxonomies, names are used to talk of organisms with others who already know the cultural significance of the organisms being discussed – but the challenge of

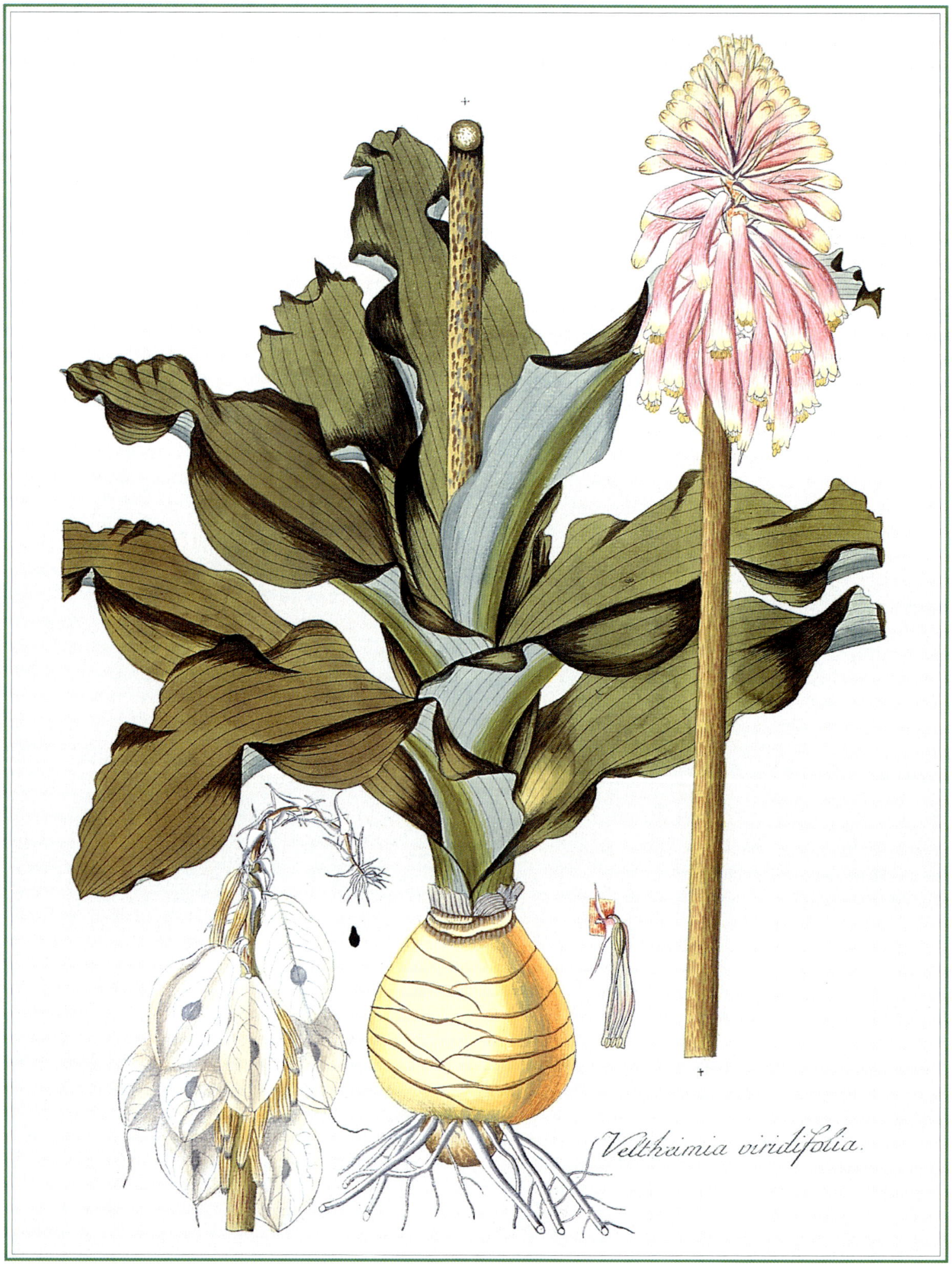

PLATE 66

Veltheimia bracteata; hand-coloured engraving from an original drawing probably by Franz von Scheidel: from Nikolaus Jacquin, *Plantarum Rariorum Horti Caesarei Schönbrunnensis Descriptiones et Icones*, vol. 1 (Vienna, 1797)

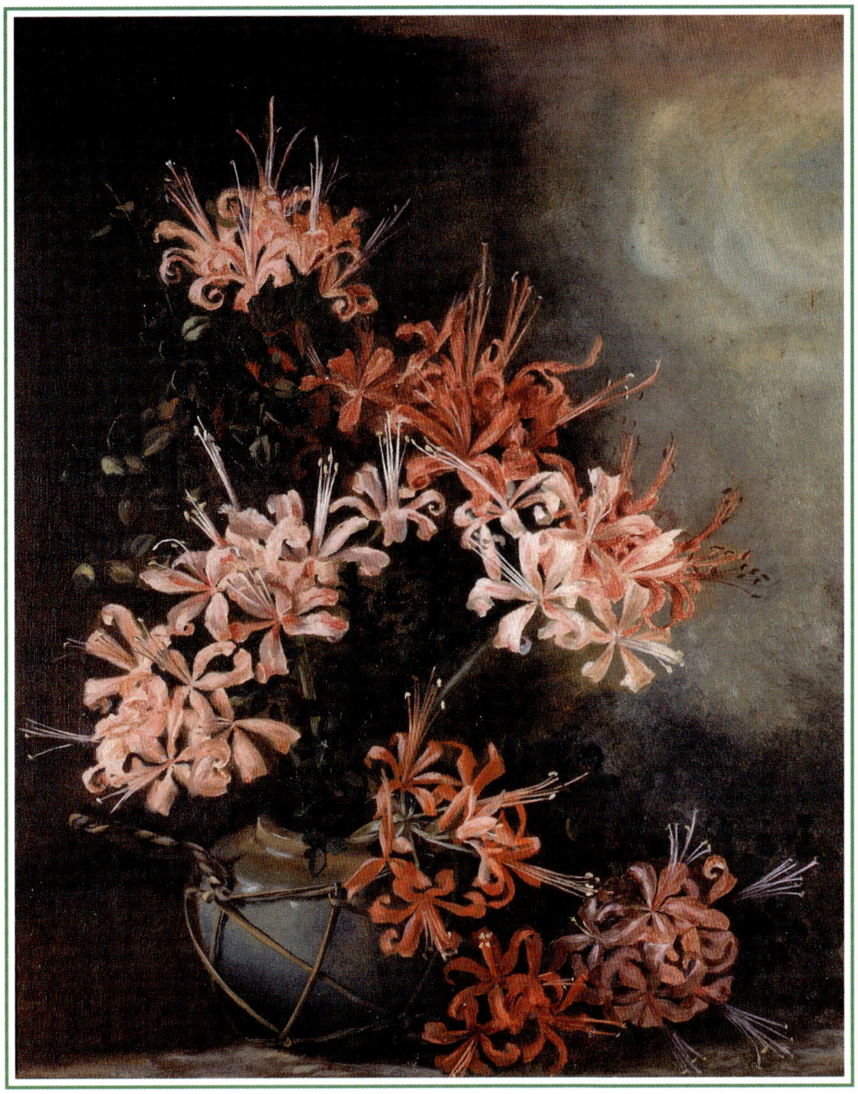

PLATE 67

[artist unknown], *Nerine sarniensis,* early twentieth century, oil on canvas, 43,5 cm x 39 cm

systematists in dealing with the vast numbers of organisms which exist is that of tracing the course of the evolution of living things and building this information into predictable classification systems. Evolution, however, is an ongoing dynamic process, with populations of plants splitting and coalescing over time and, because of this, not all plants can be attributed to neatly defined concepts with stable names.

Taxonomists carry a great responsibility not to devalue the species to a construct which is so abstract that society is no longer concerned by the threat of species' extinctions. The French anthropologist Claude Lévi Strauss (1908–) has written that each species is 'an irreplaceable treasure, equal to the works of art which we religiously preserve in museums'.[85] By revealing the true beauty and interest of plants, the challenge to art and science in the future is to continue to appeal to the aesthetic and intellectual interests of society and to promote an increased awareness of the natural world – an awareness that perhaps will enhance an understanding of the importance of all life forms and their value to future generations throughout the world.

1 Agnes Arber, *The Mind and the Eye: a Study of the Biologist's Standpoint* (Cambridge, 1964), p. 45.
2 'Systematics agenda 2000: charting the biosphere – technical report'. New York, 1994, p. 16.
3 T.H. Arnold and B.C. de Wet, eds, *Plants of Southern Africa: Names and Distribution* (Pretoria, 1993), p. 1.
4 Peter Goldblatt, 'An analysis of the flora of southern Africa: its characteristics, relationships, and origins', *Annals of the Missouri Botanical Garden*, 65, 1978, p. 384.
5 PRECIS database. National Botanical Institute, Pretoria.
6 Craig Hilton-Taylor (Pretoria, 1996), p. 1.
7 J. Loubser and P.C. Zietsman, 'Rock painting of postulated *Brunsvigia* sp. (Amaryllidaceae) at Thaba Bosiu, western Lesotho', *South African Journal of Science*, 90, 1994, p. 612.
8 (Leiden, 1605), p. 38.
9 Translated by Reverend J. Gerstner, Pretoria.
10 Agnes Arber, *Herbals: their Origin and Evolution* (Cambridge, 1912), p. 76.
11 William T. Stearn, *Botanical Latin* (London, 1973), 2 ed., p. 29.
12 John P. Rourke to D.A. Snijman, personal communication, 1998.
13 J.P. Rourke, 'Robert Brown at the Cape of Good Hope', *Journal of South African Botany*, 40, 1974, p. 47.
14 Robert Brown, 'On the Proteaceae of Jussieu', *Transactions of the Linnean Society*, 10, 1810, p. 66.
15 Kevin J. Gaston and Robert M. May, 'Taxonomy of taxonomists', *Nature*, 356, 1992, p. 281.
16 J.P. Rourke, 'Thunberg's contribution to plant systematics in South Africa', *Royal Society of South Africa*, 49, 1994, p. 157.
17 Ibid., p. 158.
18 M. Jacobs, 'Revolutions in plant descriptions', *Miscellaneous Papers, Wageningen Landbouhochschool* [Netherlands], 19, 1980, pp. 155–181.
19 Stearn, *Botanical Latin*, p. 46.
20 William Curtis, '*Strelitzia reginae*. Canna-leaved Strelitzia', *Curtis's Botanical Magazine*, 3, 1790, pl. 119.
21 J.B. Gawler, '*Amaryllis belladonna*. Belladonna lily', ibid., 19, 1804, pl. 733.
22 J.B. Gawler, '*Protea cynaroides*. Artichoke-flowered protea', ibid., 20, 1804, pl. 770.
23 D.S. Hardy, '*Aloe helenae*', illustrated by Gillian Condy, *The Flowering Plants of Africa*, 49, 1986, pl. 1934.
24 Goldblatt, 'Analysis of the flora of southern Africa', p. 379.
25 E.P. Phillips, '*Brunia stokoei*', illustrated by Stella Gower, *The Flowering Plants of South Africa*, 3, 1923, pl. 92; R.A. Dyer, '*Brunia albiflora*', illustrated by Edith Burges, ibid., 24, 1944, pl. 928.
26 J.P. Rourke, '*Glischrocolla formosa*', illustrated by Fay Anderson, *The Flowering Plants of Africa*, 40, 1969, pl. 1574.
27 I.C. Verdoorn, '*Erythrophysa transvaalensis*', illustrated by Betty Connell, *The Flowering Plants of South Africa*, 22, 1942, pl. 857.
28 H.A. Baker and E.G.H. Oliver, *Ericas in Southern Africa*, illustrated by Irma von Below, Fay Anderson, *et al.* (Cape Town, 1967).
29 E.G.H. Oliver, '*Erica*, a biodiversity record for plants?', *Yearbook of the Heather Society*, 1998, p. 2.
30 Text by A.G.J. Herre, illustrated by Harry Bolus, Beatrice Carter, Mary Page and Maisie Walgate (Cape Town, 1971).
31 Peter Goldblatt, 'Floristic diversity in the Cape flora of South Africa', *Biodiversity and Conservation*, 6, 1997, p. 361.
32 (Cape Town, 1980).
33 D. Meredith, ed., *The Grasses and Pastures of South Africa*, illustrated by Gertrude Laurence, Cythna Letty, *et al.* (Johannesburg, 1955).
34 Frances M. Leighton, 'The genus *Agapanthus* L'Héritier', supplement to *Journal of South African Botany*, 4, 1965, pp. 1–50.
35 L.E. Codd, 'The South African species of *Kniphofia* (Liliaceae)', *Bothalia*, 9, 1968, pp. 363–513.
36 I.C. Verdoorn, 'The genus *Crinum* in southern Africa', ibid., 11, 1973, pp. 27–52.
37 Ernst J. van Jaarsveld, *Gasterias of South Africa: a New Revision of a Major Succulent Group*, illustrated by Ellaphie Ward-Hilhorst (Cape Town, 1994).
38 Peter Goldblatt, *The Moraeas of Southern Africa*, illustrated by Fay Anderson, Margo Branch and Janet Klein (Cape Town, 1986).
39 Peter Goldblatt, *The Genus* Watsonia (Cape Town, 1989).
40 Peter Goldblatt, *The Woody Iridaceae* Nivenia, Klattia, *and* Witsenia, illustrated by Fay Anderson (Portland, Oregon, 1993).
41 Esmé Franklin Hennessy, *South African Erythrinas* (Durban, 1972).

42 John P. Rourke, *Mimetes: an Illustrated Account of* Mimetes Salisbury *and* Orothamnus Pappe, *Two Notable Cape Genera of the Proteaceae*, illustrated by Thalia Lincoln (Cape Town, 1982).

43 J.J.A. van der Walt, *Pelargoniums of Southern Africa*, illustrated by Ellaphie Ward-Hilhorst (vol. 1, Cape Town, 1977), vols 2, 3: J.J.A. van der Walt and P.J. Vorster (Cape Town, 1981, 1988).

44 G. Joyce Lewis, 'The genus *Babiana*', supplement to *Journal of South African Botany*, 3, 1959, pp. 1–149.

45 Marion Arnold, *Women and Art in South Africa* (Cape Town, 1996), p. 74.

46 Peter Goldblatt and John Manning, *Gladiolus in Southern Africa*, illustrated by Fay Anderson, Auriol Batten and John Manning (Cape Town, 1998).

47 Text by O.M. Hilliard and B.L. Burtt, illustrated by Auriol Batten (Johannesburg, 1991).

48 Richard Mabey, *The Flowering of Kew* (London, 1988), p. 183.

49 Frans A. Stafleu, *Linnaeus and the Linnaeans: the Spreading of their Ideas in Systematic Botany, 1735–1789* (Utrecht, 1971), p. 327.

50 Deirdré Snijman, 'A revision of the genus *Haemanthus* L. (Amaryllidaceae)', illustrated by Ellaphie Ward-Hilhorst and Fay Anderson, supplement to *Journal of South African Botany*, 12, 1984, pp. 1–139.

51 Stafleu, *Linnaeus and the Linnaeans*, p. 311.

52 *Prodromus Historiae Generalis Plantarum . . .* (Montpellier, 1689).

53 (Paris, 1789).

54 S.M. Walters, 'The shaping of angiosperm taxonomy', *New Phytologist*, 60, 1961, p. 77.

55 G. Bentham and J.D. Hooker, *Genera Plantarum*, 3 vols (London, 1862–1883).

56 Charles Darwin, *On the Origin of Species* (London, 1859).

57 H. Engler and K.A.E. Prantl, eds, *Die natürlichen Pflanzenfamilien*, vols 1–8 (Leipzig, 1887–1915).

58 William H. Harvey and Otto Wilhelm Sonder, eds, vols 1–3 (Dublin and Cape Town, 1860–1865), William T. Thistleton-Dyer, ed., vols 4–7 (London, 1896–1925), Arthur William Hill, ed., supplement to vol. 5 (2) (London, 1933).

59 4 vols in 6 pts (Cape Town and London, 1913–1932).

60 E. Warming, *Plantesamfund* (Copenhagen, 1895).

61 A.F.W. Schimper, *Pflanzen-Geographie auf physiologischer Grundlage* (Jena, 1898).

62 W. Hennig, *Grundzüge einer Theorie der phylogenetischen Systematik* (Berlin, 1950).

63 P.F. Stevens, 'Metaphors and typology in the development of botanical systematics 1690–1960, or the art of putting new wine into old bottles', *Taxon* 33, 1984, pp. 169–211.

64 Meredith A. Lane, 'Roles of natural history collections', *Annals of the Missouri Botanical Garden*, 83, 1996, p. 538.

65 Mary Gunn and L.E. Codd, *Botanical Exploration of Southern Africa* (Cape Town, 1981), p. 10.

66 Jacobs, 'Revolutions in plant descriptions', p. 165.

67 Carolus Linnaeus, *Critica Botanica* (Leiden, 1737).

68 Steve Johnson and Kathy Johnson, 'Beauty and the beast', *Veld & Flora*, 79, 1993, pp. 38–39.

69 William Jackson Hooker, '*Dryandra pteridifolia*: fern-leaved Dryandra', *Curtis's Botanical Magazine*, 63, 1836, pl. 3500.

70 Hilton-Taylor, *Red Data List*.

71 Steven Hammer, 'Twenty-five ways to look at *Haworthia*', *Cactus and Succulent Journal (U.S.)*, 70, 1998, p. 175.

72 Thalia Lincoln to D.A. Snijman, personal communication, 1995.

73 Snijman, 'Revision of the genus *Haemanthus*', p. 108.

74 2 vols (Stockholm, 1753).

75 Stafleu, *Linnaeus and the Linnaeans*, p. 307.

76 *Exoticorum*, p. 39.

77 D.J.B. Killick and E. du Plessis, eds, *The Flowering Plants of Africa*, vol. 50 ([Pretoria, 1988]), p. 4.

78 Cythna Letty, 'The genus *Zantedeschia*', *Bothalia*, 11, 1973, pp. 5–26.

79 J.P. Rourke, 'A new species of *Mimetes* (Proteaceae) from the southern Cape', *South African Journal of Botany*, 54, 1988, pp. 636–639.

80 John Rourke, '*Mimetes chrysanthus*', *Veld & Flora*, 74, 1988, pp. 143–145.

81 John Rourke, '*Mimetes chrysanthus*', *The Kew Magazine*, 11, 1994, pp. 178–183.

82 *Cape Peninsula: South African Wild Flower Guide 3* (Cape Town, 1983), 3 ed.

83 Text by Enid du Plessis (Cape Town, 1972).

84 George H.M. Lawrence, *Catalogue: 2nd International Exhibition of Botanical Art & Illustration* (Pittsburgh, Pennsylvania, 1968), p. 3.

85 Paul R. Ehrlich and Anne H. Ehrlich, 'The value of biodiversity', *Ambio*, 21, 1992, p. 220.

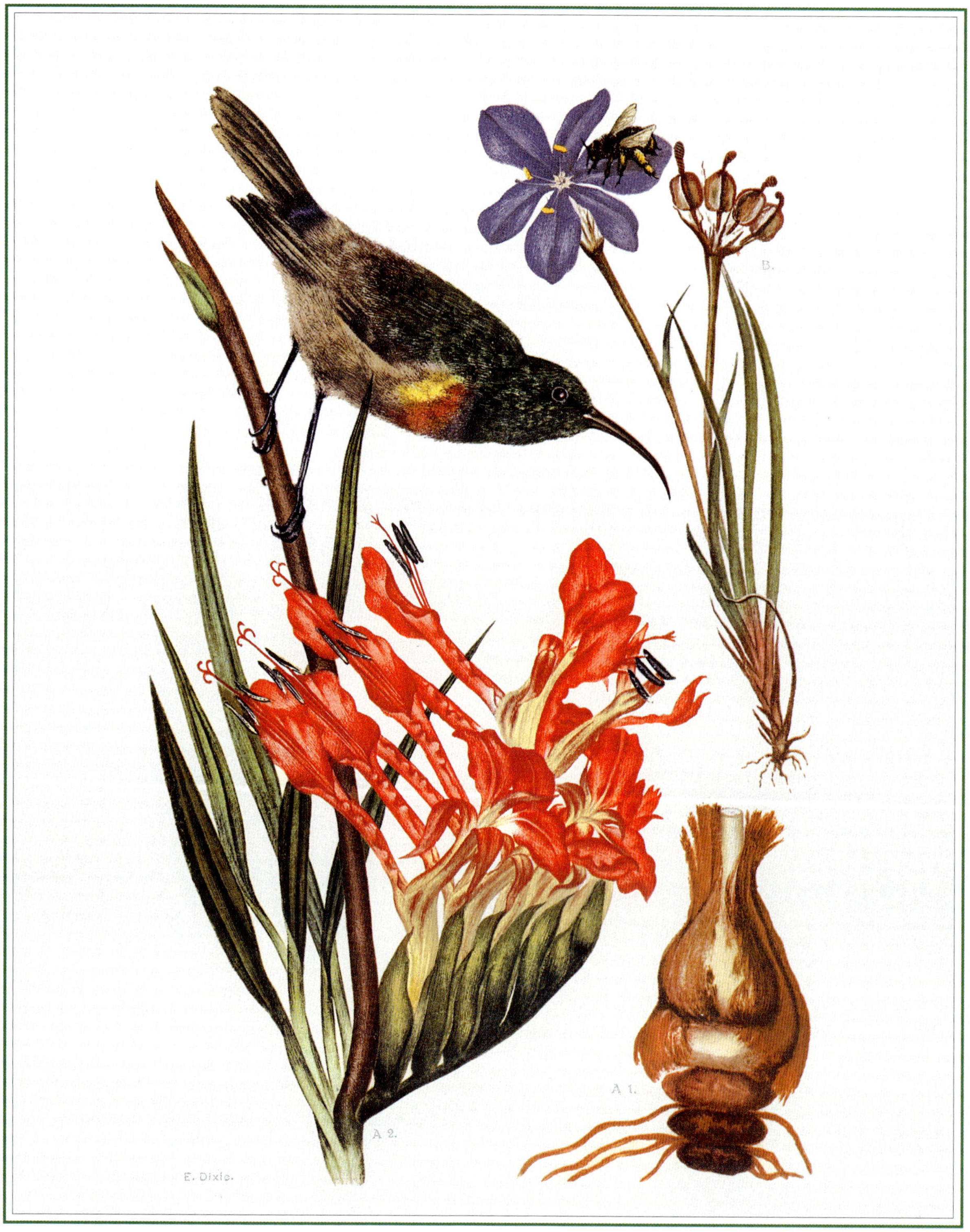

Plate 68

Ethel Dixie, A: *Babiana ringens* supporting the Lesser Doublecollared Sunbird; B: *Aristea glauca*; gouache and watercolour: reproduced in Rudolf Marloth, *The Flora of South Africa*, vol. 4 (Cape Town and London, 1915). Both the extraordinary scarlet flowers and the perch formed from a sterile extension of the flower stalk branch are extreme adaptations to bird pollination developed by the Rat's Tail Babiana, *Babiana ringens.* The short-lived blue flowers of *Aristea glauca*, by contrast, are adapted to pollination by pollen-collecting bees.

A GREATER GLORY

FLOWER FORM AND FUNCTION

JOHN MANNING AND PETER GOLDBLATT

Journeys of exploration are not the prerogative of intrepid adventurers, pitting themselves against the primitive forces of nature in their quest for rare and exotic plants or animals. While it is undeniably true that botanical explorers of past centuries endured almost constant privations and dangers in their quest for the biological treasures that would eventually grace the gardens and greenhouses of Europe, the less physical paths of understanding are still seldom trodden. More sedentary explorers can make equally exciting journeys through the jungles of confusion or across the wastes of ignorance. Little is known, for instance, about most plants' relationships with their environment and only recently have we begun to be relatively sure about their ancestry. While much of this information can be extracted only through the application of sophisticated methodologies, a surprising amount can be gleaned directly from the plant or even its image. This is possible because of the very close correlation between form and function that characterizes much of biology. In this essay we introduce some of the principles that dictate the form of flowers, and relate them to their primary function in pollination biology.

Stories without words

The most striking attribute of botanical painting is the splendid isolation in which the subject stands. Neither the intense attention to detail nor the absolute mastery of the medium distinguishes botanical art from other forms of representational art: it is the total, and quite unnatural, absence of any extraneous features that is its chief characteristic. In fact, botanical illustration differs from other representational arts more in what it leaves out than in what it puts in – it lacks a narrative or context. Plants are plucked from their surroundings, placed on an utterly blank surface, and then subjected to an intense and exclusive scrutiny of outward form. No pageant was ever so ruthlessly physical. All references to the environment in which the plant lived and evolved, and in which it carried out its daily economy of life are obliterated. The organism stands uprooted under a spotlight of scrutiny in which form alone is illuminated.

But form is the key to function and so finely are the two tuned in the natural world that the informed mind can infer a wealth of biological information from the outward appearance of a plant alone – can colour in the background, so to speak. Succulence, for example, is an adaptation to periodic water stress and indicates an arid or seasonally dry climate; while variegation or spotting on leaves, particularly in combination with a maroon or purple underside to the leaf, is an adaptation to low light intensity and indicates a forest floor habitat. This close relationship between form and function characterizes all living things but it is often most evident in plants, primarily because they are literally rooted to the spot and cannot move to accommodate changes in their environment.

There is a caveat to such sublime sympathy, however, as in all Nature's contracts, and an organism is not free to indulge the demands of the environment unconstrained. Its form and genetic makeup place some limitations on complete freedom of expression. This is one of the basic tenets of nature: organisms or individuals are not born equal even if they are born into a world of equal opportunity. Furthermore, the Economy of Nature runs on a cost-benefit budget as strict as any in the Economy of Man. Any benefit, in terms of improved survival and reproduction, must be greater than the cost of achieving that benefit if evolutionary bankruptcy, or extinction, is not to follow. In Nature nothing is ever for nothing.

At any stage of its evolution, as much in the pea as in the peacock, an organism represents a duality of operation and origin in which the demands of the environment are balanced against the constraints imposed by heredity. The garment of function is draped on the mannequin of ancestry. The form of an organism thus reflects both its relationship with other organisms and its relationship to the environment, or ecology. Flowers are particularly good examples of the compromise between capability and capacity. In fundamental details they are evolutionarily very conservative, retaining a basic form which was established early in their development. In superficial appearance, however, they are extraordinarily plastic. Nature's books cannot always be judged by their covers.

This paradox demands a very informed approach if one is to distinguish between characters that reflect deep-seated historical or genealogical influences and those more labile characters that reflect recent ecological selection. The rewards of such study are great vistas through time and space in which a precise and perfect relationship between the flower and its environment are the focus. So exquisite is this relationship that its products are, literally, timeless. Selected or hybridized garden flowers, the products of commercial selection in which the sole criterion is ultimately the pleasure of mankind, have lost this biological integrity of the flower and survive only as subjects of horticulture.

Flowers are primarily organs of attraction, structured to solicit the attentions of potential pollinators. Certainly some species rely on incidental agents such as wind as the means of pollination, in which cases the plant more or less airs its sexual apparel on the wash-line; but in the majority of species, pollination relies on the more purposeful activities of other living organisms. It is here that the flower comes into its own and achieves its greatest expression. Floral extravagance must reach its epitome in the orchids, in which no guile or wile of nature has gone unexploited. Amongst the southern African flora it is the iris family in which floral diversity has run rampant. Rich and diverse in species, it is also a family in which the pollination biology has begun to be understood in terms of form and function.

A rose by any name

The flowers of the African members of the iris family Iridaceae are particularly variable and illustrations have played a vital yet somewhat equivocal role in establishing our understanding of this complex family. The conventional method of studying plants, with its reliance on examining dry, pressed specimens preserved in herbaria, is inadequate for many of the more complex flowers. Dried flowers neither preserve the orientation of the petals and stamens, nor details such as colour, markings, scent and presence of nectar. Consequently our understanding of families like the Iridaceae, Amaryllidaceae and Hyacinthaceae followed from their popularity as garden plants and their depiction in important nineteenth-century botanical books and magazines such as *Curtis's Botanical Magazine*, H.C. Andrews's *The Botanist's Repository* and the *Flore des Serres des Jardin de l'Europe*.

Their modest size and relative ease of translocation made bulbous plants particularly popular amongst European growers, and the exotic diversity of their shapes, colours and sometimes fragrances delighted the Victorians. This led, understandably, to an exaggerated value being placed on floral form as an indicator of relationship between species. At the time, little or no thought was directed to an interpretation of these forms which so entranced botanists and growers. The direct consequence of this was that plants with apparently similar flowers were assumed to be related to one another. In the earliest classifications of these families, plants with like flowers were united in a genus, even when their underground parts (bulbs, corms, or rhizomes), leaves and fruits

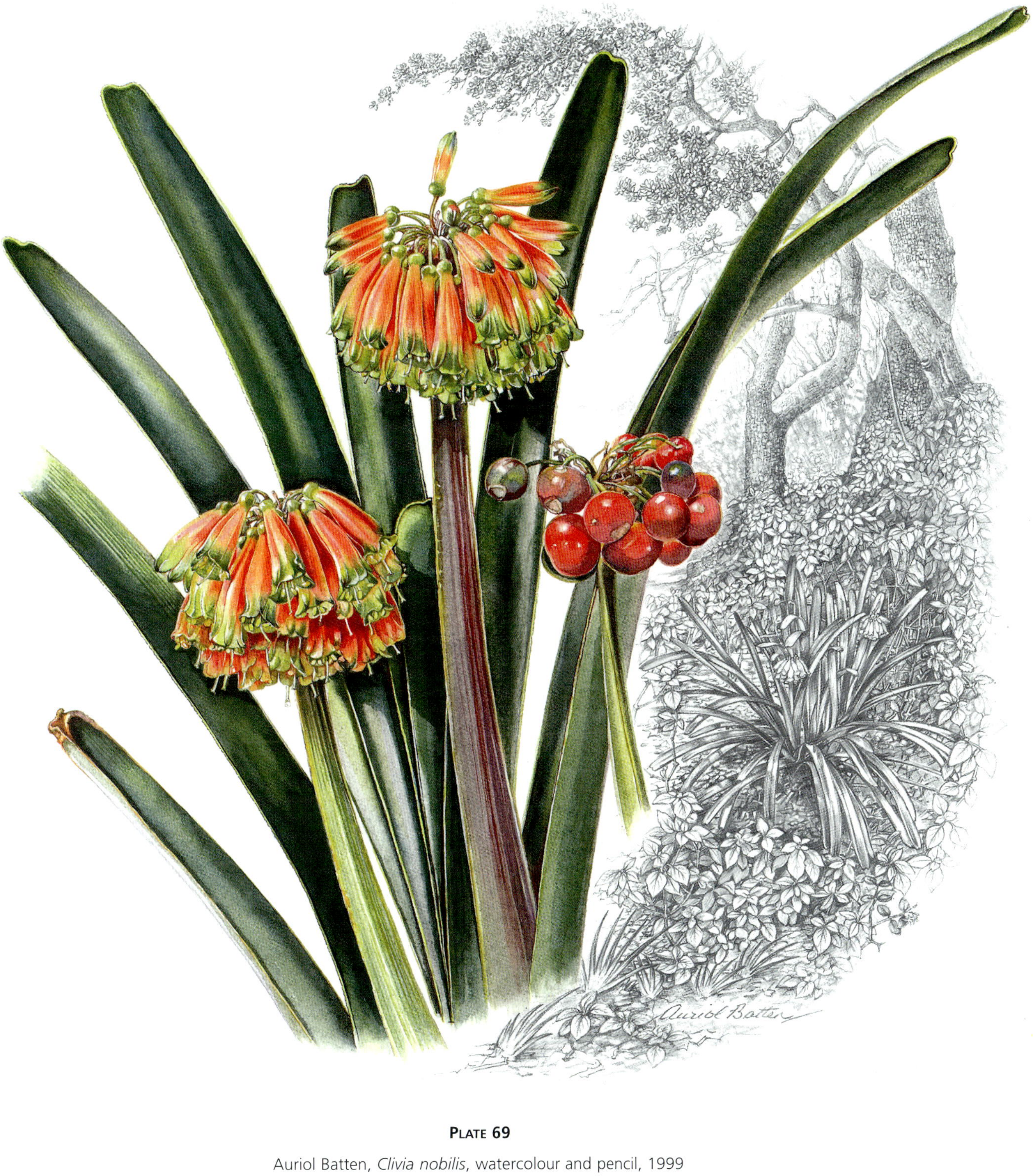

Plate 69

Auriol Batten, *Clivia nobilis*, watercolour and pencil, 1999

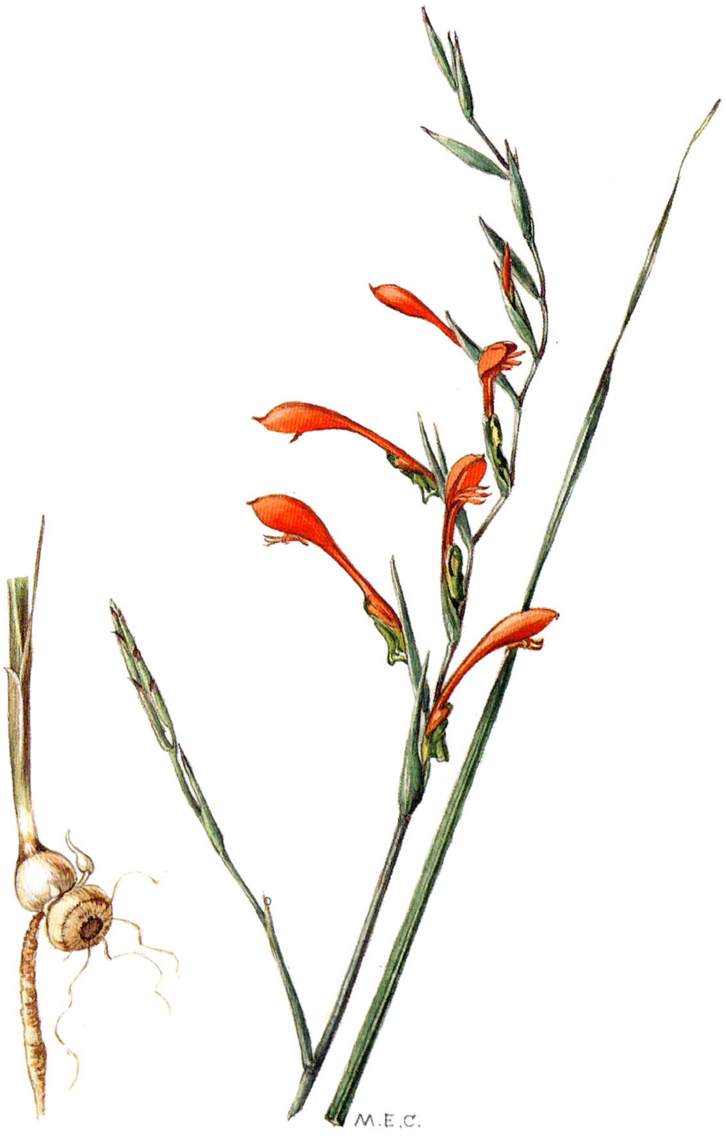

PLATE 70

M.E. Connell, detail from *Gladiolus (Kentrosiphon) saccatus,* 1959, watercolour and pencil: reproduced in *The Flowering Plants of Africa,* 1961

> The singular flowers of many bird-pollinated species confounded early botanists, leading some to classify quite unrelated species together. The genus *Antholyza*, for instance, at one stage accommodated both *Tritoniopsis caffra* and *Gladiolus saccatus*, two species which are now understood not to be very closely related.

were very different. The power of the flower has seldom been greater. This distorted grasp of ecological strategies blinded past scientists to the other side of the floral coin, and while the floral differences between related species continued to be a focus of attention, little comment can be found about the often remarkable similarity in the flowers of quite unrelated species.

It has taken some time to dismantle these early classifications and replace them with systems that reflect more accurately the evolutionary or hereditary relationships between species and genera. The genus *Antholyza* is a striking example of this process, having been established by the Swedish biologist Linnaeus (1707–1778) in 1753 for two species of the iris family which bore remarkably proportioned red flowers with an elongate floral tube which was narrow below and wide above, and from which the anthers protruded conspicuously. Later authors added to the genus various other species with more or less unusually shaped red or orange flowers until it contained a most diverse assemblage of species. It was not until the beginning of the twentieth century that it gradually became understood that such flowers were pollinated by sunbirds, and that their particular form and colour represented specific and highly specialized adaptations for this strategy. Calling a rose by any other name completely clouds the nature of its relationship to other species, dimming the light on any investigation into the origin of any of its peculiarities.

Rudolf Marloth's *The Flora of South Africa*, published between 1913 and 1932, reflects an extraordinarily holistic approach to plants and particularly to flowers (see Plate 68). Marloth (1855–1931), trained as a chemist, was especially fascinated by the physics and chemistry of colour in flowers, and the text and plates contain several references to it as well as to various alleged pollinators. Several of the latter, such as monkey beetles on the peacock lily, *Spiloxene*, and the Table Mountain Beauty butterfly on the Cluster Disa, *Disa ferruginea* (Plate 52) and Red Crassula, *Crassula coccinea* (Plate 96), have subsequently been verified – often only in recent years – as the major pollinators of these plants. Other purported associations, however, such as

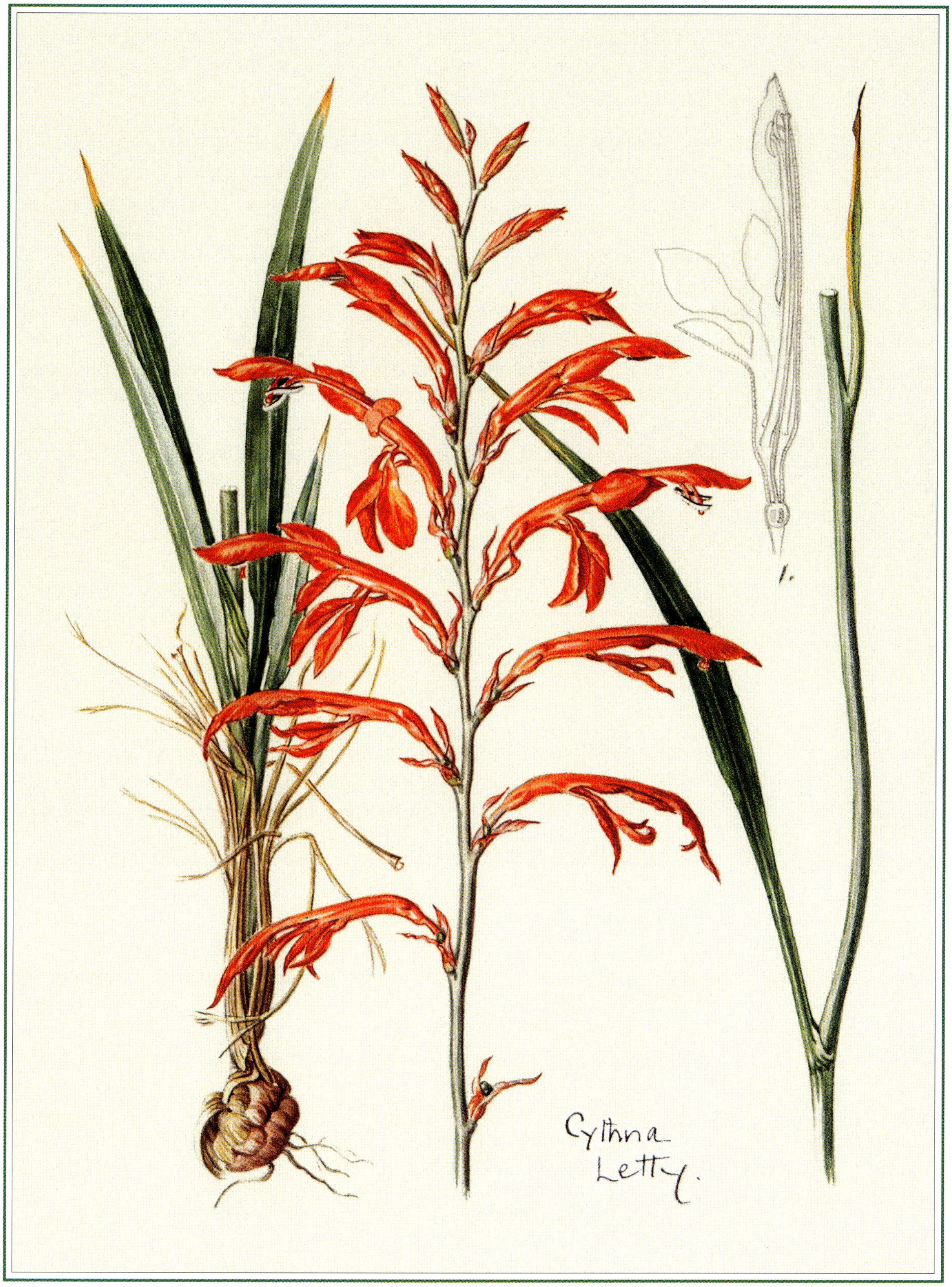

Plate 71

Cythna Letty, *Tritoniopsis (Anapalina) caffra,* 1961, watercolour and pencil: reproduced in *The Flowering Plants of Africa,* 1962

the hawkmoth visiting the kukumakranka, *Gethyllis*, are less inspired (Plate 80). Nevertheless, Marloth must be credited for his pioneering observations on the biology of Cape plants.

This increasing awareness of a real and highly functional aspect to flower shape was slow in being integrated into schemes of classification. The English botanist N.E. Brown (1849–1934), for instance, in 1932 proposed the recognition of no less than six new genera to accommodate the species previously included in *Antholyza*, each comprising either a single species or a rather heterogeneous assemblage in which apparent floral similarities were still paramount. Here matters rested for another fifty years until the philosophy of classification entered its present, phylogenetic phase. Phylogeny is the study of evolutionary relationships and the phylogenetic system of classification is founded on the simple tenet that specialized character traits having a unique evolutionary origin can, and should, be used to define lineages of species. Conversely, unspecialized features, having been present in the ancestor of that lineage and therefore potentially in each of the descendants, are useless as indicators of relationships. However handy such ancestral features may be in suggesting groupings of species, they are actually devoid of any real information about the evolutionary relationships of those species. Groups so defined, like *Antholyza*, are more specious than speciose.

A second aspect of the new school of evolutionary thought was the formalizing of the process of classification, and the explicit understanding that genera (and for that matter other higher levels of classification) must ideally be monophyletic. This means that the genera should include both the ancestor and all of the descendant species. Many aberrant species formerly placed in separate genera because of their highly distinctive appearance have subsequently been renamed in order to reflect their actual relationship with other species placed in other genera. For the bird-pollinated species of Iridaceae the paternity suits have now been resolved and diverse red-flowered plants formerly separated from their florally more conventional relatives into separate genera have been reunited with them. The Rat's Tail, *Antholyza ringens*, is now correctly seen as a specialized, bird-pollinated species of *Babiana*, a large genus better known for its brilliant blue, often scented flowers which are attractive to bees; the genus *Homoglossum* has been identified as a spurious rag-bag of unrelated species of *Gladiolus;* and *Anapalina* is understood to have harboured two separate lineages of bird-pollinated species of *Tritoniopsis*.

Functional and beautiful

Flowers adapted for pollination by African sunbirds constitute a vivid example of a specifically adapted and specialized form. All flowers are, however, essentially functional structures. Hard as it may be to accept, aesthetic considerations are meaningful only to humanity and completely irrelevant to the plant. Any flower is quite simply a suite of adaptations that has evolved to enable it to function efficiently in its environment, competing for the attention of pollinators while avoiding predation – and rewarding the pollinators without exhausting the resources of the plant. The green or vegetative parts of a plant serve the functions of photosynthesis and growth, while the flower is basically a reproductive organ, a vital link that leads to the production of fruits and seeds. The latter serve to both propagate a new generation of individuals and to disperse the species farther afield. 'Go forth and multiply' is an injunction of universal application.

The basic parts of a flower are finite in type and usually number, even if apparently limitless in variety. The treasures beyond price in the flower are the reproductive organs of the plant. In the centre are gathered the female parts of the flower, or carpels, containing the ovules and surmounted by a slender style and stigma. Attending the carpels are the male parts or stamens, containing the pollen. Surrounding this hotbed of life are whorls of accessory structures, collectively called the 'perianth', which both advertises and protects the sexual organs.

In the earliest flowers there was little or no differentiation between these accessory structures, but in the course of evolution their roles

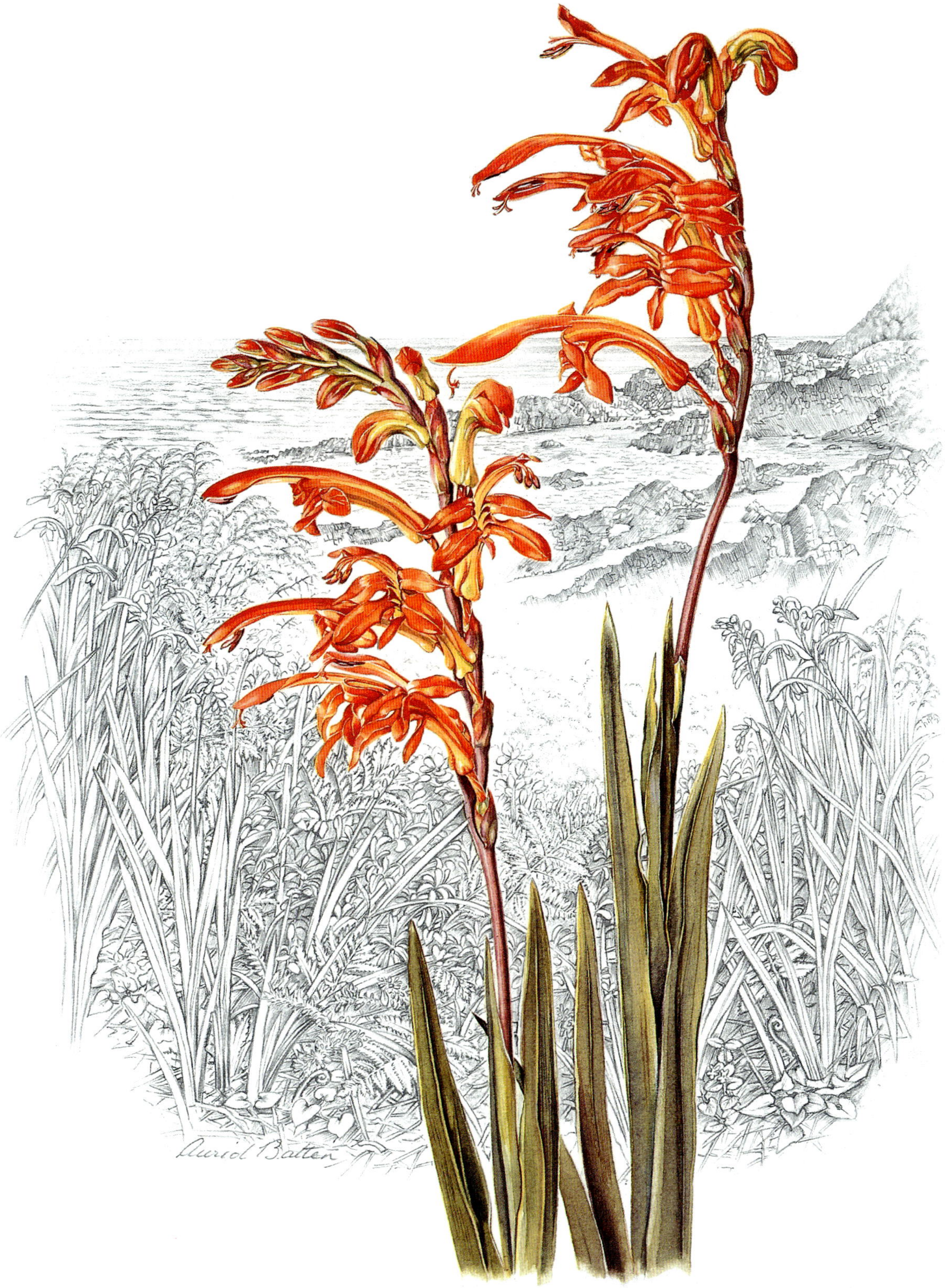

Plate 72

Auriol Batten, *Chasmanthe aethiopica, c.* 1980–1985, watercolour and pencil: reproduced in A. Batten, *Flowers of Southern Africa* (Sandton, 1986)

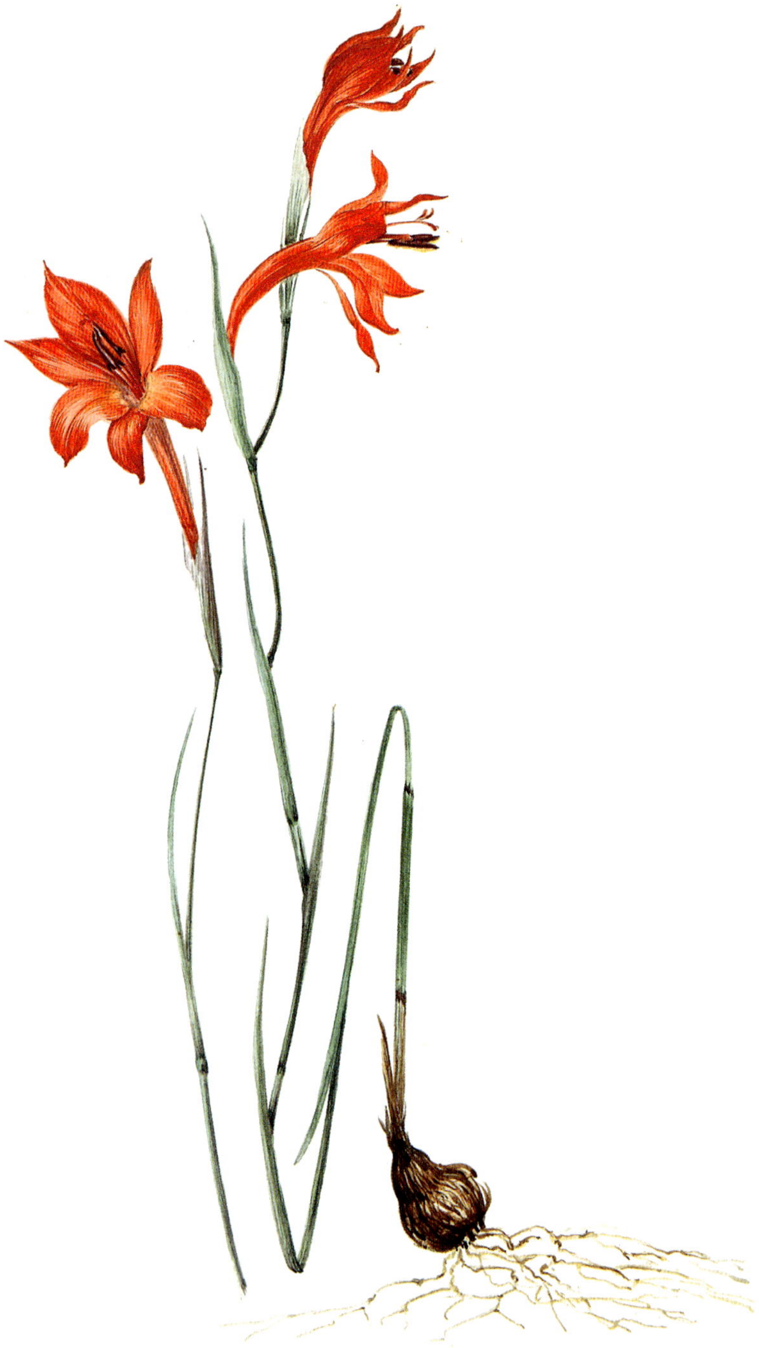

PLATE 73

Stella Gower
detail from *Gladiolus (Homoglossum) priorii*
c. 1923, watercolour:
reproduced in *The Flowering Plants of South Africa,* 1926

became refined. The outer parts are usually specialized as botanical bouncers, discouraging predators and parasites from gaining access to the inner reaches of the flower, and thereby freeing the inner parts from the necessity of playing both defensive and seductive roles. The primary function of the inner accessory parts – or petals – in all flowers which rely on a living intermediary for pollination, is to attract the attention of the go-between and encourage it to visit the flower, however briefly. During the visit, through a sleight of stamen, pollen is deposited on the unwitting visitor and any other pollen which it might be carrying from a previous visit to another flower is brushed off by the stigma. Repeated visits to different flowers will ensure that several achieve fertilization. Naturally there is a payment for services rendered, usually in the form of food. In the eyes of the visiting animal, therefore, the visual signal is an advertisement of fare, and to a pollinator a field of flowers is no more than a downtown block crammed with cafes.

Food for thought

Floral rewards usually satisfy the most basic need of a living organism – that for sustenance. At a most mundane level the reward consists of the attractive organs themselves, a case of the pollinator turning *on* the table. No flower can put up with much of this sort of wear and tear and still remain attractive to any discerning visitor, so the rewards are much more often a renewable resource, particularly nectar. This ambrosia is a solution of various sugars and sometimes also amino-acids and is a rich food source much sought after by many animals. Its secretion continues throughout the life of the flower. So great is its value in attracting pollinators that the capacity for its production has evolved numerous times within the flowering plants and almost every floral organ has been pressed into its production in one species or another.

Nectar, of course, appeals only to those animals which require or even tolerate a liquid diet. For those that prefer more solid food, some plants offer the pollen itself, which, together with nectar, is kneaded into the highly nutritious bee-bread of the honeybee. To compensate for

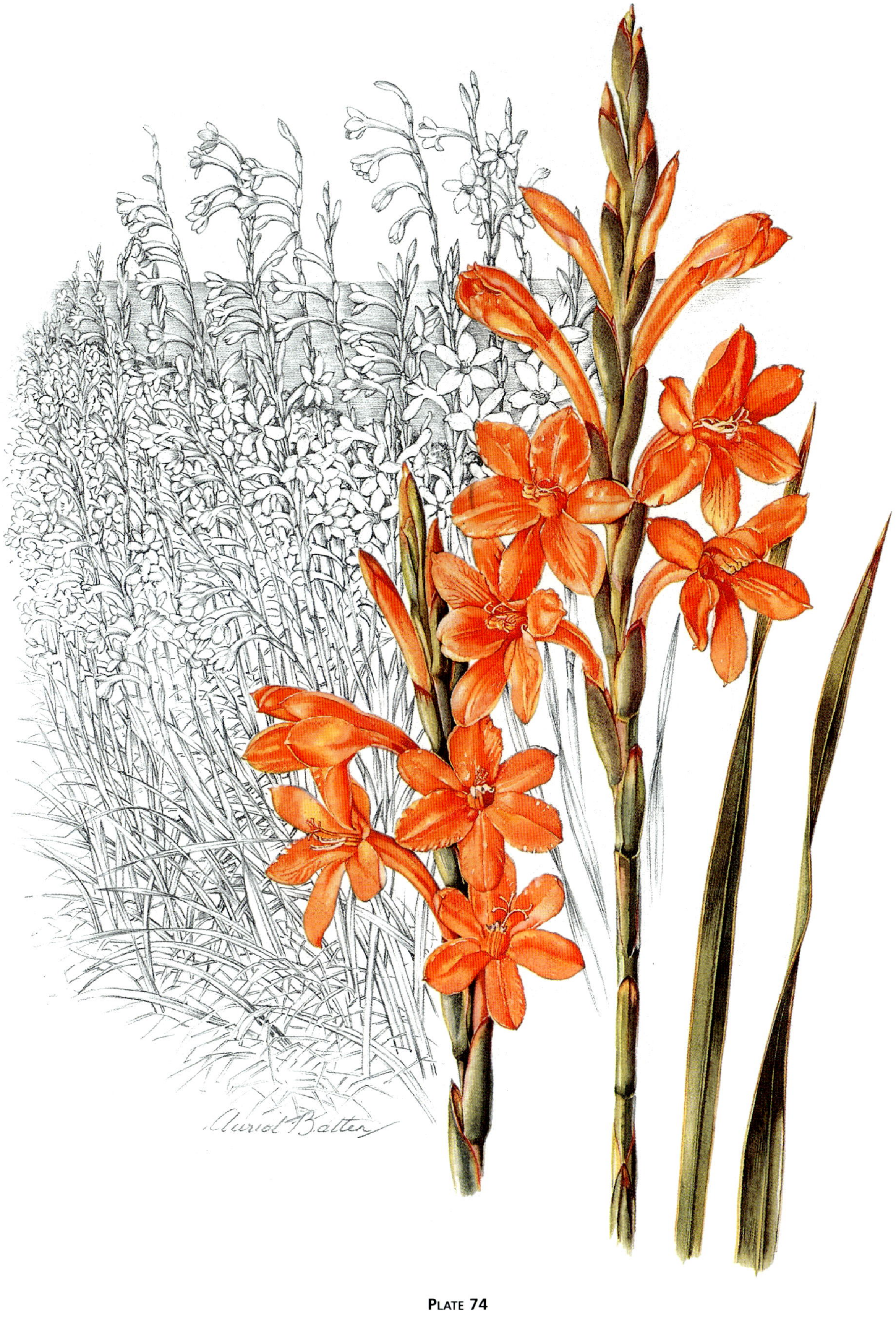

Plate 74

Auriol Batten, *Watsonia pillansii, c.* 1980–1985, watercolour and pencil: reproduced in A. Batten, *Flowers of Southern Africa* (Sandton,1986)

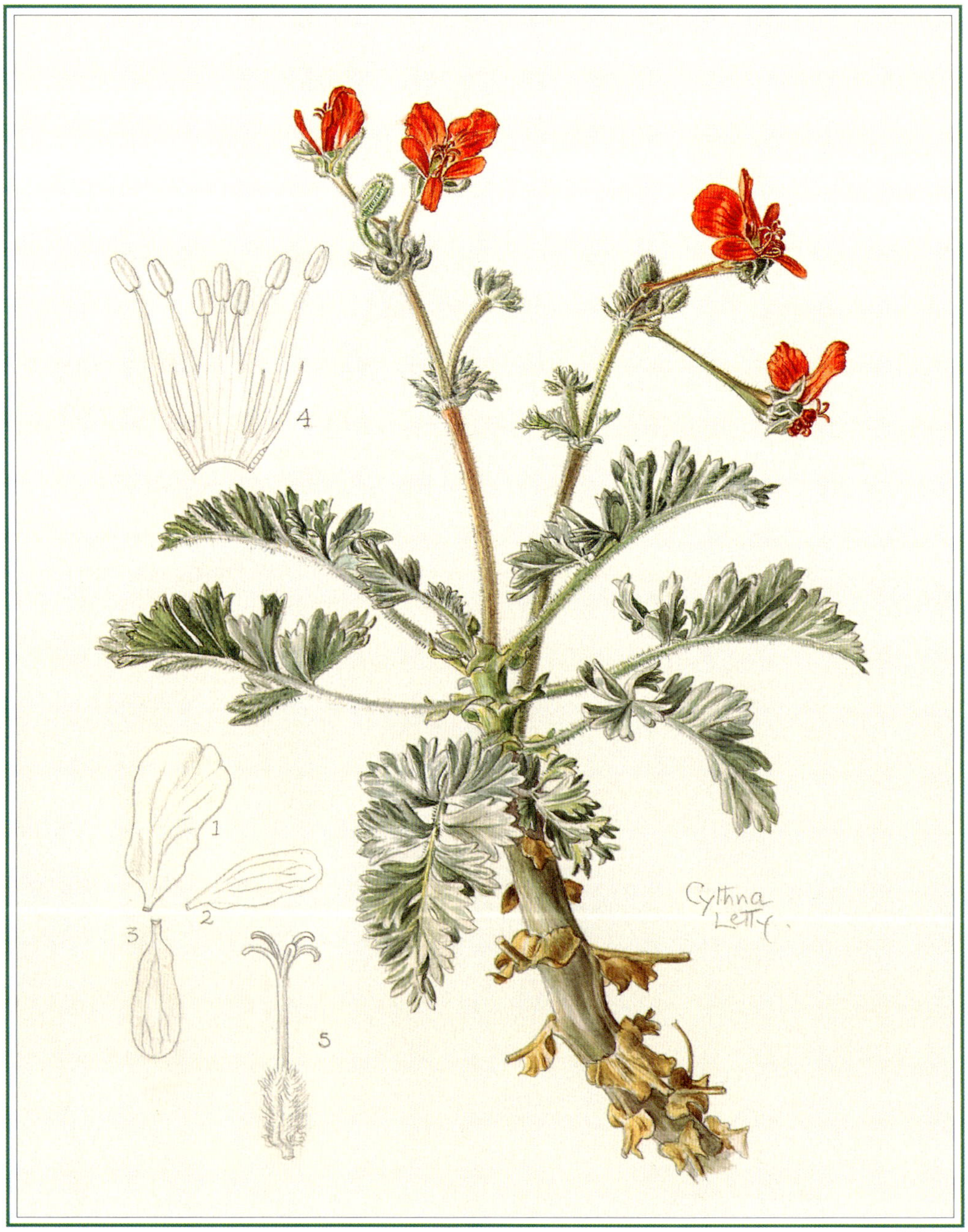

PLATE 75

Cythna Letty, *Pelargonium fulgidum,* c. 1965, watercolour and pencil: reproduced in *The Flowering Plants of Africa,* 1966

the loss of some pollen, extra quantities may be produced or it may be presented in such a way that some is placed on the body of the visitor well away from its mouthparts and where it can be transported to other flowers. While pollen may be sacrificed, the carpels themselves never constitute a reward, for loss of the offspring would negate the venture entirely.

Less commonly offered than either nectar or pollen are a range of more sophisticated products with a limited appeal to the discerning visitor. Oils or waxes used in nest construction by certain bees are secreted by some plants, but the most esoteric reward must be that purveyed by some South American orchids. These flowers secrete the components of sexual scents which are used by certain male bees to attract the females of their species. Such rarefied products are not common but several other orchids have gone further than playing the role of procurer and play the role of the prostitute instead. Their flowers are constructed so as to mimic the female bee herself, complete with her own sexual fragrance. Naive males, captivated by the scent and shape, are completely duped by these mannequins and attempt to mate with them. But if their own ambitions are frustrated, their amorous aerobics successfully consummate the sexual relations of the plant.

More usually, however, flowers that manage to attract their pollinators without providing any reward at all do so by false promises of nutrition rather than procreation. In these cases the flowers of the deceitful plant closely mimic those of a rewarding species in appearance but not in payment for services rendered. The orchid, *Disa draconis*, for example, has flowers that resemble in colour and markings the flowers of several species of *Gladiolus*, *Lapeirousia* and *Pelargonium* that grow nearby along the Cape west coast and which are pollinated by a local long-tongued fly. Like the other flowers visited by this fly, those of the disa have a long slender spur, but unlike the other flowers it is devoid of nectar. The species is capitalizing on its resemblance to nectar-producing species that are visited by the fly for this reward. The drawback to this freeloading strategy is that the deceitful mimics can never become too numerous in relation to their rewarding models. If this were to happen then the potential pollinators would begin to spurn such flowers as unrewarding, with dire consequences for the future of the species.

A great variety of agents are used by flowers for pollen transfer, some more exclusively than others. None, however, is completely reliable and some plants have therefore abandoned the need for sexual relations with another individual, either partially or entirely. While several self-fertile plants produce normal flowers and only self-pollinate in the absence of pollen from another flower, a few others have opted for this strategy completely, producing much reduced or rudimentary flowers which may never open at all. In such species the surety of setting seeds without having to rely on some external trafficker is balanced by the dangers of inbreeding. Cross-pollination is essential for the long-term survival of a species because it helps maintain the genetic diversity that allows a species to adapt to changing conditions. Self-pollinating plants sacrifice the potential for maintaining genetic diversity in return for the short-term security of reproductive success. Each strategy is appropriate under certain circumstances.

Spice of life

To us, amongst the most visual of creatures, the precise correlation between natural form and function is nowhere more strikingly evident than in the relationship between the flower and its pollinator. Although almost infinite in their variety, pollinators can be grouped into relatively few types or classes, and amongst these only a small number account for the pollination of the majority of flowering plants, particularly those with flowers of any noticeable size. Largest by far in terms of their numbers in the animal world, insects are also the most abundant pollinators.

The flowering of plants, in all senses of the expression, is intimately connected to the emergence of insects. Appearing about 150 million years ago, the first flowers were small or minute, with stamens and carpels few but variable in number. Lacking any modified petals, these flowers were probably visited by a range of short-tongued insects, as yet relatively unspecialized for the task of pollination. Various, more obviously modern forms of flowers with fixed numbers of floral parts, including petals, and even floral tubes, evolved during the subsequent fifty million years. By this time, the majority of contemporary plant families were present in the fossil record, albeit only in their most basic form.

However, at this stage two major advances had been achieved which were to affect the way in which further floral evolution and diversification were to proceed. These were the development of the floral tube through the congenital fusion of floral parts, and the change from radial to bilateral symmetry. A floral tube serves as a reservoir enabling nectar to be stored in quantities not otherwise possible, thereby both increasing the reward to a potential pollinator and limiting access to it. This is where the crux lies. Nectar at the bottom of a tube is available only to visitors with equivalently long mouthparts, whereas nectar in shallow dish-shaped flowers is available to all. The great potential for specialization between plant and pollinator which became possible with the development of a floral tube was not long in being exploited.

The massive extinction of animals which took place around sixty-five million years ago and which saw the demise of the dinosaurs, seems not only to have had little adverse effect on flowering plants but actually presaged an extensive explosion of evolution which has culminated in the diversity we see today. This second radiation is even more closely linked with insects than the first, and the number of specialist insect-pollinators – butterflies and moths, long-tongued flies and bees, which are today's most important flower visitors – proliferates in the fossil record at about this time.

While the main consequence of the development of the floral tube was an increase in the variety of pollinators, the major consequence of the development of bilateral symmetry was an increase in the variety of flowers. Bilateral symmetry meant that no longer did all the members of each of the floral whorls have to be identical, thereby opening up endless avenues for evolutionary exploration: some petals could become specialized for attraction and others for protection, or some could be lost entirely, and likewise for the stamens and even carpels. In a radially symmetrical, or star-shaped flower, there are several equally direct ways to reach the centre where the pollen or nectar rewards are held, and no certainty which of them will be used by any individual visitor. In a bilaterally symmetrical flower, however, the entrance is specifically defined by both sepal orientation and markings. Because of this, anthers and stigmas can be aligned in all sorts of ways in order to place and pick up pollen from very specific parts of the visitor's body.

While insects were both the impetus for the development of the primitive flower and a strong driving force for its later elaboration, the evolutionary radiation of many insect groups – particularly those of moths and butterflies, many flies, and bees – was only possible once the flower had evolved. This intimate evolutionary relationship, in which it is impossible to isolate cause and effect, epitomizes the process of co-evolution.

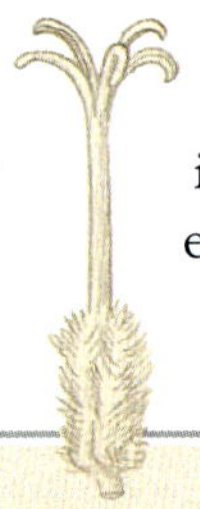

A brief examination of any regional flora, or even a walk in the wild, will reveal two important principles, each the result of this co-evolutionary process but seen from a different perspective. Firstly, many plant families or

Flowers adapted to pollination by birds are often characterized by their orange or red colour, tubular shape and reduced lower petals. Similar flowers occur throughout the tropics and sub-tropics in a plethora of families. The orange trumpets of the Cape Honeysuckle, *Tecoma (= Tecomaria) capensis* (family Bignoniaceae), for instance, are remarkably similar in general appearance to the flowers of many species of the southern African Iridaceae, especially species of *Chasmanthe*, *Gladiolus* and *Watsonia*, and the blooms of *Clivia nobilis* recall those of many other bird-pollinated plants, including *Aloe* and *Lachenalia*. Even the occasional species of *Pelargonium* has become specialized for bird-pollination.

The preponderance of red colouring among bird-pollinated flowers is not because of an innate preference by birds. While both birds and insects have good colour vision, insects are unable to perceive shorter wavelengths of light and are thus colour-blind to red. This fact has been exploited by bird-pollinated plants to ensure that their flowers are invisible to insects which might damage or rob them of their nectar. It is no coincidence that fruits which rely on avian or mammalian agents for seed dispersal are green when unripe, and thus difficult to distinguish from the foliage by both birds and insects, but red or yellow when ripe, thereby becoming selectively visible to potential dispersers while remaining invisible to insect predators.

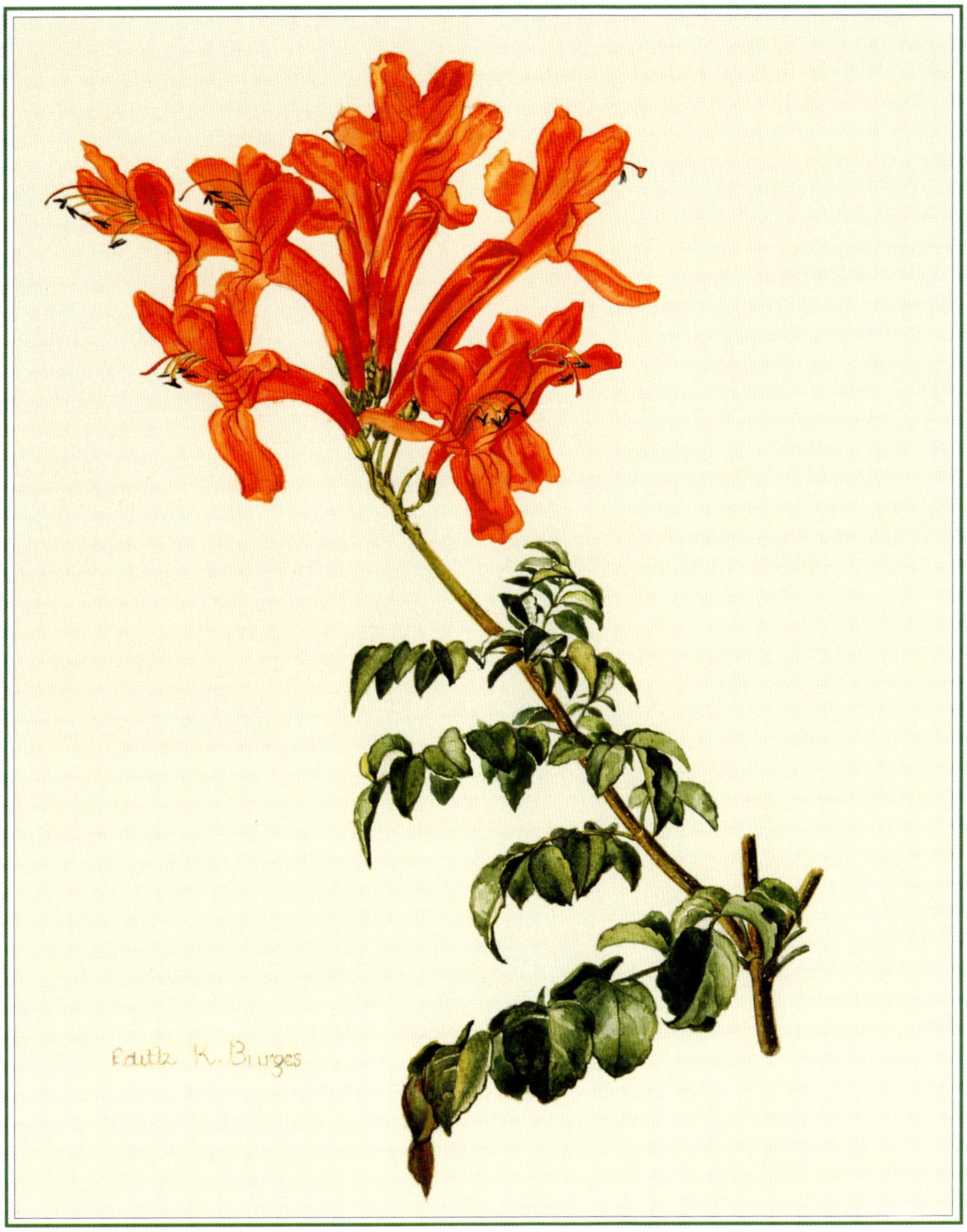

PLATE 76

E.K. Burges, *Tecoma capensis, c.*1940, watercolour and pencil: reproduced in *The Flowering Plants of South Africa,* 1943

even genera exhibit a diversity of flower form, clearly the result of related plant species specializing for pollination by different types of pollinator. Within the genus *Gladiolus*, for example, flowers range from star- or bowl-shaped, through a diversity of more or less two-lipped forms to highly asymmetrical, tubular types, and the variation within the whole family is substantially greater even than this. The second striking thing is the similarity between the flowers of species from quite different genera or families, in this case the result of unrelated species specializing for pollination by the same type of pollinator. Some species of *Gladiolus* have flowers which are so unusual in the genus that they were originally described in another, and can only be correctly placed in *Gladiolus* by examining details of their seeds and chromosomes, which are unaffected by the influences being exerted on the flowers. This absolute interdependence between floral form and pollinator function is the lodestar which links the appearance of a flower with the identity of its pollinator and enables the one to be inferred from the other.

Flowers and their pollinators

FLORAL FANFARE

We have already briefly described the striking, trumpet shape of the flowers of many bird-pollinated plants. Although present almost throughout the world, bird-pollinated flowers – amongst the most distinctive of all floral shapes – are absent from Europe and northern Asia, and the importance of bird-pollination was only fully recognized in the early decades of the twentieth century. The flowers of bird-pollinated plants are usually bright red (see Plate 68), sometimes yellow or greenish, often with little or no markings of any sort but sometimes with the petals tipped green (see Plate 69) or black. The floral parts are often thick-textured and sturdy enough to survive probing by the bird, and the anthers are mostly held in a position to brush against the feathers of the head or throat as the bird pushes its beak into the flower to reach the nectar (see plants illustrated in Plates 68 and 77). Bird-pollinated flowers are always unscented and typically produce large quantities of dilute nectar, a case of watering the more expensive vintages because of the energy costs of secreting sugars. The bird-pollinated *Protea repens* owes its traditional name 'sugarbush' to its copious nectar which was collected and boiled down to a sweet syrup by Cape residents in the last century.

Although bird-pollination is widespread in the tropics and subtropics, each of the continents supports a different suite of bird, and often flower families adapted to this mode of pollination. The main nectarivores in the New World are the hummingbirds and honeycreepers (families Trochilidae and Coeribidae); in Africa, the sunbirds (family Nectariniidae) and sugarbirds (family Meliphagidae); in Australasia, the honeyeaters (family Meliphagidae) and brush-tongued parrots (family Loriidae); and in the Hawaiian Islands, the Hawaiian honeycreepers (family Fringiliidae). This is a clear indication that bird-pollination arose independently in each of these faunal regions.

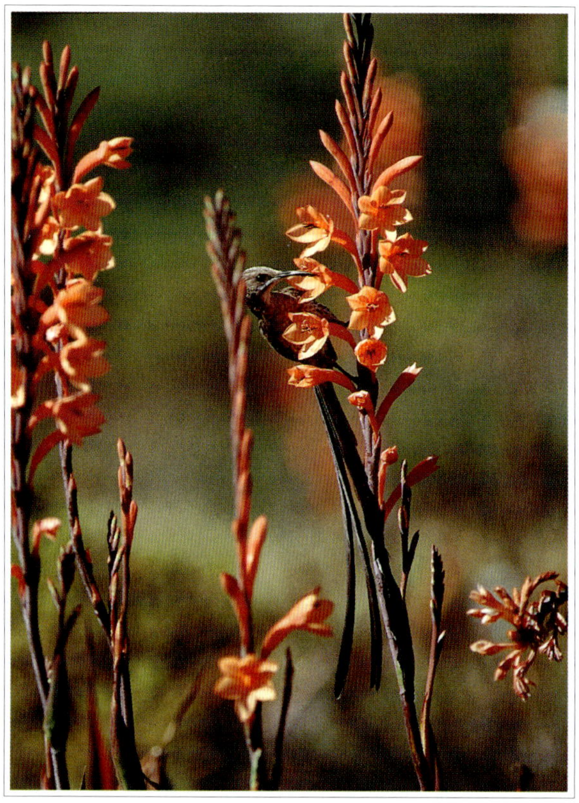

PLATE 77

Cape sugarbird visiting *Watsonia tabularis (Photographer:* Colin Paterson-Jones, Cape Town)

There is always a greater diversity of bird-pollinated plants than of pollinating birds, though some plant genera or families are more typical of one region than another. Africa without aloes just would not be Africa, and almost all the species in this large and successful genus are bird-pollinated. Similarly, bird-pollinated members of the iris family are rarely encountered outside Africa, yet are quite common in southern Africa and have evolved repeatedly here.

Their size and vibrant colours have made many bird-pollinated species welcome additions to subtropical and tropical gardens. Think bold and beautiful and the flowers that spring to mind are sure to include many that are adapted to pollination by birds. The coral tree (*Erythrina:* Plate 5), strelitzia (Plates 106, 120–126), Cape honeysuckle (*Tecoma capensis:* Plate 76), aloe (Plates 32, 108, 116), and red-hot poker (*Kniphofia*) are typical examples.

DANCING IN THE DARK

The vast majority of moths are active at night, which deprives them of many of the visual cues available to day-flying insects. Consequently, bright colours and complex markings are not a feature of moth-pollinated flowers, and white or other pale shades predominate as these are more easily perceived at low-light intensities (Plates 30, 43). A fringed or broken outline is also a common feature of such flowers as it increases the optical area occupied by the bloom. Many other species have abandoned visual attraction entirely and are dull green or brown, even mottled or camouflaged. This is thought to render them inconspicuous to either herbivores or other nectar-feeders which might rob the flower of its nectar without pollinating it.

In another variant of this strategy certain moth-pollinated plants close during the day and may even lose their scent. At night, however, they open and begin to release their scent. The heady wafts of fragrance produced in the early evening by many night-blooming plants herald the start of business. Even more wonderful than these clock-watchers is the Cinderella of flowers, *Gladiolus liliaceus*. Inconspicuous during the day, the large lily-like flowers are mottled brown and tan, and quite odourless. With the fall of dusk, however, an almost magical transformation occurs as the flowers change colour to a pale lilac or mauve and become beautifully perfumed. At dawn these evening embellishments are lost again and the flowers revert to their drab daytime state.

More than most other flower types, moth-pollinated flowers rely on scent for attraction. The scents are often either heavy and lily-like, or spicy and redolent of clove, unlike the sweet or floral fragrances produced by some bee-pollinated flowers. The specialized mouthparts of moths and butterflies are unsuited to anything other than a liquid diet, but well adapted to sucking nectar from floral tubes. Another characteristic of moths and butterflies, however, poses some problems for the flower. The loose scales which cover the body and wings of a moth are an unstable surface unsuitable for carrying pollen between flowers. The only part of the insect which is not covered in scales is the proboscis. It is a characteristic of moth-pollinated flowers, therefore, to have the anthers hidden in the mouth of the tube or only shortly protruding from it, where they contact the base of the insect's mouthparts. Sticky nectar residues help the pollen grains to adhere.

A great variety of moths from many families are involved in pollination. The larger flowers, however, are mostly pollinated by members of the families Noctuidae and Sphingidae. Noctuid moths are medium-sized and often dull in colour, with mouthparts normally no more than 40 mm long. Sphinx moths are mostly large-bodied fast-flying insects with narrow wings and

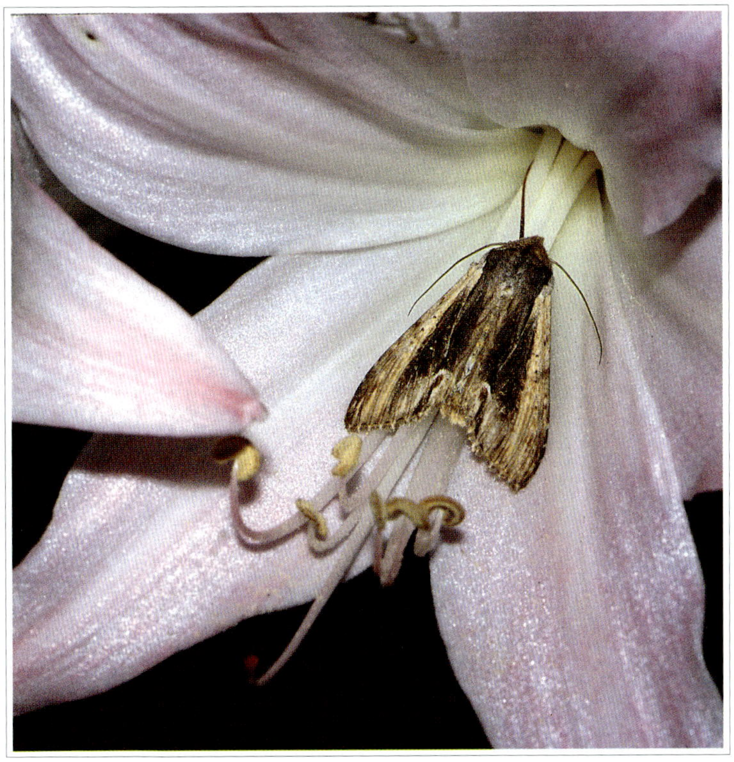

PLATE 78

Cucullia extricata moth visiting *Amaryllis belladonna*
(*Photographer:* Colin Paterson-Jones, Cape Town)

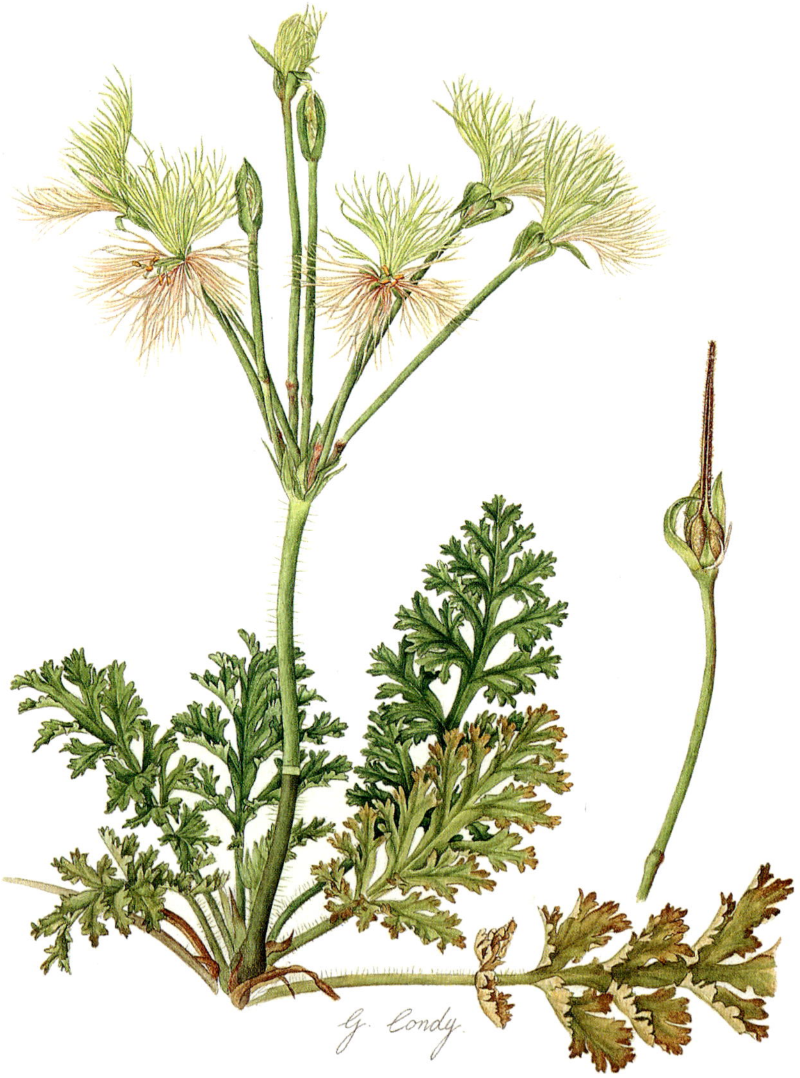

PLATE 79

Gillian Condy, *Pelargonium schizopetalum,* watercolour: reproduced in *The Flowering Plants of Africa*, 1991

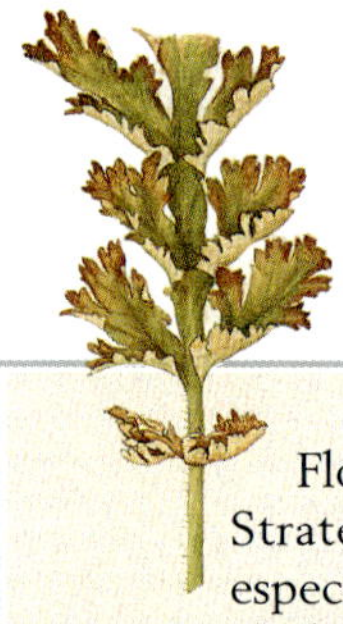

Flowers adapted to pollination by moths rely mainly on scent to attract their nocturnal visitors. Strategies for improving their visibility at low light intensities include a pale colouration, especially white, and a fringed or broken outline.

A slender floral tube concealing the nectar is also characteristic of moth-pollinated flowers. The flowers of *Pelargonium schizopetalum* exhibit many of the features typical of moth-pollinated species. The similarity of the flowers of the kukumakranka (*Gethyllis afra*) to various moth flowers, however, is quite specious. Bursting directly from the parched ground in mid summer, the almost impossibly fragile flowers of *Gethyllis* last only a few days, during which time they exude a wonderful perfume. Their colour, fragrance and slender floral tube all conspired to convince the botanist Rudolf Marloth that the pollinator had to be the large Convolvulus Hawkmoth, and so he directed that it be depicted visiting the flowers of *Gethyllis* on the illustration in his *Flora of South Africa*. More careful examination, however, would have revealed the implausibility of this association: the floral tube in *Gethyllis* is solid, serving as a stalk rather than as a nectar reservoir and the flowers actually tend to close at night. The true pollinators are in fact pollen-collecting bees and not moths at all. Appearances can be deceiving.

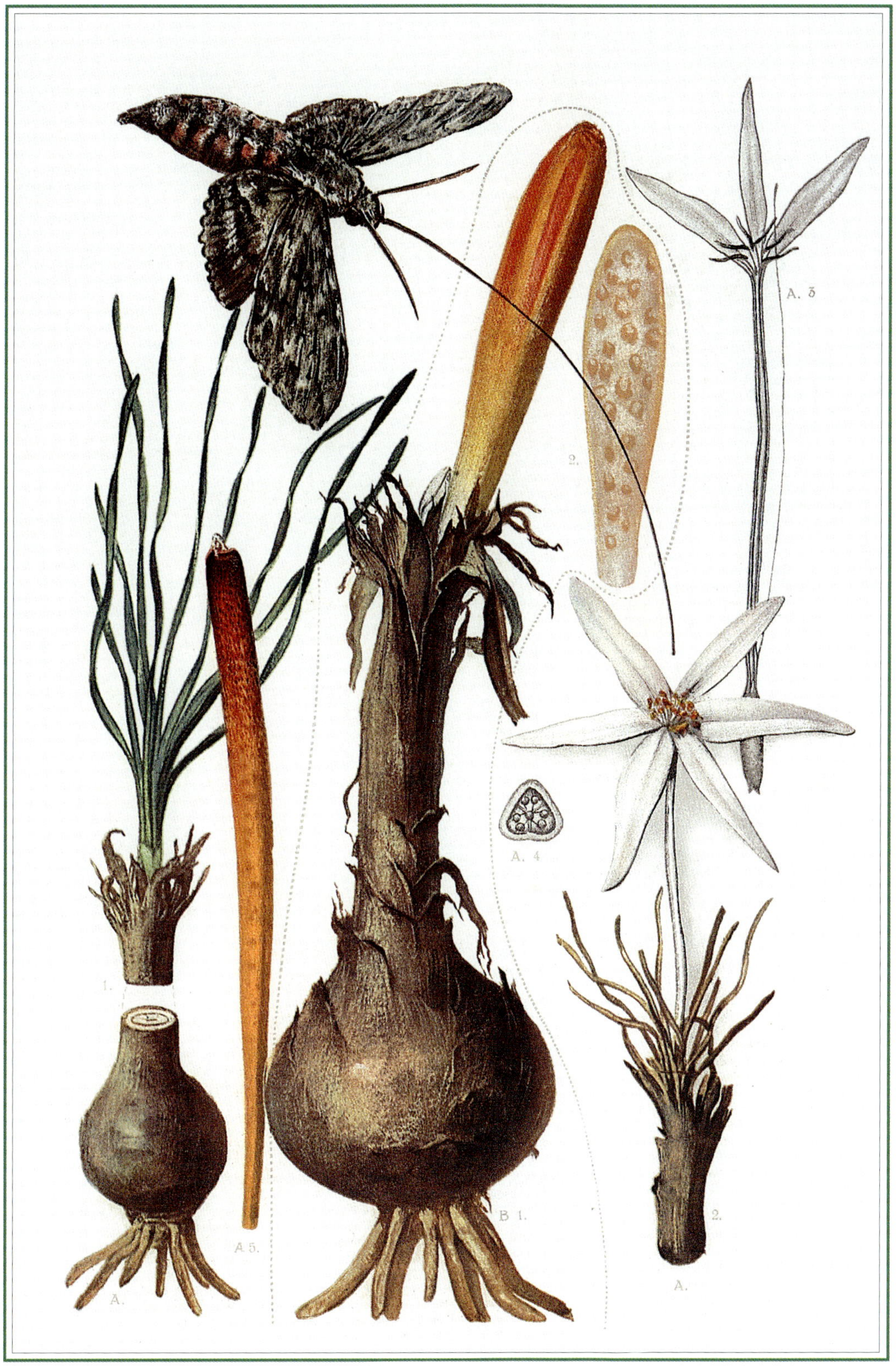

PLATE 80

Esther Smith, A: *Gethyllis afra* and the Convolvulus Hawkmoth, *Agrius convolvuli*; B. *Gethyllis ciliaris*; gouache and watercolour: reproduced in Rudolf Marloth, *The Flora of South Africa*, vol. 4 (Cape Town and London, 1915)

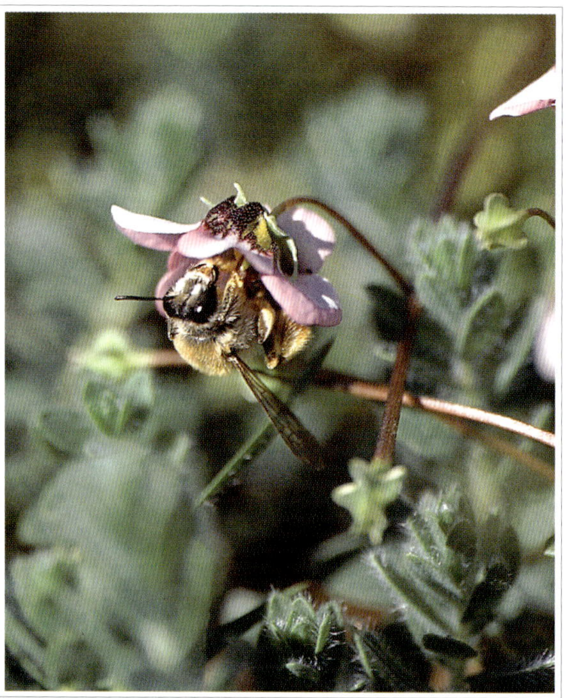

PLATE 81

Rediviva bee visiting *Diascia cardiosepala (Photographer:* John Manning, Cape Town)

may have rather longer mouthparts. The large Convolvulus Hawkmoth, for instance, has a proboscis up to 110 mm long.

DARE TO BE DIFFERENT

Bees, unlike almost all other insect or animal pollinators, are dependent on floral products for sustenance at all stages of their life-cycle. In other insects and in mammals, pollen or nectar either form only part of the diet or are consumed for only part of the life-cycle. In addition, the larval food requirements in bees are met by the parental generation, which collects sufficient nectar and pollen to sustain the larva and stores it alongside the egg. Because of this, individual adult bees spend more of their time visiting flowers than any other flower visitor. This means that they have to be both efficient at it and versatile enough to handle different flowers if they are to remain above the bee bread-line.

It comes as no surprise then, that bees are by far the most important pollinators of plants, nor that bee flowers are highly variable in their appearance and in the rewards they offer. Flowers can be star-shaped, tubular or variously two-lipped, often marked with contrasting lines or spots which will guide the insect towards both the reward, and the anthers which will dust the insect's body with pollen (see Plate 82). These markings, although usually visible to us, are in some cases visible only to the insects, which can see further into the ultraviolet spectrum than we can. The flowers may have keels which conceal the anthers or flag-like petals for attracting the insects, they may be unscented or scented, and they may last several days or only a few hours. The rewards offered are most commonly pollen and nectar, but in other cases may be shelter, oils, pheromones or even the promise of sex itself. Bee flowers may be almost any colour except red, a hue which bees, like most insects, see poorly or not at all.

Flowers in which pollen is the reward are usually star-like, with prominent, yellow anthers in the centre (Plates 64, 65, 119). The petals are often white or yellow, colours which are highly reflective and attractive to a diversity of bees. Offering pollen as a reward has several potential dangers to the plant and the ways in which flowers have overcome these are wonderful in their variety. Because it is a non-renewable resource, pollen flowers cease to be rewarding to visitors once all of their pollen has been removed. Bees are quick to realize when this has happened and will then switch their attentions to other flowers.

For the flower two options are open: increase the rate of turnover, or retain the pollen in the anthers for longer. Several genera of

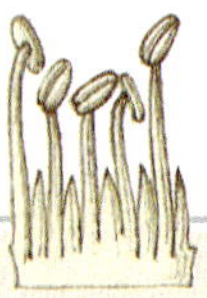

Bees are by far the most assiduous visitors to flowers because they depend on floral products at all stages of their life-cycle. Although bee flowers are extremely variable in shape and colour they are more often distinctly two-lipped than other flowers. The bold markings on many of these flowers act as landing lights, guiding the insect directly to the centre of the flower and facilitating a speedy transit time.

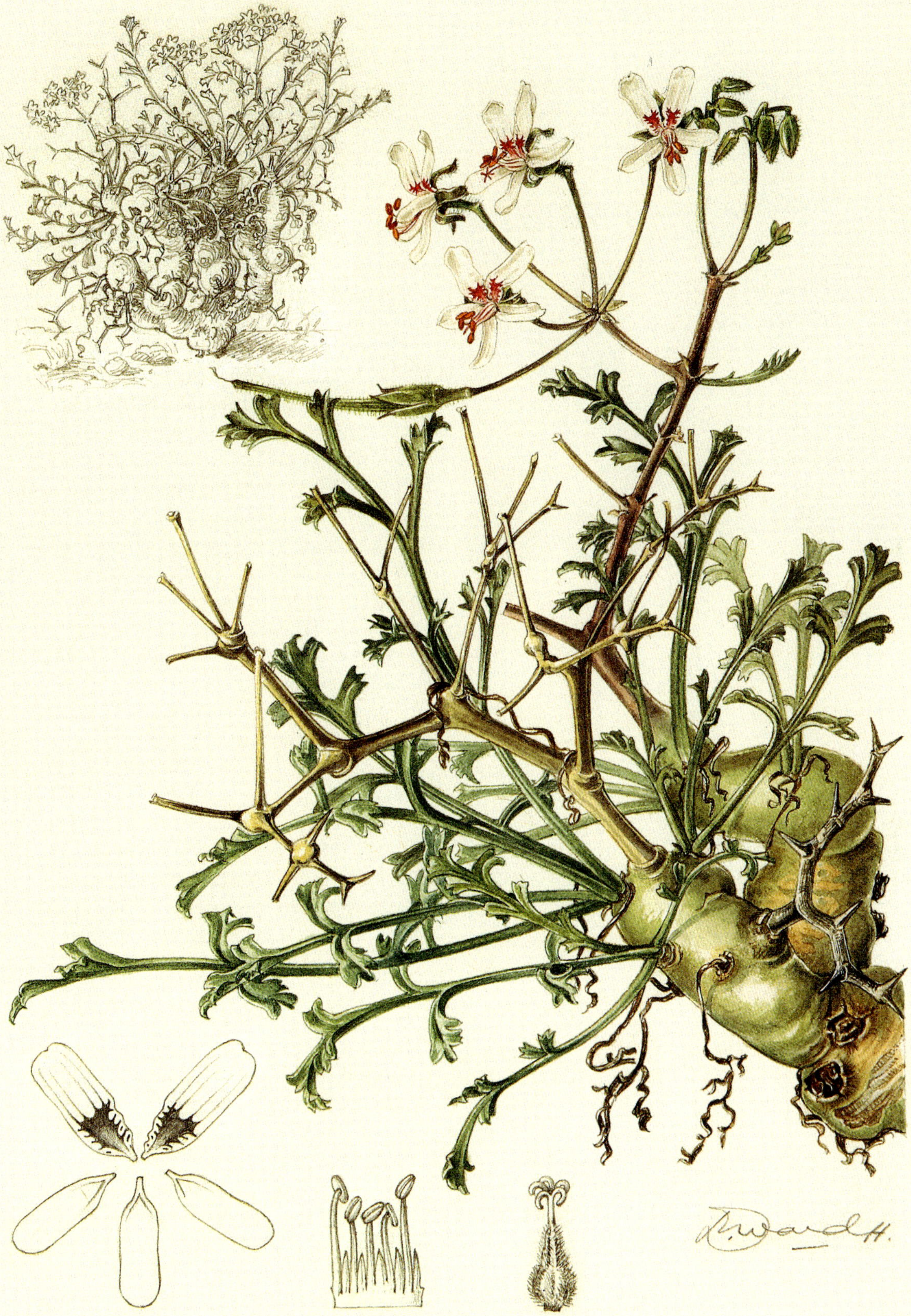

Plate 82

Ellaphie Ward-Hilhorst, *Pelargonium crithmifolium, c.* 1975, watercolour and pencil: reproduced in *The Flowering Plants of Africa*, 1978

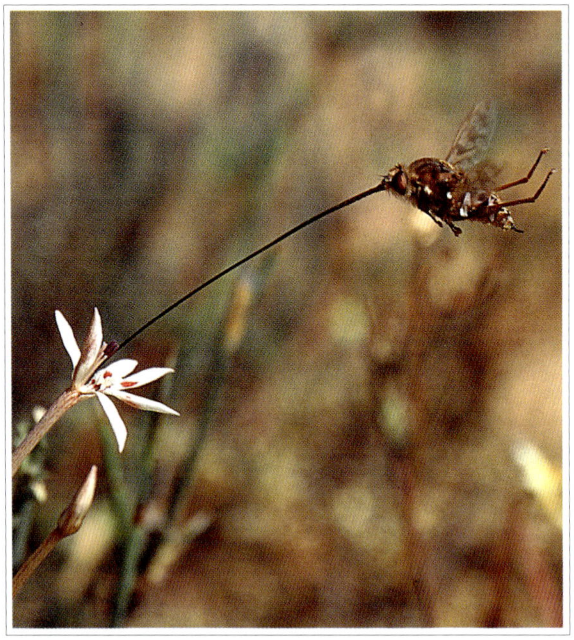

PLATE 83

Moegistorhynchus longirostris fly visiting *Lapeirousia anceps* *(Photographer:* John Manning, Cape Town)

plants have adopted the first strategy and produce flowers which last only a few hours before withering. These disposable flowers are usually delicate in texture and precisely timed to open synchronously each day. Often this is in the late afternoon, by which time the resources of the day offered by other species have been exhausted, but it is not uncommon to find morning flowers growing with afternoon flowers, one species opening as the other is closing. To increase the display, many of these flowers also bear tufts of pollen-mimicking hairs near the anthers.

Several unrelated plant genera have adopted the second option and retain their pollen in the anthers. In these species the anthers do not open to display the pollen, but remain sealed except for small pores at the apex or base. Pollen can only be extracted from these little floral saltcellars by shaking or buzzing them. Visiting bees which are able to accomplish this, grasp the anthers tightly before vibrating their wings at a high frequency, whereupon pollen is driven out of the anthers in a cloud. In this way each flower is good for several puffs before its rewards are exhausted, and even then this is not apparent until the bee has alighted and tried to extract it.

Flowers which offer nectar as a reward, face none of these potential hazards because the nectar is replenished daily for the life of the flower. In nectar-producing flowers the anthers are usually inconspicuously coloured or even concealed so that the pollen is picked up inadvertently by the visiting bee. The longer life of these flowers justifies their more elaborate construction, which is often also necessary in order to protect both the nectar and pollen from potential thieves.

In southern Africa the bees most commonly involved in pollination are the solitary or colonial long-tongued bees (family Apidae) and leaf-cutter bees (family Megachilidae). The latter are specialized for collecting pollen from pea flowers and have short mouthparts. They play a relatively minor role in pollinating the flowers of other families. The long-tongued bees, however, have longer mouthparts, up to fifteen mm long, and include honeybees, large carpenter bees and several other medium-sized, robust bees which visit a range of flowers. Honeybees are exceptional in forming large hives which require adults to be active throughout the year. The majority of other bees are solitary and exist as adults for only a few months of the year. Because solitary bees are unable to survive as adults during the cold season, in the south-western Cape, for example, the only bees active during the winter months are colonial honeybees. Early-season flowers thus have to rely on honeybees alone for pollination.

TO THE POINT

It has only recently become apparent that southern Africa is very rich in specialized nectar-feeding flies. The flies belong to several families but all have mouthparts drawn into slender, siphon-like structures which are used to suck up nectar. All, incidentally, also have rather unusual life-cycles, with the adults free living but the larval stages parasitic, feeding on various other insects or spiders. The most numerous are the furry bee-flies (family Bombyliidae), a common sight in the drier parts of the region where they can be seen sunning themselves on stones or hovering above flowers. It is the horseflies (family Tabanidae) and the tangle-veined flies

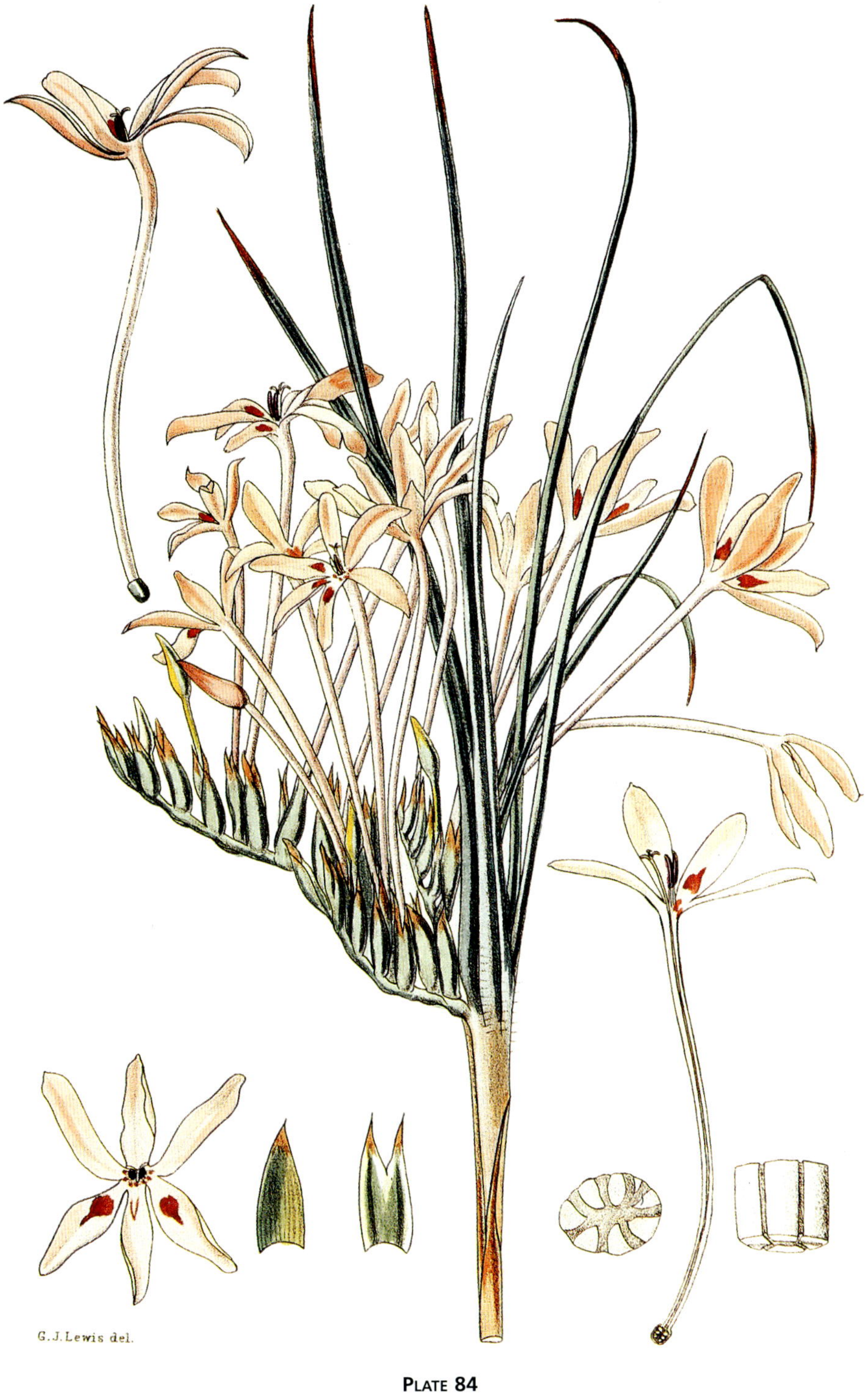

Plate 84

Gwendoline Joyce Lewis, *Babiana brachystachys,* 1933, watercolour and pencil: reproduced in *The Flowering Plants of South Africa*, 1934

More than two hundred species of southern African plants rely on long-tongued flies for their pollination. Over a dozen species of fly are involved, and the flowers they prefer are all either purple to magenta with white markings or cream to pink with maroon markings and always have a long, slender floral tube.

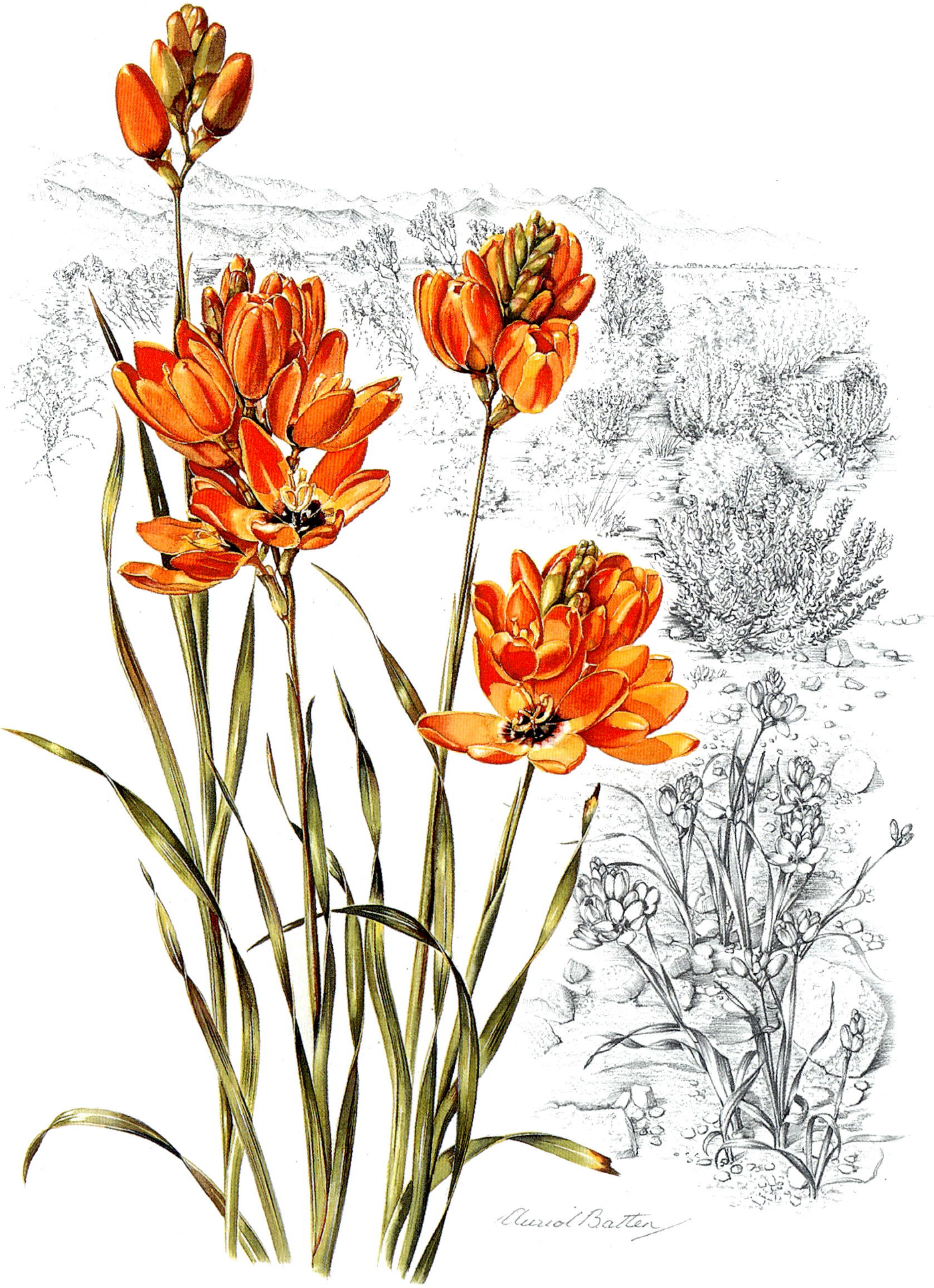

Plate 85

Auriol Batten, *Ixia maculata*, c. 1980–1985, watercolour and pencil: reproduced in A. Batten, *Flowers of Southern Africa* (Sandton, 1986)

Glorious colours and glittering markings characterize flowers adapted to pollination by monkey beetles. The dark, sometimes iridescent, centre typical of these flowers mimics the glossy wing-covers of the beetles and attracts passing individuals to the flowers with the promise of congenial company. Amongst the most brilliant of southern Africa's floral gems, several of these species are now widely cultivated.

(family Nemstrinidae), however, which are the most important in terms of floral evolution and speciation. Bee-flies, for all that they are specialized nectar-feeders, have relatively short mouthparts – never more than ten mm long – and as such form part of a somewhat heterogeneous group of floral visitors that includes many butterflies. The larger tangle-veined flies and horseflies, with mouthparts more than twenty mm long (Plate 83), however, comprise a unique category of pollinators which is highly developed only in southern Africa.

Flowers adapted for pollination by these deceptively fearsome-looking insects with their needle-like mouthparts are characterized by slender, elongate floral tubes in distinctive colours, either pale pink to cream in colour with dark pink to red markings, or magenta to purple with cream markings. Only in the last few years has this remarkable pollination strategy become understood, and the reason for the close floral similarity of many, often unrelated, southern African flowers become apparent.

Altogether about a dozen species of long-tongued fly are involved in this class of pollination, mostly in the south-western winter-rainfall parts of the region. The most distinctive group of flowers specialized for this pollination strategy is that pollinated by the two long-tongued nemestrinid flies which occur in Namaqualand as far south as Clanwilliam. These flies between them are responsible for the pollination of almost thirty different species of plants, all in the families Iridaceae and Geraniaceae, and all with brilliant carmine, magenta or purple flowers marked with cream. The most spectacular of the flies, however, is undoubtedly the one with the longest mouthparts, *Moegistorhynchus longirostris*, a beautiful russet fly with chequered abdomen and wings and a proboscis even longer than its name. This almost fabulous insect, with mouthparts up to four times as long as its body, is restricted to the Cape west coast and interior, from Namaqualand to the Cape Peninsula.

BULL'S EYE

Another group of pollinators that has been largely overlooked until the past few years are the scarab beetles in the tribe Rutelinae: subtribe Hopliini, commonly known as 'monkey beetles'. These somewhat comical beetles are often densely hairy and the males may have their hind legs grotesquely enlarged. The larval stages of the beetles are subterranean and feed on the roots of various shrubs, while the adults forage on floral parts, mostly pollen. Their importance as pollinators, however, results from another, more exotic use of flowers. Monkey beetles use a variety of flowers as sites for courting and mating, literally as singles clubs and nuptial couches. During the rough and tumble which ensues, the beetles' hairy bodies become liberally covered with pollen.

The beetles favour flat- or dish-shaped flowers and are commonly found on daisies but also on the flowers of a range of other species, particularly those with dark centres or large dark markings on the petals near the centre of the flower. These markings may be highly effective mimics of the beetles, often reproducing the glittering highlights or metallic sheens on their wing covers and sometimes even the pollen which becomes deposited on their bodies. Beetle flowers may be white, yellow, orange, red, mauve or blue. The anthers and pollen are

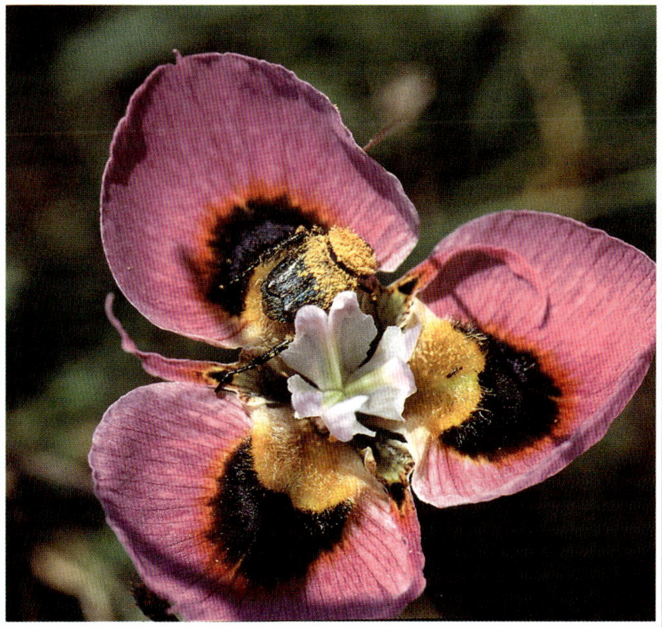

PLATE 86

Anisonyx monkey beetle visiting *Moraea villosa*
(*Photographer:* John Manning, Cape Town)

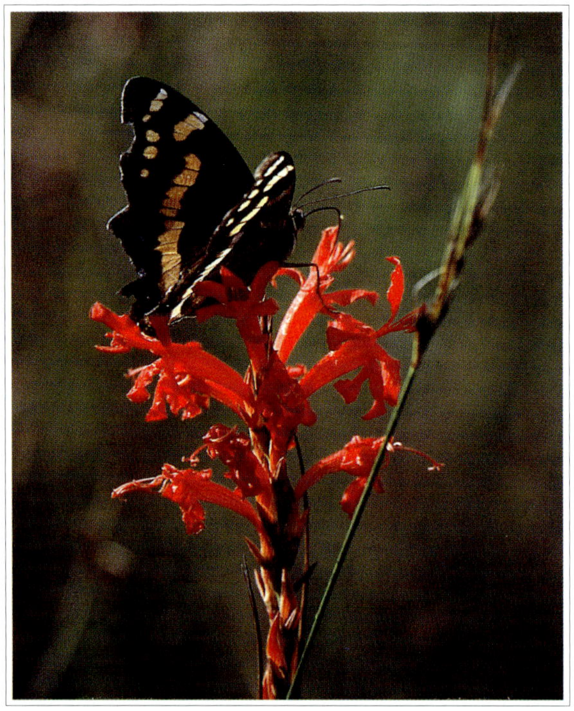

PLATE 87

Aeropetes tulbaghia butterfly visiting *Tritoniopsis burchellii* (*Photographer:* John Manning, Cape Town)

usually dark, even black, and probably quite unattractive to pollen-collecting bees. Because monkey beetles are not nectar feeders, floral tubes are usually reduced or absent and nectar production is largely or completely suppressed. In genera that normally have flowers with a floral tube, the tube may be extremely narrow and blocked by the style, or it may be closed at the mouth by the fused stamens. This is the case in many species of *Ixia* in which the tube has assumed the role of a flower stalk instead of a nectar reservoir.

Big, brightly coloured and boldly marked: the criteria for successful beetle-pollination could come from a plant-breeder's wish-list, and the results of Nature's breeding programmes have not been a disappointment. Several beetle-pollinated species are now widely cultivated and enjoyed, many in their natural state. These include the nemesia (*Nemesia strumosa*), chincherinchee (*Ornithogalum thyrsoides*), and *Ixia*, *Tritonia* and *Sparaxis* cultivars (derived from the hybridization of several beetle-pollinated species [Plate 21]).

Unlike some of the other specialized pollination systems, beetle-pollination is not often an exclusive system. Beetle flowers by their very nature have their pollen exposed above the disc- or bowl-like perianth and may thus be inadvertently shared with bees which collect pollen, or flies which seek nectar, sometimes present in tiny amounts. Indeed, some species may take advantage of more than one class of pollinator, thus hedging their bets in the pollination stakes. Exclusively beetle-pollinated flowers are often unusually coloured, like the large-flowered *Romulea* species of the western and northern Cape, which produce dazzling carpets of ruby-brilliant flowers which are difficult for other insects to see (Plate 89).

SEEING RED

Flowers adapted to butterfly pollination are relatively uncommon outside the tropics but in the southern African flora some twenty species of plants are specifically adapted to pollination by the Table Mountain Beauty butterfly, *Aeropetes tulbaghia*. This large brown butterfly with bright yellow banding on the wings occurs along the eastern seaboard from the Zimbabwe highlands to the Cape fold mountains. Unlike most insects, it has an inborn affinity for red flowers, which is so strong that hikers wearing red clothing are often surprised by exploratory visits by the butterfly.

The adults emerge after mid summer and are a conspicuous feature of the Cape mountains and Drakensberg in February and March. They fly rapidly across the slopes, alighting abruptly on various red flowers. In the summer-rainfall region these are often species of *Kniphofia* (which are also pollinated by birds) but also a couple of species of gladiolus. In the winter-rainfall region the diversity of plants adapted to this pollinator is far greater, and this co-evolution can be thanked for giving us some of our most beautiful bulbous species. These include the red disa (*Disa uniflora*), new year lily (*Gladiolus cardinalis*), George lily (*Cyrtanthus elatus*), and Guernsey lily (*Nerine sarniensis*). All told, some fifteen species of Cape plants are dependent on this butterfly for pollination. Although a few plants are visited by both *Aeropetes* and sunbirds,

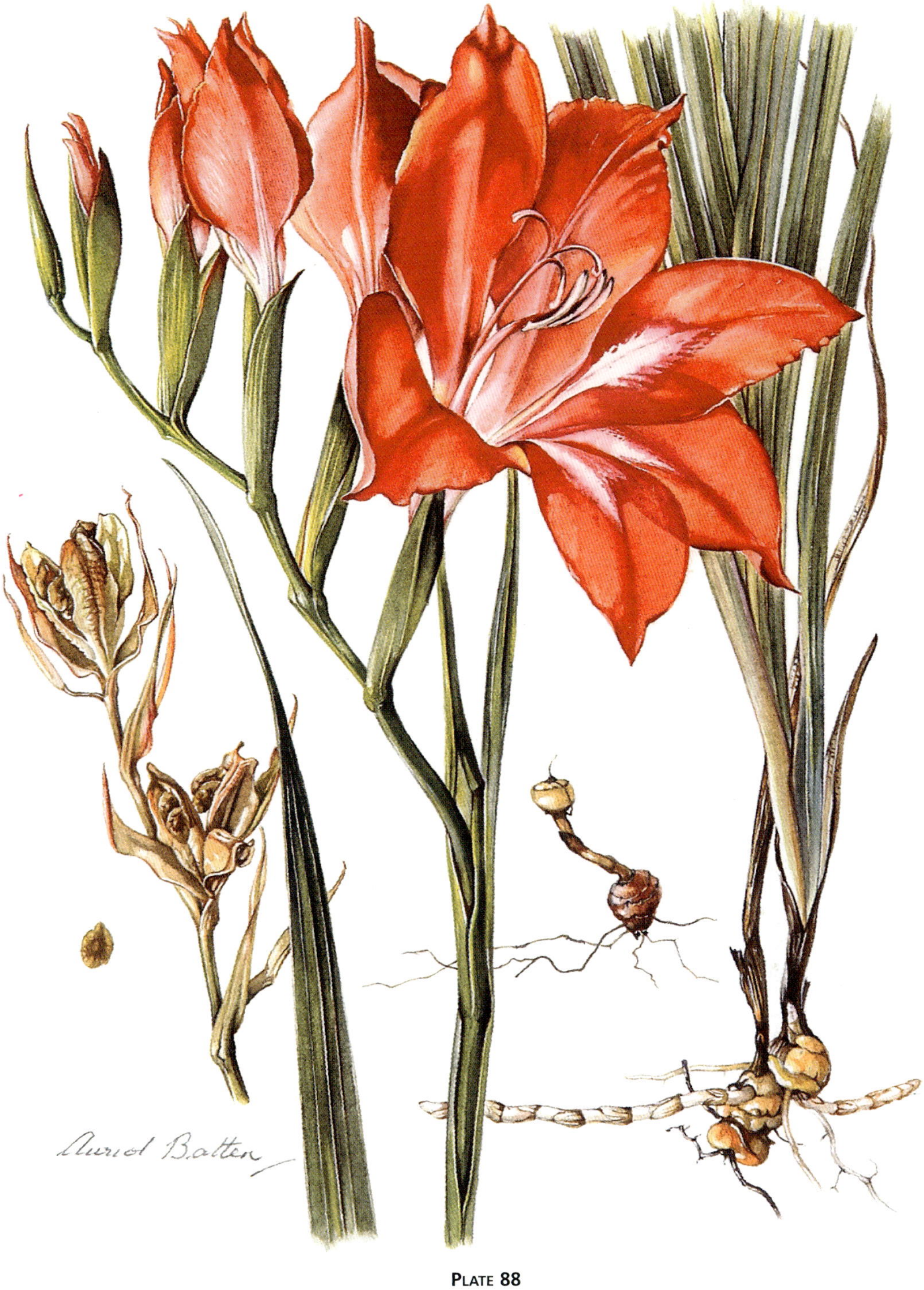

PLATE 88

Auriol Batten, *Gladiolus sempervirens*, watercolour: reproduced in P. Goldblatt and J. Manning, *Gladiolus in Southern Africa* (Cape Town, 1998)

Several of southern Africa's brightest flowers have literally rolled out the red carpet for their pollinator. The butterfly *Aeropetes tulbaghia* is passionate about red and in order to attract the insect, during the summer months when the butterfly is on the wing, some twenty plant species produce brilliant scarlet flowers.

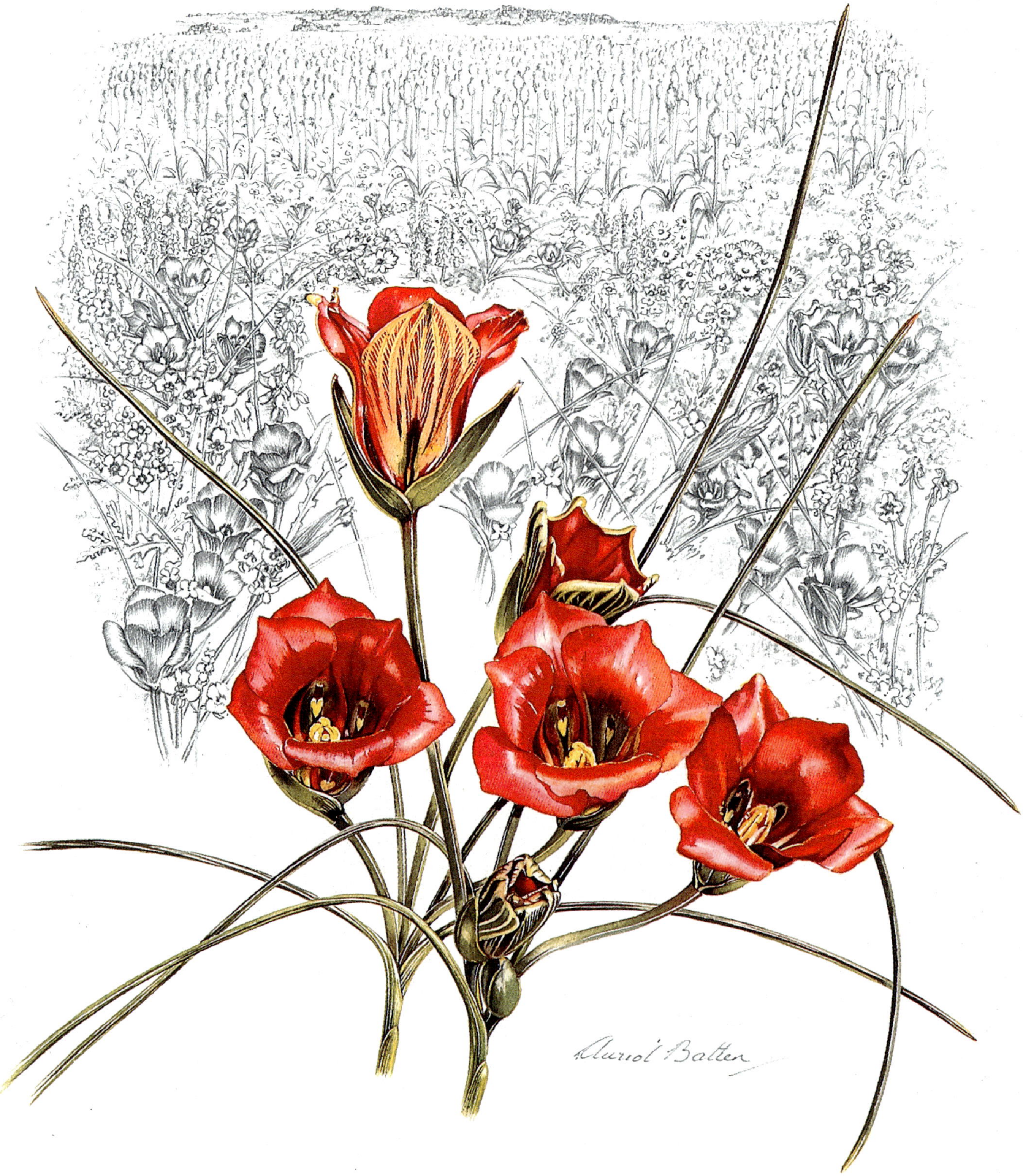

PLATE 89

Auriol Batten, *Romulea monadelpha, c.* 1980–1985, watercolour and pencil: reproduced in A. Batten, *Flowers of Southern Africa* (Sandton, 1986)

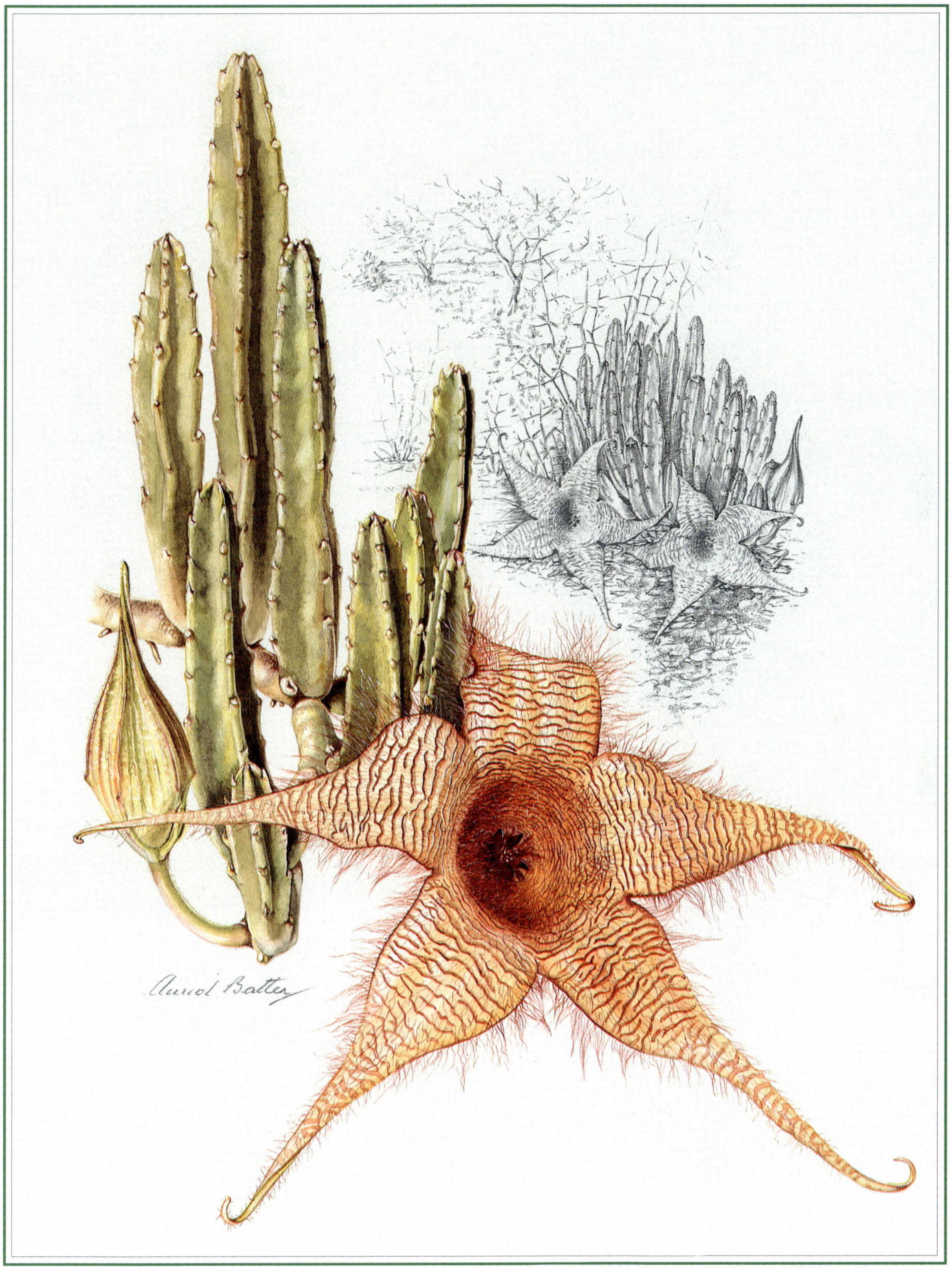

PLATE 90

Auriol Batten, *Stapelia gigantea, c.* 1980–1985, watercolour and pencil: reproduced in A. Batten, *Flowers of Southern Africa* (Sandton, 1986)

flowers specifically adapted to pollination by *Aeropetes* can be distinguished by their narrower tubes, often larger flowers with spreading sepals and the very precise shade of red, a glorious scarlet flecked with gold highlights.

The emergence of *Aeropetes* in the summer months, regardless of the rainfall regime, has meant that it is available to pollinate only those Cape species which are able to flower during what is there the dry season. The plants have achieved this either by growing in habitats which remain wet through the summer – places such as stream-sides or damp upper slopes which are often covered with cloud – or by flowering without their leaves, which are produced later.

FROM THE GUTTER

Possibly the least appealing of the pollination strategies relies on Nature's undertakers to do the job. The pollinators are house flies and carrion flies, in the families Muscidae and Calliphoridae. These dull- or metallic-coloured flies have short, lapping mouthparts and are attracted to decaying organic matter as a rich source of nutrients, either for themselves or in which their larval stages can mature. Flowers pollinated by carrion flies are dull in colour, usually brown or maroon, often mottled and

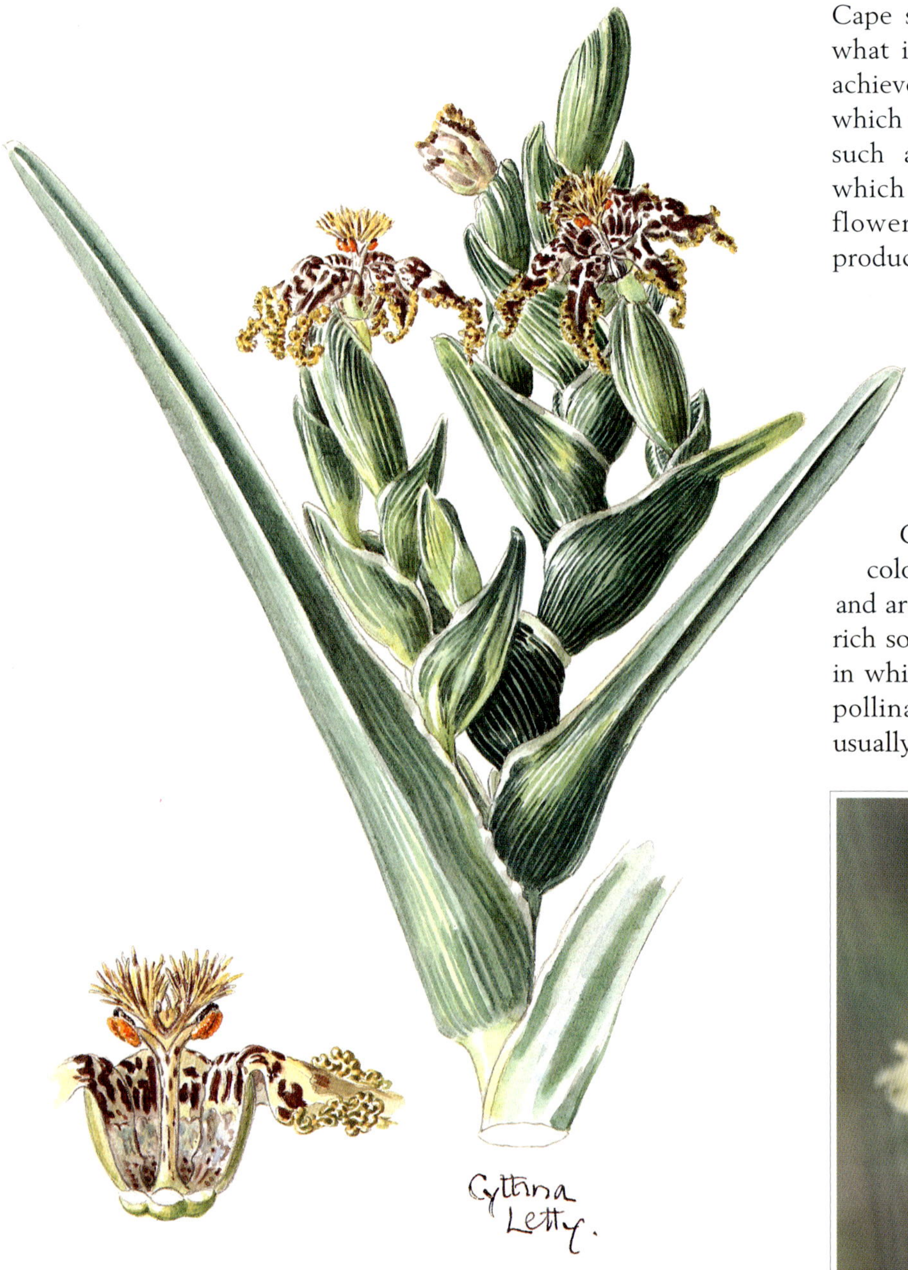

Plate 91

Cythna Letty, *Ferraria crispa,* 1955, watercolour and pencil: reproduced in *The Flowering Plants of Africa,* 1959

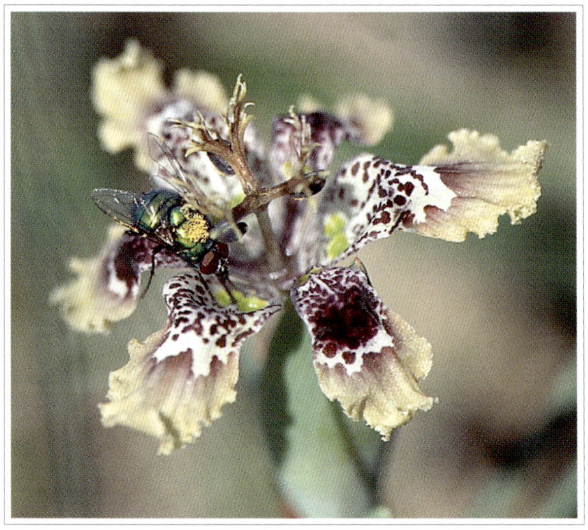

Plate 92

Lucilia carrion fly visiting *Ferraria densepunctulata* *(Photographer:* John Manning, Cape Town)

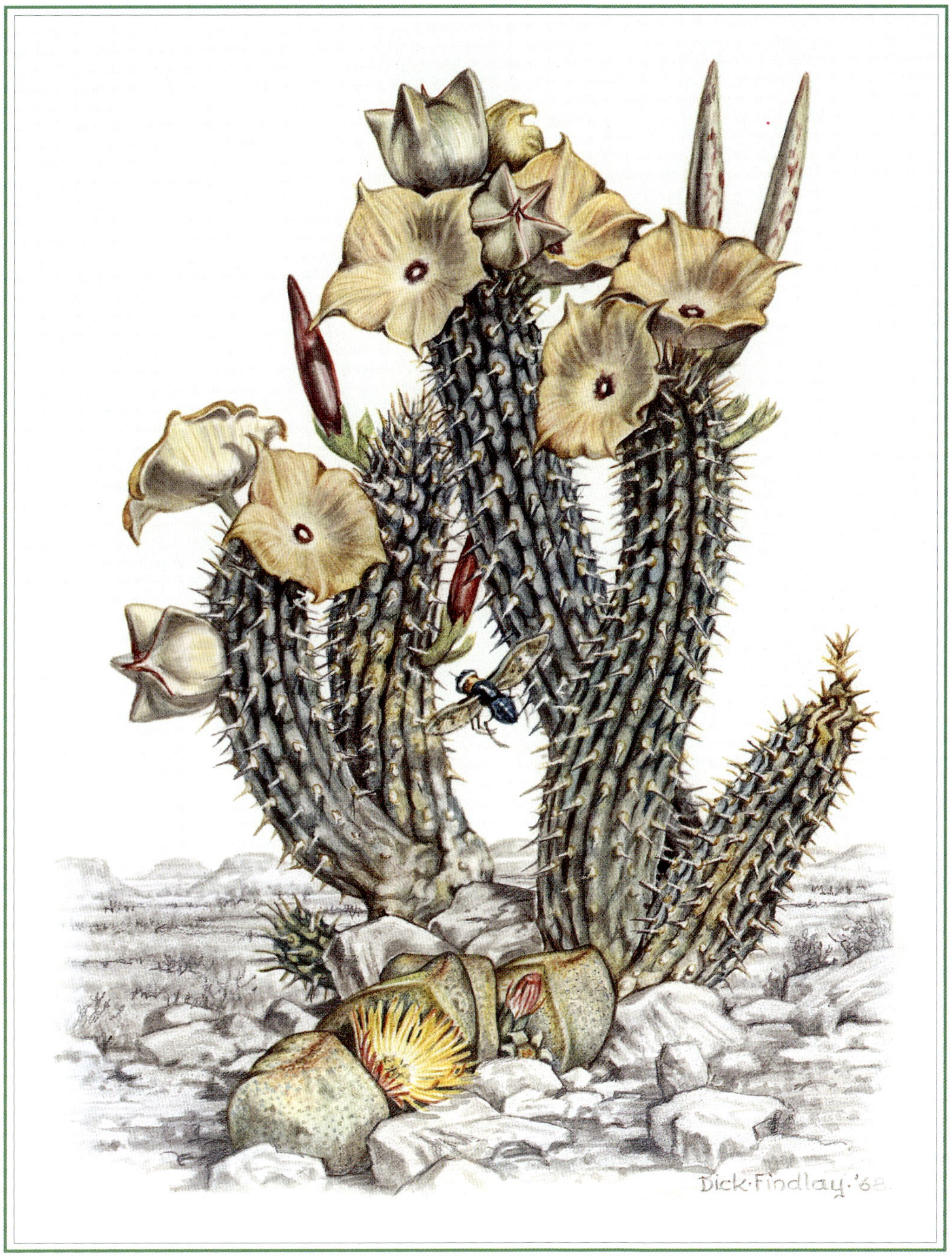

PLATE 93

Dick Findlay, *Hoodia gordonii* (= *Hoodia bainii*) and *Pleiospilos bolusii*, 1968, watercolour and pencil

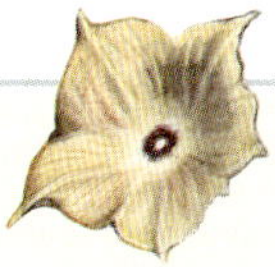

Carrion flies and house flies are too ubiquitous to be ignored as pollinators, and their ghoulish associations are only unpleasant to us. Many members of the milkweed family produce flowers with a scent that has an almost repulsive resemblance to carrion – so convincing that they are visited regularly by these flies.

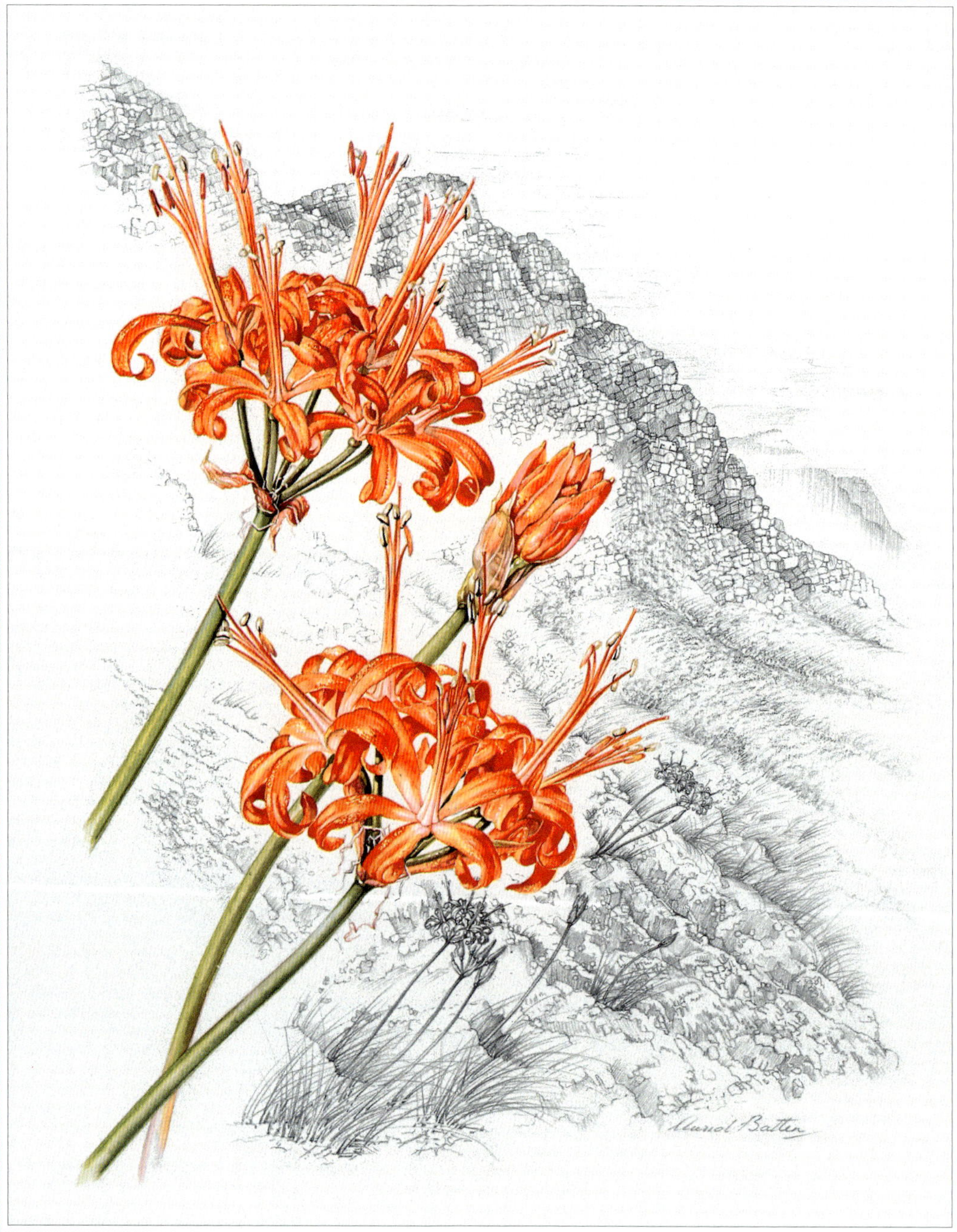

PLATE 94

Auriol Batten, *Nerine sarniensis, c.* 1980–1985, watercolour and pencil: reproduced in A. Batten, *Flowers of Southern Africa* (Sandton, 1986)

sometimes with a textured or hairy surface. Nectar may be offered as a reward but in many cases is not. Odour is the prime attractant, ranging from relatively innocuous musty smells to highly repellent stenches of decay. Carrion fly pollination occurs in rather few families, most commonly amongst the stapelias, in which it has been refined and perfected. The flowers of some of these plants reproduce the textures, lurid colours and sickening stench of decaying animal hide or organs with gruesome fidelity. The deception may be so complete that the flies are actually induced to lay their eggs in the flowers. Less offensive are the flowers of the iris-like genus *Ferraria*, which resemble the ruched forms of sea anemones and give off faintly musty but inoffensive odours. Flies visiting these flowers are rewarded by abundant nectar.

Several other classes of pollinator exist, each with its inimitable imprint on the form and behaviour of the flowers which it pollinates. One of the most important of these is bats. These mammals are invaluable in both pollination and seed dispersal in New and Old World tropics and subtropics, and in Africa are responsible for the pollination of the baobab (*Adansonia*) and sausage tree (*Kigelia*). Nearer ground level, small mouse-like rodents are pollinators of several of the Cape proteas. Other highly specific systems with a regional bias include marsupial pollination in Australia, oil-collecting bees in southern Africa, and yucca moths in North America.

Handsome is as handsome does

One of the more striking things about these pollination strategies – even more than their variety – is that throughout the world they remain recognizable. Although each region will have different pollinators and different plants, the essential characteristics of each system are the same. It cannot, in fact, be otherwise. Since form follows function, it is inevitable and natural that organisms performing similar functions will come to resemble one another.

Because of this, it is often quite easy to infer the type of pollinator for which a particular flower is adapted. The flowers of *Babiana brachystachys*, for instance, show the characteristics that are associated with pollination by long-tongued flies: cream or pale pink flowers marked with maroon and with a long, slender tube more than twenty millimetres long. A more intimate knowledge of the local fauna can even make it possible to infer the exact species or, at least, group of species that comprise the primary pollinators. Returning to our example, the natural distribution of *Babiana brachystachys* and the prodigious length of the floral tube indicate unequivocally that the pollinator is the fly *Moegistorhynchus longirostris*.

Sometimes it is possible to infer quite a lot about the ecology of the plant species merely from its appearance. Take the Guernsey lily (*Nerine sarniensis*) as an example. A botanical painting might show a head of brilliant scarlet flowers appearing from a leafless bulb. The flower colour and shape enable one to infer that the plant is pollinated by the butterfly *Aeropetes tulbaghia*. From this, even the habitat may be painted in, for *Aeropetes* favours rocky slopes. The leafless habit at flowering indicates that this plant is flowering in the dry season. Since *Aeropetes* is on the wing in the summer and autumn, this indicates that *Nerine sarniensis* is a plant of the winter-rainfall region. Some quite intimate information on the lifestyle of the plant can thus be gained from a more informed look at its flower, and the educated viewer can begin to sketch in the background around the portrait.

Gilding the lily

Because a single plant species is mostly unable to provide sufficient nectar or pollen to supply the daily needs of its pollinator throughout the season, a single pollinator will visit several different plant species during the course of its daily rounds. Such groups of species are known as 'guilds', so called because they co-exist in an ecosystem and use a particular resource in a similar way. A suite of bird-pollinated species at a particular site would constitute a bird-pollination guild. The magenta or purple-flowered species in Namaqualand pollinated by long-tongued flies likewise comprise a long-tongued fly-pollination guild, in this case more particularly the *Prosoeca peringueyi* pollination guild. A visit to any natural site will invariably reveal several guilds of flowers, collectively providing for the needs of their pollinators.

The number of different guild members that occur in any particular locality is not limitless. There are very precise constraints on the maximum number of species in a guild which can co-occur at a site. These constraints are purely economic and relate directly to the rate of successful pollination. As guild members are added to a particular site, so the likelihood that the appropriate visitor will call at that site increases. This is beneficial to both visitor and plant – but only to a point. As the number of

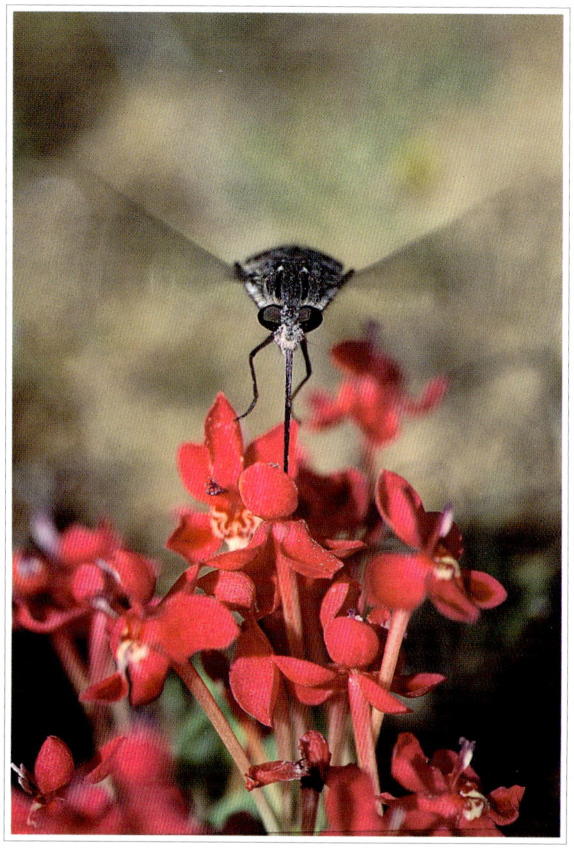

PLATE 95

Prosoeca peringueyi fly visiting *Lapeirousia silenoides*
(*Photographer:* John Manning, Cape Town)

guild members increases, so the likelihood that pollen will be transferred directly from one individual of a species to another of the same species decreases. This is because the pollinator does not distinguish between guild members and so will move indiscriminately from one species to another, in the process not only depositing the precious pollen load onto an inappropriate stigma but also clogging that stigma so that even were the correct pollen to be deposited there at a later stage, it would be unable to germinate.

Since any suitable site, in Namaqualand for instance, will support five or even seven different species of plants pollinated by the same long-tongued fly, some sort of solution to this dilemma must exist. Strolling amongst the granite outcrops on the Kamiesberg near Leliefontein will reveal several species belonging to the local long-tongued fly guild, bringing back to these rocks some of the fiery colours which attended their formation. Shimmering pink skirts of *Hesperantha latifolia* may spread over the damper flats at the base, while gravelled washes amongst the boulders are often aflame with *Lapeirousia silenoides*, interrupted here and there by taller gouts of *Pelargonium incrassatum*. In crevices in the rocks, tufts of *Babiana dregei* glow a bright azure, and in seeps the rare little *Xenoscapa uliginosa* sparkles amongst the dark mosses. In other localities the flowers of *Romulea kamisensis* may be glowing like embers in the washes, or plants of *Babiana curviscapa* can be found scattered about like live coals in drier, sandy places.

A closer examination of the flowers at any site reveals that each of the species, for all their overall similarity, has a very different arrangement of the stamens and style. These differences relate directly to where the pollen will be deposited on the visiting pollinator. In one species the stamens may be very short, depositing pollen on the face of the visitor; in a second, they may be longer and arching, depositing pollen on its back; in a third, the pollen may be placed on the underside of the insect. In this way the pollinators can carry the pollen of several species simultaneously without mixing them up, and several species of a guild can reap the benefits of co-existence, avoiding the potential loss in reproductive success.

The way in which the pollinator's body is so rigidly partitioned by guild members reflects the extraordinary value of this strategy in promoting reproductive success. There is a second, less obvious consequence. Because the number of plant species that can belong to a guild at any particular site is limited by the necessity for preventing pollen mixing and stigma clogging, the local distribution of guild members is determined by whether or not another guild member with the same stamen position has already occupied a particular site. Plant distribution is thus not totally determined by the availability of suitable physical habitats but also by the availability of suitable biological niches.

Paradise lost

When a guild of plant species depends for its immediate or ultimate survival on a particular organism, this organism assumes an importance which extends beyond its own survival. Any highly specific pollinator constitutes such a *key-stone* species, without which the entire biological arch will crumble. Long-tongued flies, sunbirds and *Aeropetes* are all keystone species, for without them whole guilds of plant species will fail to reproduce successfully.

How important these keystone species can be to the survival of other species has been brought vividly to our attention during our research on the pollination biology of the family Iridaceae. Long-tongued flies are often rare and hard to find in the field. With perseverance, however, flies may be sighted where their favourite flower resources grow. At some sites which support good stands of the appropriate plants, however, we have never seen the flies, despite repeated visits over several years. At these sites the plants fail to set seed, and multiply vegetatively only.

Certain other members of the guild that use the same fly for their pollination are able to set seed in the absence of the pollinator by self-pollination, but this is at the expense of the genetic diversity that cross-pollination ensures. How long they can maintain their populations in the absence of cross-pollination remains to be studied, but continued inbreeding cannot be good for the population.

The fact remains that destruction or disturbance of the habitat affects not only the plants themselves but also their pollinators. Conserving the ground on which the plant grows is a short-term strategy to conserving the species unless the area that its pollinator requires to complete its life-cycle is also protected.

In many instances, conserving the area around the plants will include this, but in the case of long-tongued flies so little is known about their biology that we cannot even guess whether this applies. Too often the fragmentation of habitats through farming or industrialization restricts plant species to the small pockets of natural vegetation which remain. Unless these pockets can also support viable populations of the pollinators, whole groups of species are started down an inexorable track to extinction.

More than skin deep

The fact that botany and other biological sciences developed in Europe and North America has had a profound impact on the way in which biological systems were perceived in the past. Europe, especially northern Europe, and North America have relatively depauperate floras and faunas largely derived from recolonization of the areas from the south following the retreats of the glaciers. The boundaries between the species are in most cases well defined. These biological characteristics coincided perfectly with eighteenth-century philosophical constructs dominated by a Creationist theology which postulated the immutability of organisms. Species boundaries were closely and clearly circumscribed by biologists of the time, and lo!, Nature by and large actually fitted in. Once boundaries were erected around species, relatively larger morphological differences between groups of species led to the recognition of genera, a simple extension of the process of grouping like organisms into ever larger groups. Classification by perceived similarity of appearance was the norm for over two hundred years, until the mid twentieth century.

The exploration of the tropics and other regions with older or more actively evolving floras in the nineteenth century, had begun to reveal the inadequacy of this methodology in accounting for the degree of variation that exists in many parts of the world. Colonization and increased possibilities for travel enabled scientists to make a more intimate acquaintance with species in the wild. Here they could appreciate the functional aspects of many of the structures that were used in classification.

The recent explosive development of the sciences combined with a revolutionary change in the philosophy of classification have led to a reappraisal of the so-called 'typological' method of classification based on overall similarity of form. Today biologists look for classification systems that reflect the evolutionary relationships between species. The goal of modern biologists is ultimately to understand the variation patterns in the organisms they are studying and to interpret them in a historical, or evolutionary, context.

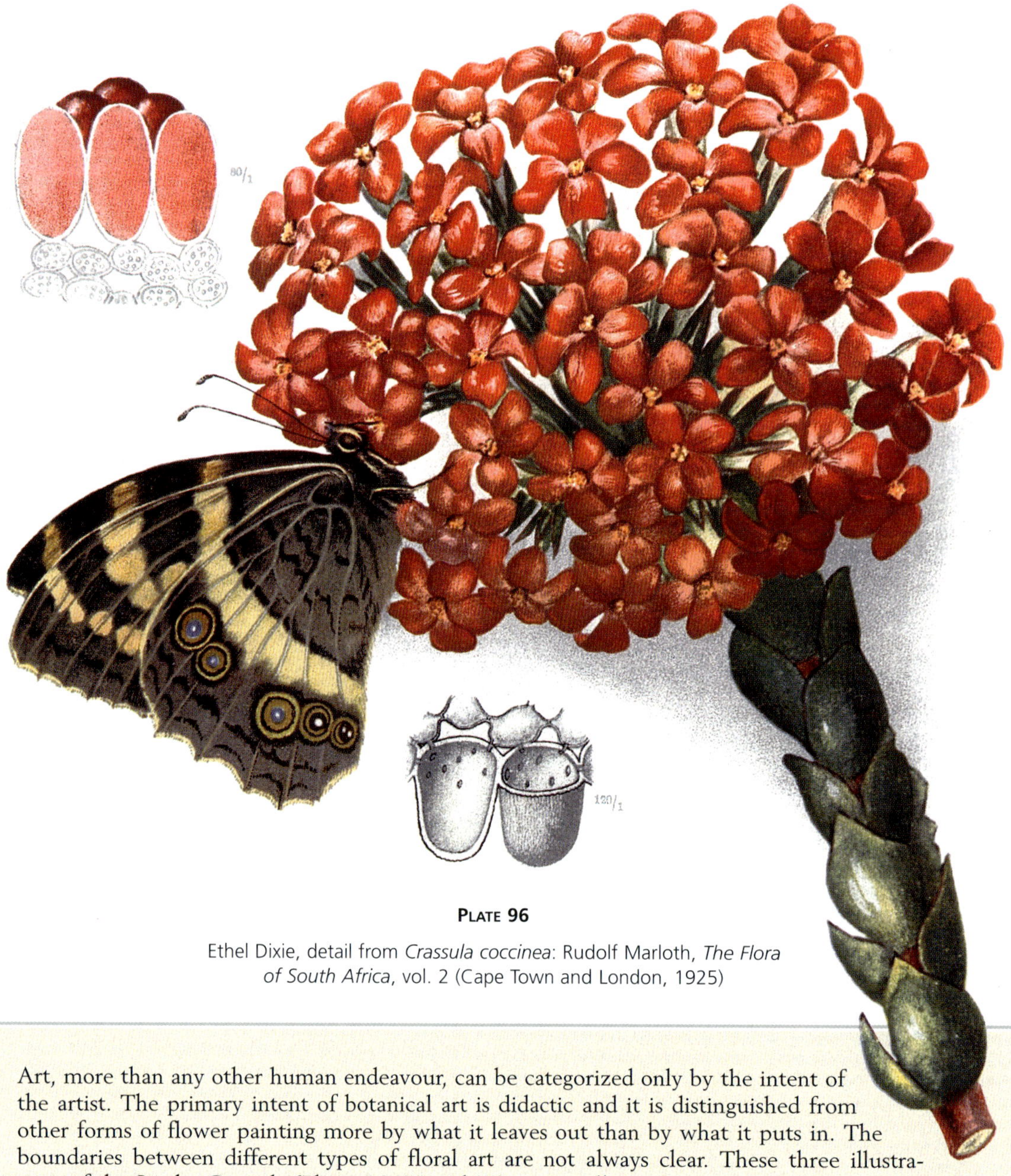

PLATE 96

Ethel Dixie, detail from *Crassula coccinea*: Rudolf Marloth, *The Flora of South Africa*, vol. 2 (Cape Town and London, 1925)

Art, more than any other human endeavour, can be categorized only by the intent of the artist. The primary intent of botanical art is didactic and it is distinguished from other forms of flower painting more by what it leaves out than by what it puts in. The boundaries between different types of floral art are not always clear. These three illustrations of the Scarlet Crassula (Plates 96, 97 and 98) are equally accurate in their representations but were contrived from three disparate outlooks with different intents.

The watercolour by Thwaits is aggressively clinical. It isolates the crassula as a single specimen for close examination and morphological comparison.

The image by Dixie, although similar in style to that of Thwaits, highlights the insect pollinator for which this flower is adapted as well as cellular details of the petals. This immediately reveals the plant to be a living organism and places it in both a functional and evolutionary context not evident in the other images.

Although contemporary with the illustration of Thwaits, the still life by Morstatt uses the crassula as one element of a summer bouquet clearly gathered on the upper slopes of Table Mountain where all the depicted wildflowers occur. This image more than the other two reflects the response that nature evoked in the artist.

Similar form does not always reflect evolutionary or genetic relationships, and a conflict between the typological classifications based on perceived relationships and more recent evidence of actual relationships has become apparent in many plant groups. Understanding the function of various modifications in flowers, leaves and other organs has gradually led to a thorough reassessment of the relationships between species. The need for such reassessments is especially evident when conflicting relationships are suggested by different organs, say a particular flower type versus an underground organ.

Analysing the evolutionary relationships between species allows us to gain insights that not even the fossil record can provide. It allows us, as it were, to sit in on the actual evolution of characteristics and species. Using these methodologies we can at last begin to understand in very specific terms that most perplexing question in biology, the origin of the species. Our analysis of the genus *Lapeirousia* is an example of how these types of analyses can improve our understanding of the processes of speciation. But what does it tell us about the evolution of the flower?

The genus *Lapeirousia* contains about forty species of small spring-flowering geophytes, mostly from semi-arid areas in the winter-rainfall part of southern Africa. The different species have quite discrete ranges and habitats and exhibit a remarkable range of flower types. We constructed a phylogeny, or evolutionary tree, for the genus and used this to follow the evolution of various ecological strategies, particularly pollination strategies. From this study it appears that the impetus for speciation in the genus was often an adaptation to a different habitat, basically a different soil type. As the daughter population became better adapted to its particular habitat, interbreeding with the original populations became more and more deleterious.

One way of reducing the potential for interbreeding between two species, or incipient species, is to utilize different pollinators. This is exactly what seems to have happened in the case of the genus *Lapeirousia*: as two species diverged in ecology, so their flowers became adapted to different pollinators. And although the most obvious differences between different species of *Lapeirousia* are often in their flowers, it seems that the primary step in speciation was a shift to a new habitat and not a change in pollination strategy. These results illustrate one of the great paradoxes of the flower: what may be essential in defining a particular species, often may not

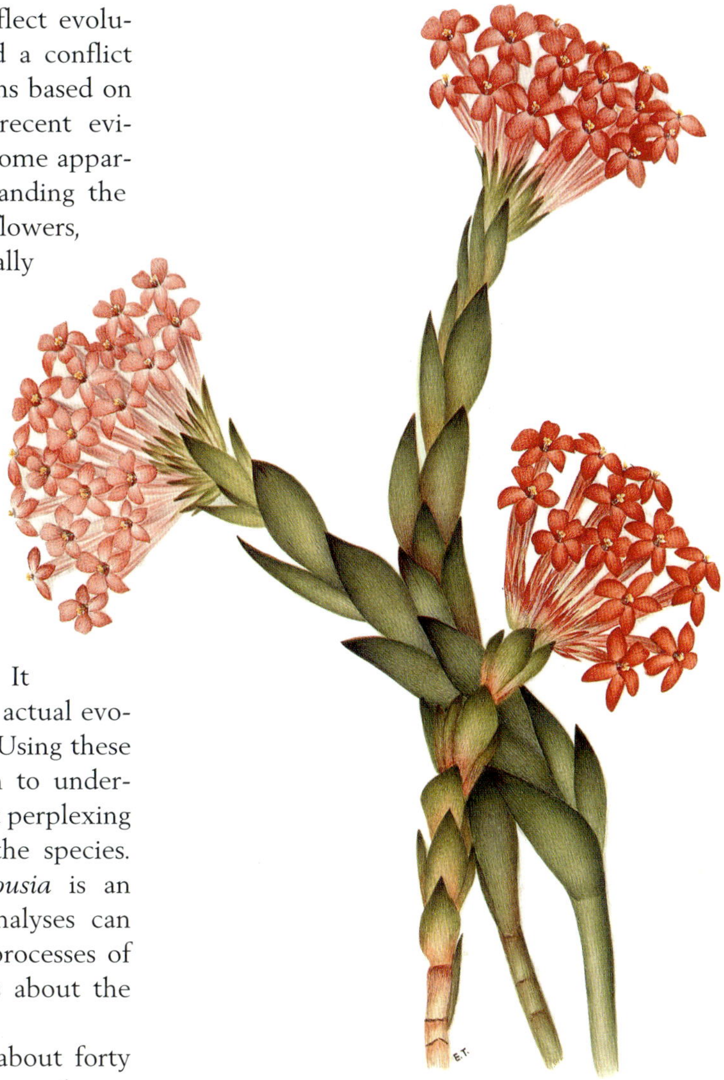

PLATE 97

Emily Thwaits, *Crassula coccinea*, c. 1890, watercolour, 30 cm x 20 cm

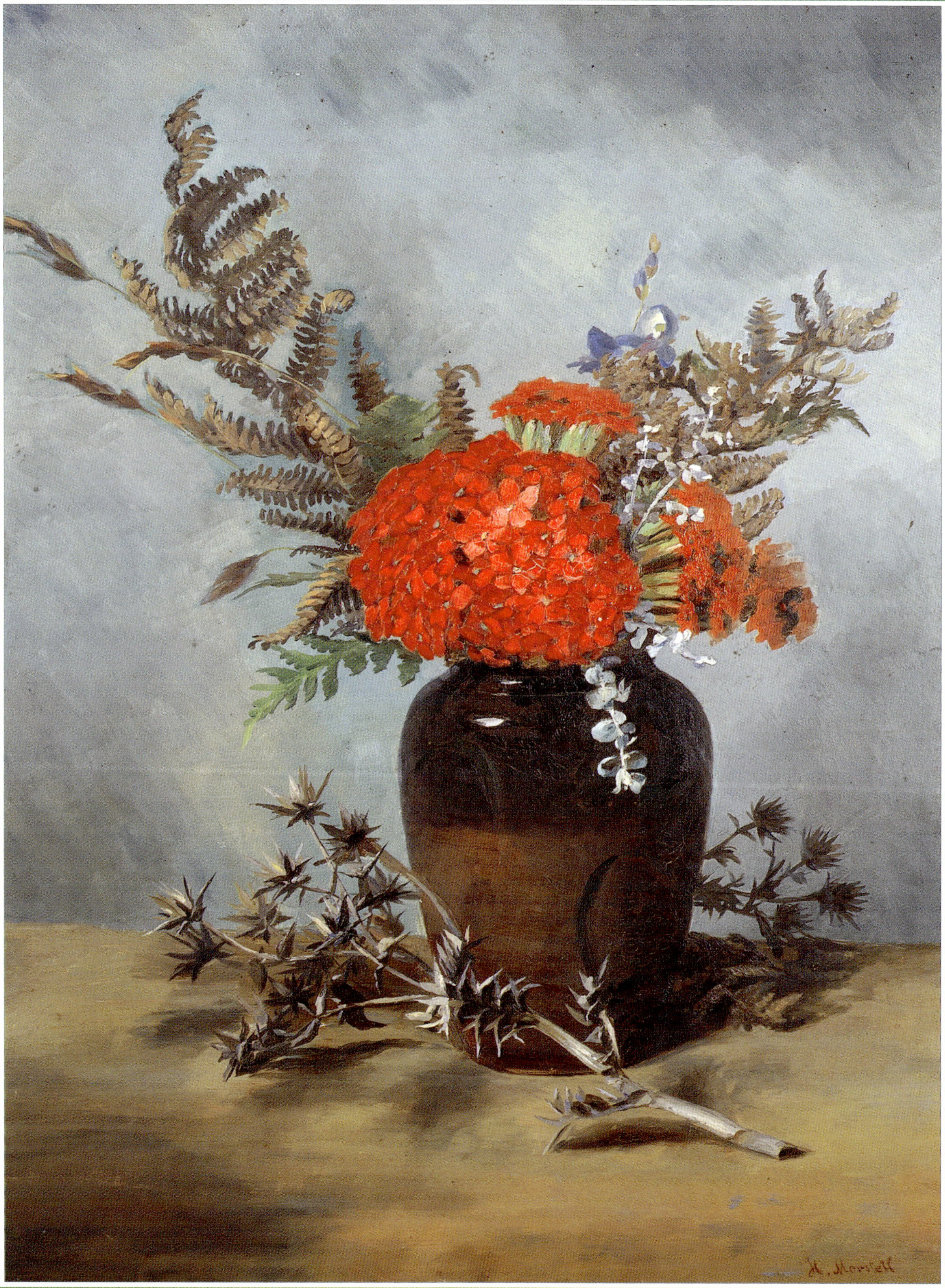

PLATE 98

H. Morstatt, *Crassula coccinea,* c. 1890, oil on canvas, 60 cm x 44 cm

necessarily be the stimulus which led to the evolution of that species.

Each according to its means

The flowers that we see around us are not randomly selected and accumulated. They are the result of directed forces acting on defined resources. If individual floral types are the result of adaptation to individual pollinator classes, then the collection of floral types present in any plant community is the result of adaptation to the local pollinator community. Many of Namaqualand's distinctive spring flowers owe their appearance to the presence of a long-tongued fly species with a preference for long-tubed red or purple flowers. In the western Cape the repeated evolution of disc- or bowl-shaped flowers with contrasting dark markings was stimulated by the unusual preponderance of monkey beetles. These are only two examples of the influence that pollinators can have, not just on plant species but on the floral characteristics of entire communities.

For further evidence on how strong this influence can be, we need look no further than the flowers themselves. Each starts with a different set of characteristics from which it must develop the features which suit it to a particular pollinator. The hairs which are such an integral part of the deception practised by flowers that rely on carrion flies for pollination are abundantly evident on the petals of *Stapelia*, but are an evolutionary hurdle for species of *Ferraria*. In the latter genus, a similar effect has had to be invoked by different means, namely, the crisped petal margins and the finely fringed style crests. Conversely, the long floral tube necessary in flowers pollinated by birds or long-tongued flies is easily developed from the existing floral tube in the family Iridaceae, but is an evolutionary impossibility for the family Geraniaceae, which has free floral parts. Species of pelargonium that rely on birds or long-tongued flies for pollination have therefore relied on the fusion between one of the sepals and the flower stalk to produce the necessary nectar-containing tube. Each flower represents a compromise between the demands of the pollinator and the constraints of its underlying structure.

These compromises are, however, never final and the forces that have shaped the flowers we see around us are relentless in their impact. It is only because natural organisms change on a scale so much vaster than the life of the individual that species appear to be indifferent to time. Botanical portraits are, ultimately, snapshots in the life of a species. It is essentially in its intent that botanical art differs from other forms of floral representation (Plates 96–98). Its primary intention is to inform about form. Unlike other types of representation, which permit or encourage a much greater freedom of interpretation, botanical art is therefore imbued with a timelessness reminiscent of the species itself. Adding the more active (or interactive) dimension of ecology immediately transforms the object into a subject, and we are drawn inescapably into the picture.

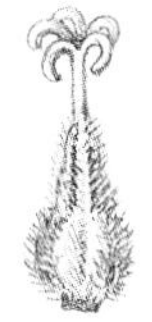

FURTHER READING

Goldblatt, P., Bernhardt, P. and Manning, J.C., 'Pollination of petaloid geophytes by monkey beetles (Scarabeidae: Rutelinae: Hopliini) in southern Africa,' *Annals of the Missouri Botanical Garden*, 85, 1998, pp. 215–230.

Goldblatt, P. and Manning, J.C., 'Phylogeny and speciation in *Lapeirousia* subgenus *Lapeirousia* (Iridaceae: Ixioideae)', *Annals of the Missouri Botanical Garden*, 83, 1996, pp. 346–361.

Johnson, S.D. and Bond, W.J., 'Red flowers and butterfly pollination in the fynbos of South Africa', in M. Arianoutsou and R.H. Groves, eds, *Plant-Animal Interactions in Mediterranean-Type Ecosystems*. Amsterdam: Kluwer Academic, 1994, pp. 137–148.

Manning, J.C. and Goldblatt, P., 'The *Moegistorhynchus longirostris* (Diptera: Nemestrinidae) pollination guild', *Plant Systematics & Evolution*, 206, 1997, pp. 51–69.

Manning, J.C. and Goldblatt, P., 'The *Prosoeca peringeyi* (Diptera: Nemestrinidae) pollination guild', *Annals of the Missouri Botanical Garden*, 83, 1996, pp. 67–86.

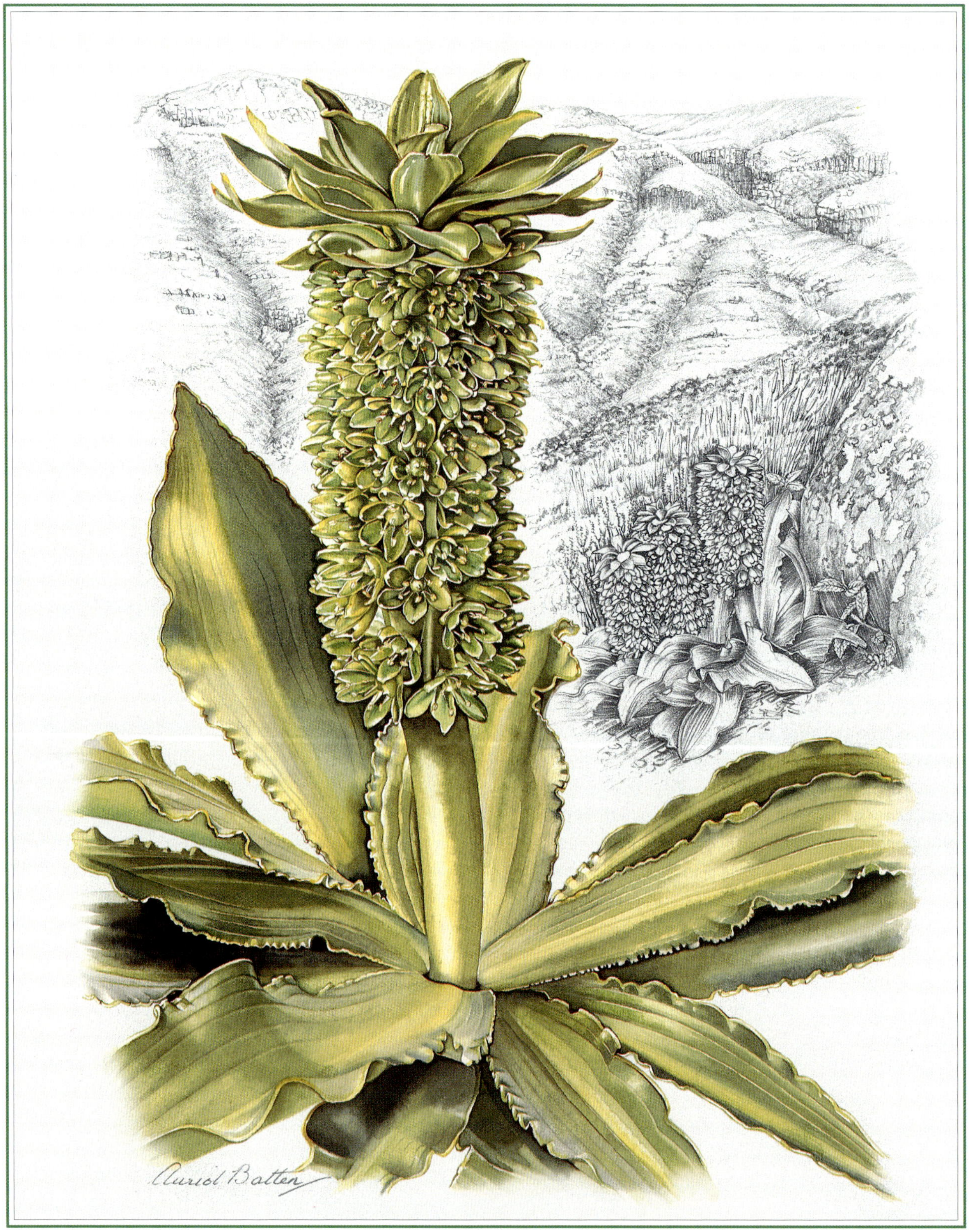

Plate 99

Auriol Batten, *Eucomis autumnalis*, 1986, watercolour and pencil: reproduced in A. Batten, *Flowers of Southern Africa* (Sandton, 1986)

PETALS AND STIGMAS
REFLECTIONS ON PLANT PORTRAITURE

MARION ARNOLD

Botany: 'living portraits of plants'

Contemporary plant portraiture, a hybrid genre with a complex history, is located within both botany as a scientific discipline, and art as a transformative visual practice. Possessing aesthetic and documentary attributes, southern African botanical art makes a significant contribution to the long history of 'scientific' plant representation, which commences centuries before plant science became separated from medicine in the sixteenth century.

During the Renaissance a significant change in representation occurred when, in 1530, Otto Brunfels began publishing *Herbarum Vivae Eicones (Living Portraits of Plants)*, an aptly named book, for the commentary was accompanied by Hans Weiditz's realistic drawings. Unusually, Brunfels credited his artist, commenting that the 'renowned painter Johann Weiditz . . . has provided illustrations of such artistry for the engravers that some have said they are real plants!'[1] Weiditz's coloured drawings (translated into an edition of woodcuts) were products of direct observation. Their realism modernized botanical imagery and set a precedent which saw medieval fantasies, initially derived from classical sources, being replaced by empirically based plant portraiture – an art of accurate description and interpretation of the particular.

The Brunfels herbal, produced at a time when affluence and humanist philosophy motivated patrons to commission their individual likenesses, was followed by Leonhart Fuchs's book, *De Historia Stirpium* (1542), illustrated by Albrecht Meyer. Whereas Weiditz portrayed what he saw – specimens with irregularities and flaws – Meyer idealized his plants. Thus the first modern botanical artists established the two paradigms that still characterize plant portraiture – realism and idealization, the depiction of the particular or the typical.

The emergence of 'botanical art' from the realm of plant portraiture was prompted by Linnaeus's (1707–1778) contribution to systematic science. He recognized the importance of illustration and, in *Philosophia Botanica* (1751), he states that illustrations must be of natural size and show the parts of the plants.[2] Linnaeus focused the demands that botany made on the image, and his theories were then demonstrated visually by artist Georg Ehret (1708–1770) who had met Linnaeus in 1736 at Hartekamp, the Dutch estate of wealthy banker and garden-lover, George Clifford (1685–1760). Impressed by Ehret's work, Clifford commissioned him to produce drawings for Linnaeus's descriptions of the rare plants in the Hartekamp gardens, the combination of text and illustration being published in 1737 as *Hortus Cliffortianus.*[3]

Ehret contributed twenty drawings to the thirty-four plates of *Hortus Cliffortianus* and he included *Gladiolus foliis linearibus*, now *Gladiolus angustus* (Plate 100), recently reinterpreted by Fay Anderson (1931–) and John Manning (1962–).[4] Thus the first significant publication of modern 'botanical art' includes an example of an important southern African genus. According to Linnaeus, 'Ehret did in the beginning absolutely not want to paint the stamina,

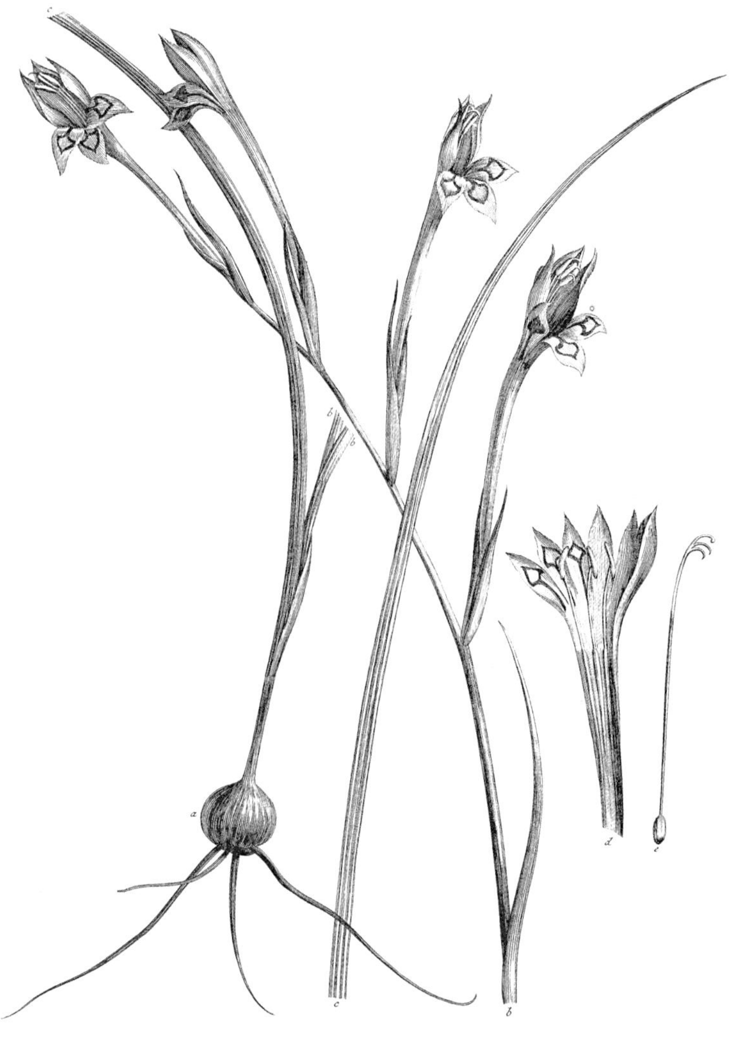

PLATE 100

Georg Ehret, *Gladiolus foliis linearibus* (= *G. angustus*), engraving: *Hortus Cliffortianus* (Amsterdam, 1737)

pistilla, and other small parts, as he argued that they would spoil the drawing; in the end he gave in, however, and then he liked this kind of work so much that thereafter he observed the most minute and inessential particulars.'[5] Under Linnaeus's guidance, Ehret did drawings of dissections (seen to advantage in the *Gladiolus foliis linearibus* details) and he published a diagram in 1736 to promote the Linnaean system.

Hortus Cliffortianus is important because it contains the first 'pictured' examples of the Linnaean system of plant classification. In drawing attention to the flowering parts, and the *flower* as the locus of sexuality, Ehret established new principles for botanical illustrations. Generated by scientific texts, illustrations became images that document and support verbal classification and identification with visual evidence, often utilizing two different visual languages – diagrammatic stylization and naturalism, the conceptual and perceptual signifiers.

Destined to appear as visual texts interacting with written discourse, botanical illustrations are circumscribed by the expectation that imagery will be informative, descriptive, accurate, and mimetic, but not necessarily beautiful. Indeed, however useful they may be as visual references, not all botanical illustrations satisfy the eye as 'art'. The *Crassula retroflexa* (Plate 101) from *Plantarum Selectarum Icones Pictae*[6] by Nicolaas Meerburgh (active 18th C.) is awkwardly drawn and hand coloured. The line lacks decisiveness, hesitant cross hatching makes an inadequate attempt to render tone, and the plant structure is wholly unconvincing as form in space. Nonetheless, in this and other more accomplished images by early travellers in southern Africa, there is sufficient schematic information in the salient features of the work to establish plant identity (see Plates 16, 17 and 20).

Botanical illustrations gain artistic credibility when they fulfil and transcend the imitative requirements of science. Auriol Batten's (1918–) depiction of *Eucomis autumnalis* (Plate 99) demonstrates an exhilarating understanding of watercolour as a medium requiring discipline and audacity, and displays a convincing knowledge of plant character. Controlling her washes, working freely on large areas and with crisp delineation of details, Batten manipulates a limited palette with assurance, imbuing a range of greens with density in shadow areas and evoking the fall of light on stiff leaves and frilly leaf edges. Looking at her painting, one sees two realities: the likeness of a sculptural plant, and a wonderfully accomplished watercolour bearing the evidence of its making process. Batten's knowledge of botany and formal control of visual language and medium, create an informative and affective image. In short, she *interprets* her subject and produces a complex portrait, not merely a likeness. Her paintings reveal how illustration becomes art, even when defined by scientific parameters.

The scientific nature of *illustrations*, produced specifically for botanical publications or for herbarium sheets, can be distinguished from images in a florilegium, the type of book described by Stearn as '*consisting largely or entirely of pictures of flowers*'[7]. The famous florilegia, featuring works by artists such as Pierre-Joseph Redouté (1759–1840) (Plate 122) are essentially aesthetic, not scientific in intent. Although these naturalistic images are products of accurate, detailed observation, they are not intended to provide classificatory information but to function as catalogues of important plant collections. While illustrators commissioned by botanists often have to portray plants lacking striking attributes, florilegium artists are free to choose subjects imbued with dramatic form and colour, and need not include scientific detail. The latter can indulge their own virtuosity and sense of spectacle.

The great florilegia paintings provide excellent examples of plant portraits since they prioritize the decorative appearance of plant models while showing due concern for accurate resemblance. They do, however, raise questions about the definition of 'botanical art'. In fact these opulent images are 'botanical' only in a generic sense; more correctly they are 'horticultural art'. Many florilegia testify to the age of discovery, the import of exotic species into Europe, and prevailing horticultural practices. As sumptuous flower portraits, the studies reveal the desire of wealthy nature lovers to use art to immortalize the ornamental plants they owned and grew in their gardens. These images differ from the genre of flower painting only because they are produced as plates for books, rather than as commodities for display in art exhibitions.

In considering botany as a site of meaning for plant portraits one must remember that the image is more than a likeness of a plant: it is also part of a historically determined relationship

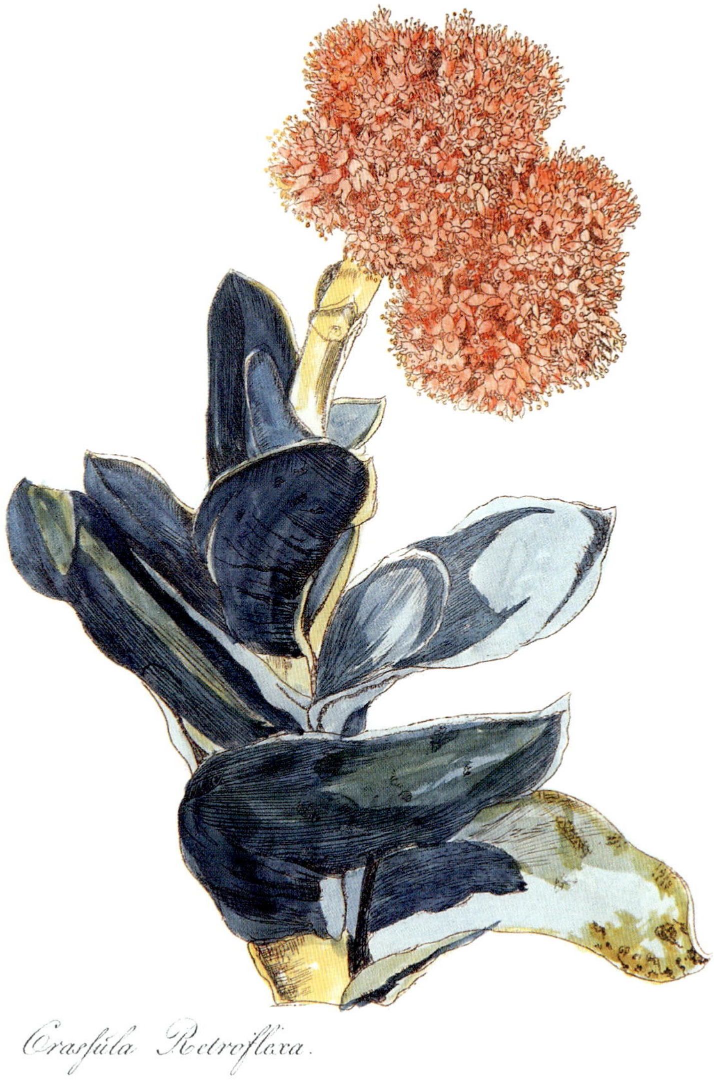

PLATE 101
Nicolaas Meerburgh, *Crassula retroflexa*,
hand-coloured engraving:
Plantarum Selectarum Icones Pictae (Leiden, 1798)

between image and text. Illustrated botanical publications require the services of two contributors: the botanist who writes the plant description, and (assisted by engraver or lithographer) the artist who produces images which support the process of identification and offer evidence of personal observation and style. In many instances, however, the text/image relationship is weighted in favour of the writer: the artist is not seen as a collaborator of equal stature with the botanist, but rather as a subordinate player whose sole purpose is to serve science in printed and published form.

Art and the mark of disgrace

Flowers are unashamedly about sexuality, a fact that greatly disturbed Victorian art critic, John Ruskin (1819–1900). Although he painted acutely observed studies of natural forms himself, he refused to pry into the sex life of flowers with a microscope, observing primly, 'with these obscene processes and prurient apparitions the gentle and happy scholar of flowers has nothing whatever to do'.[8] His remark has been well heeded by legions of amateur painters, whose interest in flowers ends with the application of a watercolour wash to pencilled outlines of leaves and petals.

As a sex organ, a flower stigma plays a crucial role in facilitating fertilization and the continuation of a species. But the term has another meaning – it is a mark of disgrace. Within botany, the stigma contributes positively to species reproduction but in the High Art world plant portraiture has been stigmatized and marginalized, despite (or perhaps because of) its popularity with the public at large.

There are historical reasons for the negative reception of plants in general and flowers in particular. When the art academies of Europe were established in the seventeenth and eighteenth centuries, History Paintings (grand, narrative compositions based on subjects from history, literature, mythology and religion) were considered the pinnacle of achievement. The rites of man took precedence over the delights of Nature; still-life and flower painting were relegated to the lowest ranks in the hierarchy of genres.

The production of History Painting required formal training – knowledge of drawing, perspective, anatomy, composition, art history, and literature. It aimed to tease the mind, stir the imagination and convey moral dictums about the human condition. In the philosophical climate of The Age of Reason, Art demanded demonstrations of visual and moral rhetoric through large-scale renderings of human drama and passion, and it required portraits to evoke greatness, grandeur and dignity. Art had to provide more than the sensual gratification abundantly offered by small still-life and flower paintings – such works were considered to require no more than an ability to observe and render form realistically. In other words, History Painting defined itself through inventiveness, elaborate design, and the exercise of reason, while plant portraiture was considered to rely only on visual acuity and manual dexterity.

There is, of course, no way of proving that Sir Joshua Reynolds (1723–1792), who founded the British Royal Academy in 1768, created portraits of human models that are inherently greater than the plant portraits of his contemporary, Franz Bauer (1758–1840)[9] (see Plates 9 and 120). What is obvious is that the Reynolds and Bauer studies possess utterly different frames of reference. It is also interesting to note that science became institutionalized in Britain before art: The Royal Society of London for Improving Natural Knowledge (usually known as 'The Royal Society') was founded in 1662, over a century before the Royal Academy.

The institutionalized Art of the European Academies promoted its cultural authority by looking back to classical antiquity and the Renaissance for philosophical and stylistic justification, but Science concerned itself with extending knowledge and formulating theories relevant to a contemporary world. While scientists located new formulae to present new information, academicians were bent on entrenching tradition through fiction, not facts, and they failed to recognize the analytical and conceptual underpinning that characterized specialist imagery such as botanical art, bird paintings, or medical drawings.

Despite the anthropomorphic focus of seventeenth- and eighteenth-century art, flower imagery was visible in northern Europe as the flower-piece, a sub-category of still-life painting. During the golden age of Dutch art, flower-pieces by artists such as Jan Breughel (1568–1625) described floral beauty and were highly valued in a materialist society that esteemed portrayals of wealth and possessions in preference to religious scenes. Complex and highly contrived, flower-pieces recorded the spectacle of natural forms and colours, often

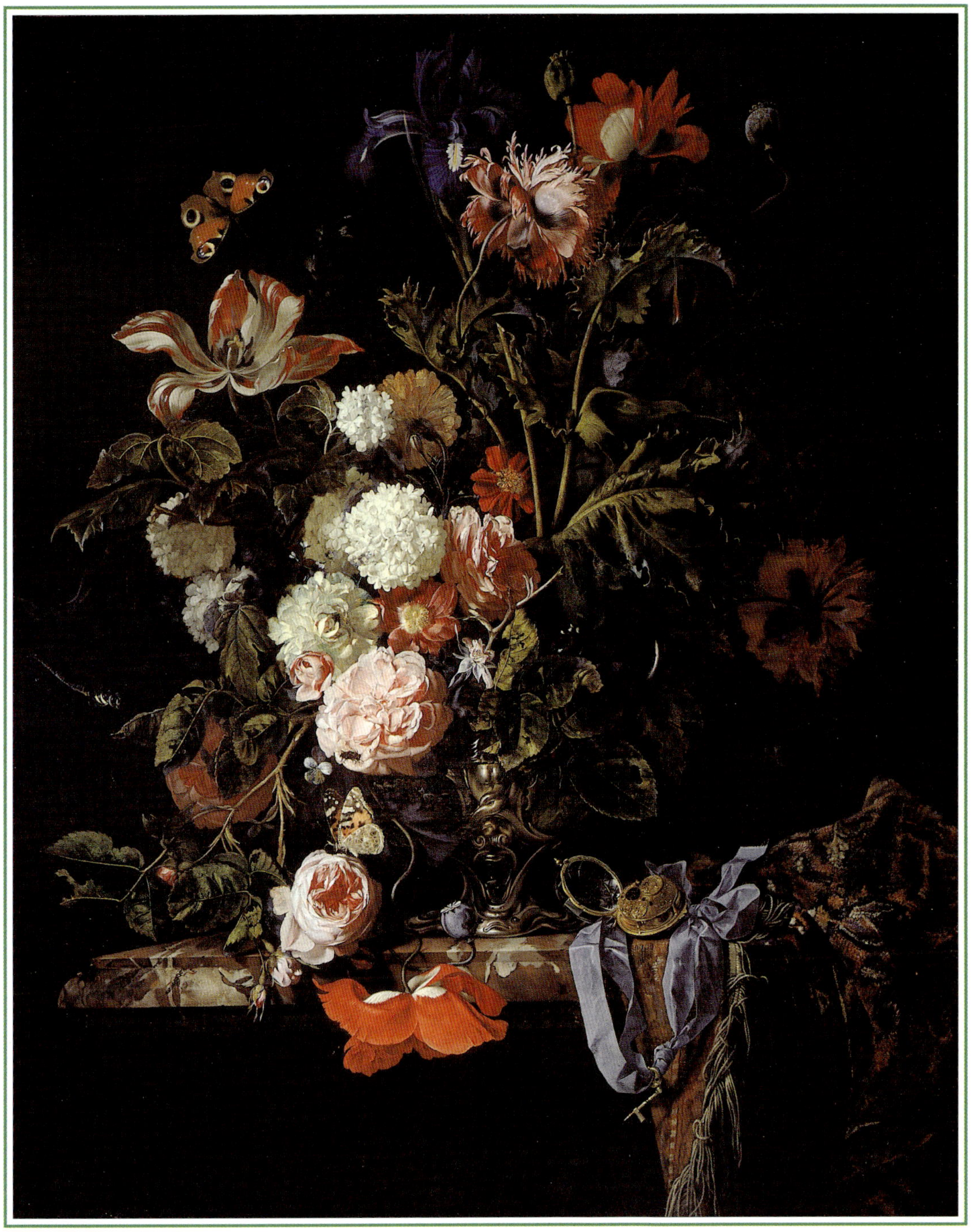

Plate 102

Willem van Aelst (1625/6–c.1683), *Vase with Flowers*, 1660, oil on canvas on wood, 83,8 cm x 66,7 cm

Dutch flower-pieces are not just displays of *trompe l'oeil* virtuosity: facts are combined with symbols. Vases of cut flowers, and objects such as watches or candles, allude to the ephemeral nature of time. These decorative, moralizing works spawned generations of less evocative and alluring offspring – flower paintings devoid of the power to preach, but with a residual sentiment that renders them attractive to viewers devoid of much knowledge about either art or botany.

juxtaposed with objects of lush textures and reflective surfaces. These paintings are elaborate, detailed descriptions of commodities, amongst them the flora collected globally to be cultivated and hybridized in Europe, and they celebrate Holland's mercantile wealth and horticultural achievements. The commodification of flowers resulted in tulipmania – speculation on tulip bulbs – which gripped Holland in the mid seventeenth century and which relied, to some extent, on images of tulips to proclaim the extraordinary diversity of colours and patterns that resulted from genetic accidents produced by a virus.

Many southern African artists of note, such as Irma Stern (1894–1966), used flowers in still-life paintings. Stern's *Still Life with Gladioli* (1958) (Plate 103), an expressionistic painting of flowers easily identifiable as gladioli, is more than a decorative statement of vigorously applied, strong colour; it speaks of the hybridization process that makes these large flowers unlike the delicate blooms of indigenous species (Plates 70, 73, 88). But despite the fact that flowers within still-life painting possess meanings related to commercial enterprise or ecological concerns, flower painting has come to epitomize popular, light-weight, purely decorative art, the antithesis of arcane and serious avant-garde imagery.

Botanical art publications tend to introduce discussion on Dutch flower-pieces apologetically[10] but fine-art publications seldom include botanical imagery at all. This is partly attributable to the stigma that has been attached to illustration. During the Modernist period, 'illustration' was a contentious term. Although it ought to be understood as a genre of art, it has often been used adjectivally, in a derogatory sense, to imply that 'illustrative' images – because they relate to texts – lack the visual autonomy so highly prized by Modernist theory. Postmodernism, with its enthusiasm for intertextuality, has rehabilitated illustration. Moreover, 'fine art' is no longer regarded as the most significant category of visual imagery. In the age of capitalism and technology, advertising, the graphic novel and comics have all contributed to positive reassessments of 'illustration', but so far this has not had any major repercussion on botanical illustration.

Botanical art and illustration have been allowed to remain in the realm of Science, under-researched by art historians, undervalued by art theorists and critics, uncollected by art institutions, and seldom exhibited in art galleries.

Although plant portraiture, by and large, is still lodged in Science, feminist research has contributed to a reappraisal. Revising the canon of history, feminist art historians rediscovered the superb illustrations of Maria Sibylla Merian (1647–1717), the accomplished flower-pieces of Rachel Ruysch (1664–1736) and works by a host of Victorian women painters. It is also noteworthy that The Royal Academy had two women founder members – the Neoclassical painter, Angelica Kauffmann (1740–1807), and Mary Moser (1744–1819) whose *oeuvre* was largely comprised of flower paintings. It is now obvious that the hierarchy of art genres, constructed to favour male artists, reinforced the obscurity of many women artists – including those working in southern Africa – who created intimate studies of nature.[11] Thus revisionist art historians, focusing on the art object as a site of social documentation, have shifted attention to a neglected genre where, as we peel back the petals, we discover that the stigma carries complex meanings.

Grey territory

Within Art, the genre of landscape has traditionally enjoyed more credibility than flower painting, partly because some radical changes in form and content were effected through landscape representations. The grey territory between Art and Science, or flower painting and botanical art, is nowhere more clearly demonstrated than in landscape painting.

Initially the outdoor background for human action, landscape painting gained autonomy when seventeenth-century Dutch artists proudly recorded the appearance of newly independent Holland. Their landscape paintings were thus related to geography – to specific places and regions. In the early nineteenth century the Romantics used landscape to evoke spiritual and transcendental states, and then the

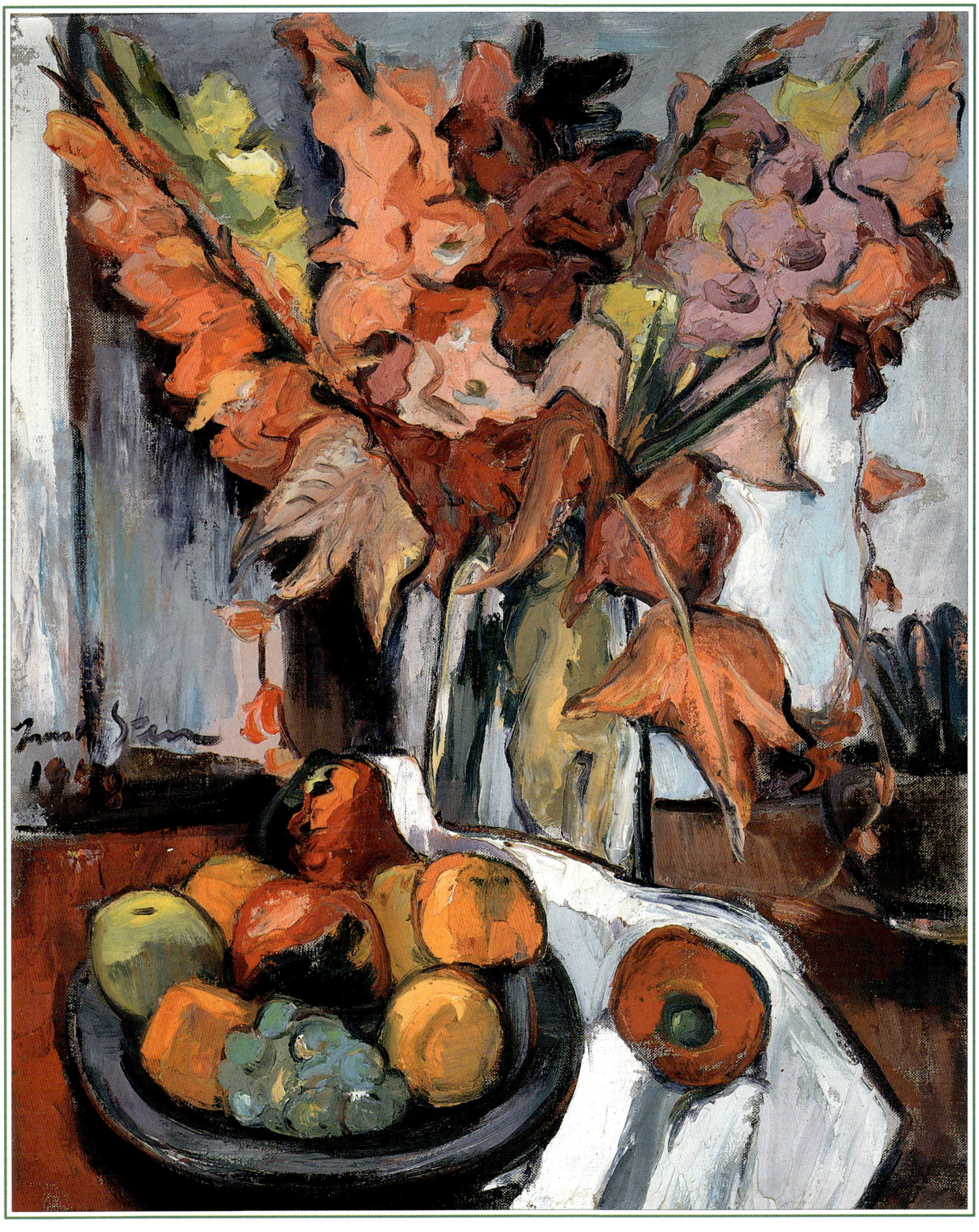

PLATE 103

Irma Stern, *Still Life with Gladioli,* 1958, oil on canvas, 86 cm x 67,5 cm

Impressionists, conscious of their modernity, revolutionized pictorial language by developing new methods of paint application to denote the present moment and transitory play of light on natural forms.

One of the most extravagant productions of floral landscapes – *The Temple of Flora,* issued from 1799 to 1807 – exposes the claims made on the floral landscape by both art and botany. Conceived by Robert John Thornton (1768?–1837) as a florilegium, the plates for the work are part of an ambitious project. At the outset Thornton planned a multi-volume botanical publication called a 'Philosophy of Botany', and he announced in 1797 that the concluding portion of the work would be a *'New Illustration of the Sexual System of Carolus von Linnaeus'.*

Thornton's enthusiasm for modern botany and his desire to effect an alliance between botany and art were not, however, realized in the plates of his extravagant undertaking. He conceived the idea of representing plants against fanciful backgrounds that would evoke atmosphere. The melodramatic illustrations are devoid of scientific detachment and there is very little that is literary in the explanatory flowery prose and self-conscious poetry that accompanies the plates. Nevertheless, the fusion of romantic sensibility, fact and fiction, provide a fascinating glimpse of the eighteenth century's contradictory impulses.

Most of the artists who worked on Thornton's project came from the fine-art world and earned their livings as landscape and portrait artists, not as botanical illustrators. The plants they depicted are, however, recognizable portraits of tulip, lily, rose, stapelia and other striking flowers. The southern African *Strelitzia reginae,* the queen flower, or bird-of-paradise flower – introduced to England in 1773 – was included in *The Temple of Flora* as a compliment to Queen Charlotte to whom the folio was dedicated. This striking flower appears twice: in *Cupid Inspiring the Plants with Love* (Plate 104) by Philip Reinagle (1749–1833) and *The Queen* by Peter Henderson (active 1799–1829).

Reinagle's painting (translated as a coloured stipple engraving partly finished by hand) depicts the boy Cupid shooting his arrows towards the foreground clump of strelitzias while behind him stretches a picturesque tropical paradise. The image is one of romantic fancy, ridiculous in a literal sense, and the antithesis of scientific veracity. And yet it is not devoid of scientific significance. If the image is understood as allegorical, Cupid's role as god of love relates to Thornton's eighteenth-century conception of nature.

Thornton believed in an ontological continuum where, as he wrote in *The Philosophy of Medicine,* 'all is combination, relation, affinity, connection'.[12] He was greatly influenced in his philosophy of nature by Erasmus Darwin (1731–1802) whose long poem, *The Botanic Garden,* explores the theme of the shared characteristics of humans and plants, and advances the proposition that plants possess 'feeling'. The first part of Darwin's poem, 'The Loves of the Plants', anthropomorphizes plant sexuality and thus it is not strange at all that Thornton should have commissioned Reinagle to depict Cupid inspiring plants with love. Furthermore, although Reinagle's image is allegorical, it carries a message of interconnectedness not far removed from late twentieth-century ecological concerns and systems theory. *The Temple of Flora* paintings certainly promoted the idea that parts of a plant, laid out as stark images in neutral space, are limited as signifiers for the complex relationships of the natural world. An anomaly in its time, *The Temple of Flora* bankrupted Thornton, but – despite its excesses – it remains a landmark of botanical publication, a passionate tribute to plants.

As indicators of philosophical and aesthetic attitudes to nature, nineteenth-century landscape paintings expressed the conflict that arose when subjectivity and objectivity were positioned in binary opposition. In southern Africa, travelling artists were confronted with the difficulties of expressing feelings of wonder generated by the landscape of the Dark Continent, and the need to make accurate records of new and strange forms. They tended to adopt picturesque conventions and naturalism to serve their purposes.

Thomas Baines (1820–1875), immensely curious about the world, produced his best work

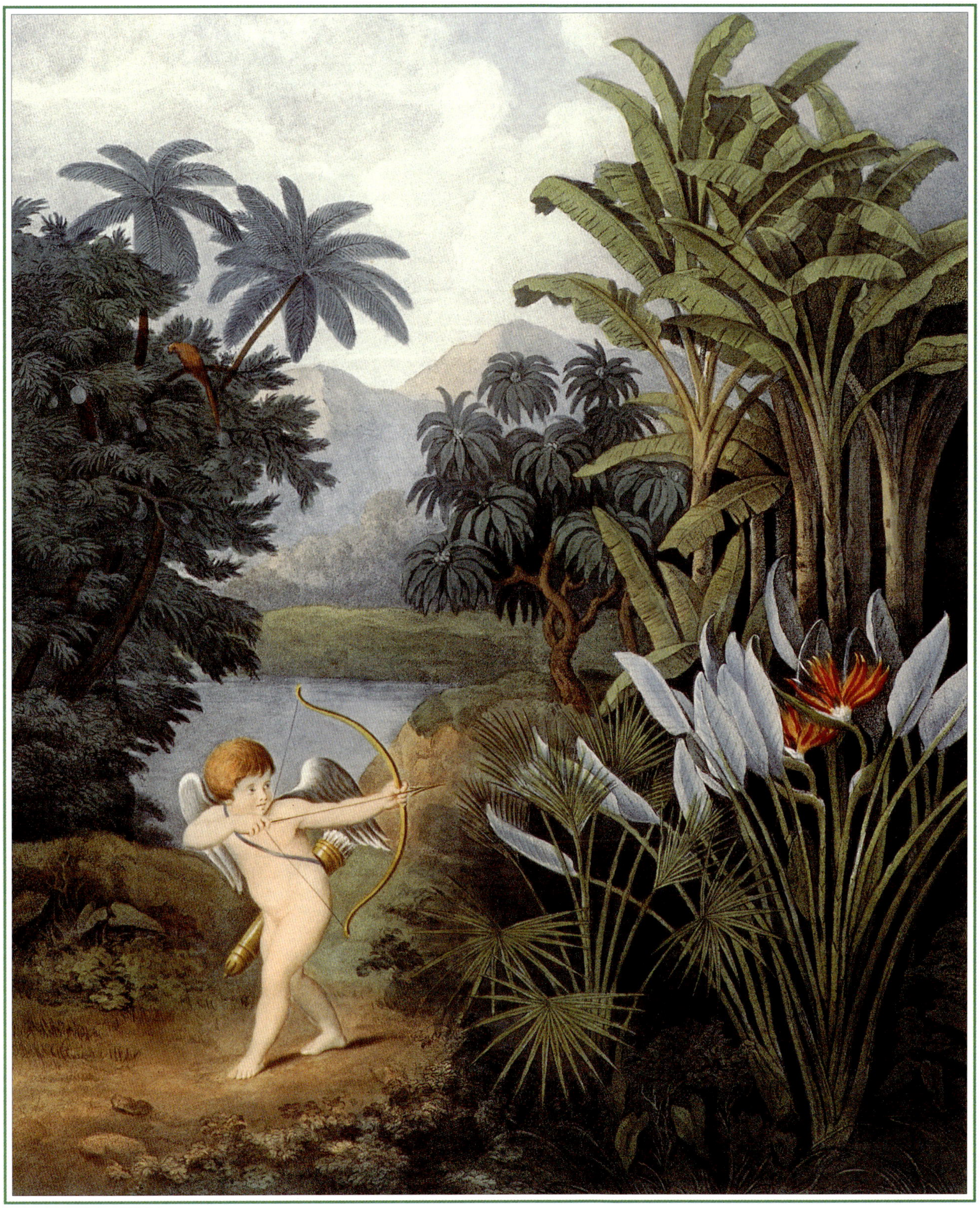

PLATE 104

Thomas Burke, *Cupid Inspiring the Plants with Love*, engraving from original painting by Philip Reinagle: *New Illustration of the Sexual System of Carolus von Linneaus; . . . the Temple of Flora, or Garden of Nature . . .* (London, [1799]–1807)

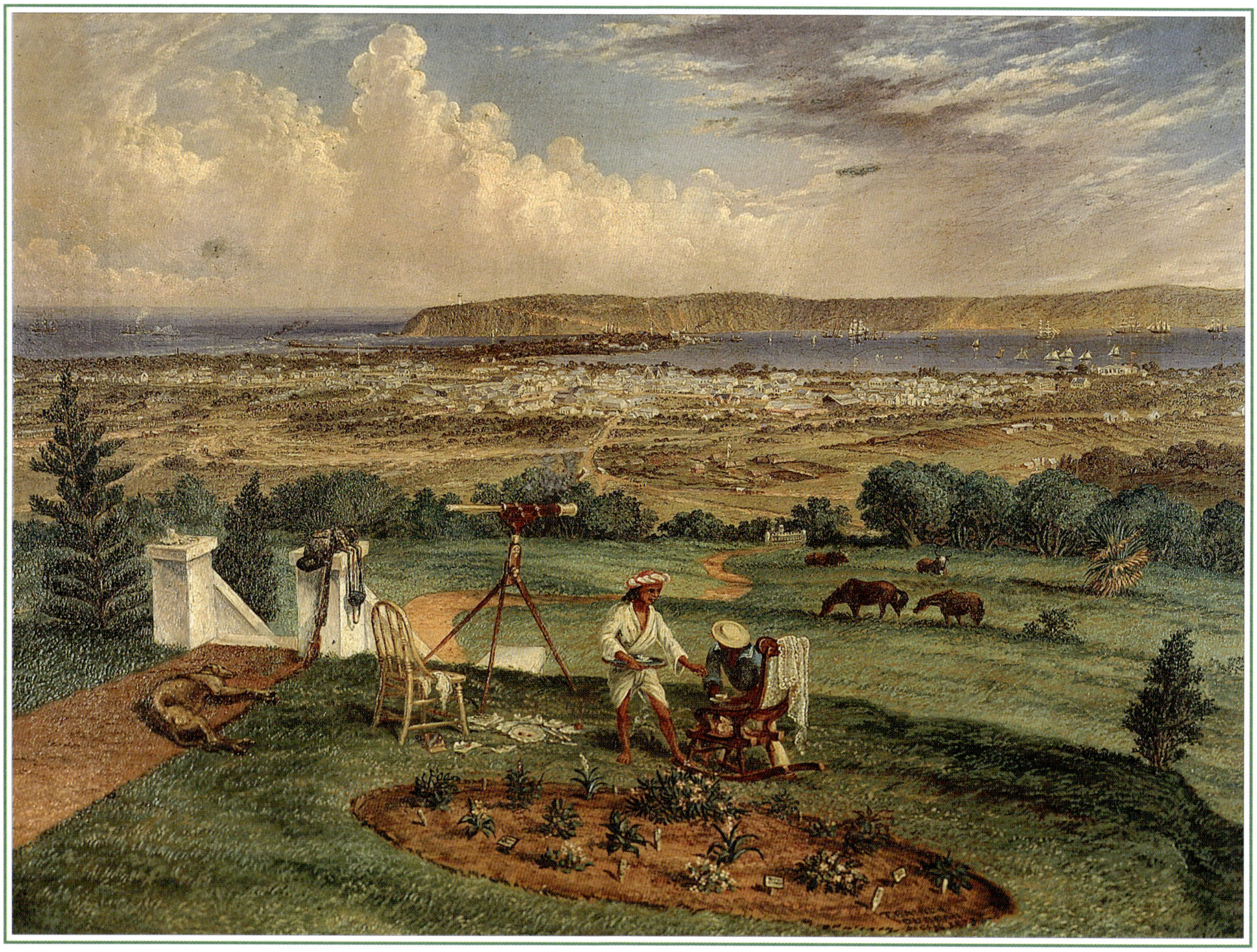

PLATE 105

Thomas Baines, *Durban from Mr Currie's Residence, Berea 1873*, 1873, oil, 45,5 cm x 60 cm
(Local History Museums' Collection, Durban)

When Baines painted this landscape, the Durban Botanic Gardens (close by, to the left of this view) had just come under the direction of William Keit who, a year later, became H.W. Currie's son-in-law. Currie's Fountain, feeding from a spring at the foot of the gardens, commemorates mayor Currie's irrigation works of 1878 an innovative system that saved Durban in a time of drought.

when his romantic imagination was stirred by the spectacle of space and the natural elements, but he also paid attention to detail, sketching animals, birds and plants for subsequent inclusion in his oil paintings. A fellow of the Royal Geographical Society, Baines had a well-developed interest in natural history and he collected plants. Some of these plants were dispatched by Baines to Kew Botanical Gardens in England to Sir William Hooker (1785–1865) – with whom he corresponded – and later to his son Dr Joseph Hooker (1817–1911). Baines must be considered as a naturalist rather than a scientific botanical illustrator but his watercolour sketches are accurate records of plants.

Baines's interest in botany finds expression in a landscape painting that places plants in the significant foreground zone. *Durban from Mr Currie's Residence, Berea 1873* (1873) (Plate 105) is a colonial landscape replete with telling detail. In the distance is the developing port of Durban depicted thirty years after the annexation of Natal in 1843; in the foreground is Henry William Currie's garden with a circular flowerbed containing neatly labelled plants.

This synthesis of macrocosm and microcosm is a complex socio-political statement and amongst the issues it represents is the importance accorded to the development of gardens in the Victorian British Empire. Natal was South Africa's garden colony and Durban contained one of the region's major botanic gardens. As imperial institutions, developed for the systematic cultivation of plants, botanic gardens were highly valued in Britain. They helped to sustain what has been defined as 'a sense of botanical nationalism'.[13] In 1873, when Baines painted Currie's garden, indigenous plants were still sent to Kew Gardens for naming and classifying, and Baines draws attention to the importance of taxonomy in the labels placed adjacent to the plant specimens. He also chooses to inscribe his signature in the space of the flowerbed, thus identifying himself with this area of the pictorial content.

Baines was not the only painter of the southern African landscape to concern himself with the close and distant views of plants. A fascination with plants in the environment characterizes the work of Marianne North (1830–1890).[14] Just as Thornton's *Temple of Flora* poses problems for writers on botanical art, so too does North's oeuvre of 832 oil paintings, executed over a period of thirteen years as she travelled the world documenting the floral landscape. North arrived in Cape Town in August 1882 and during her nine-month visit to the Cape and Natal, she produced more than a hundred works. These are now housed in the Marianne North Gallery at Kew, with the studies of her other travels where they constitute an idiosyncratic installation conceived decades before Claude Monet's famous waterlilies series at the Orangerie, Paris, which also draws attention to the relationship between plants and their habitats.

Challenging Victorian norms for women, North was a determined and courageous traveller who produced her paintings *in situ*, under difficult conditions.[15] Atypical of the delicate studies expected from gentlewomen, North's vigorously brushed and strongly coloured plant portraits show scant regard for botanical conventions, but are deeply felt responses to nature (Plate 106). They disconcerted Wilfrid Blunt who, in his seminal text, *The Art of Botanical Illustration*, commented, 'Botanists consider her primarily as an artist, although she discovered and portrayed species new to science . . . but artists will hardly agree, for her painting is almost wholly lacking in sensibility'.[16] This summary dismissal of North is unjust. It probably results from Blunt's discomfort with North's robustly sensual oil painting, which he denigrates as 'gaudy',[17] rather than from her integration of plants into landscapes, something to which Blunt did not take exception in his discussion of *The Temple of Flora*.

Landscape, as space and place, provides flora for the scrutiny of both taxonomists and ecologists. Marianne North's habitat studies constitute the first collection of paintings produced in southern Africa to foreground flowering plants, but not the last. Two of the region's foremost plant-portrait artists, Thalia Lincoln (1924–) and Auriol Batten, created work that relates plants to environment.

Lincoln's collaboration with John Rourke

resulted in the publication of *Mimetes*,[18] its closest relative being *The Temple of Flora* – not because it bears the remotest resemblance to it in the interpretation of plants – but because it is equally iconoclastic in challenging boundaries between landscape and botanical art. This massive volume (measuring 66 x 49 centimetres and weighing 9 kilograms) is the product of Lincoln's dedicated pursuit of her vision, realized in her unorthodox image-making process, choice of technique and materials, and stubborn attention to detail during the printing process.

Lincoln's method of working in coloured pencils, permits her to superimpose layers of colour on petals, leaves and stems, giving each form extraordinary intensity and luminosity. *Mimetes hottentoticus* (Plate 107) conveys all the structural data required by a taxonomist, it presents the ecologist with information on habitat, and it offers the viewer a plant portrait replete with personality and a metaphysical life force.

In Lincoln's drawing, elongated, bright red flower styles, tipped with yellow and black, protrude from hairy cream petals and punctuate cool pale-blue space, while silvery leaves shimmer in the clear light. The wonderfully intricate rhythms of the robust shrub contrast with a simplified panoramic view of mountains descending to the distant ocean. The image is both very specific and very timeless. To portray the mimetes, Lincoln climbed mountains, studied her plants *in situ*, and collected and took back to her studio material she needed. Recalling her first experience of *Mimetes hottentoticus,* Lincoln commented, 'It was like looking at gold and silver with touches of green and red and in the sunlight the leaves shone brilliantly'.[19] Her works convey this sense of visual delight. They are products of her determination to 'get in touch with the soul of the plant',[20] an artistic objective of little concern to purely scientific illustrators whose main intention is to render diagnostic features present in the external appearance of plants, not to ponder on and evoke inner life force.

In 1986 another milestone book featuring the floral landscape was published – Auriol Batten's *Flowers of Southern Africa*. Like *Mimetes*, it challenged evaluations of botanical illustration. Batten, a trained botanist and artist, produced both text and images, and thus defined her own objectives. On her choice of plants for illustrations she observed, 'My reasons for choosing them were various and they were selected in most cases for their great beauty.'[21] In order to sustain this emphasis on the aesthetic, she developed a schema that fuses botanical and landscape traditions (Plates 72, 74, 85, 89, 90, 94, 99, 125).

Batten explores plant structure and character in one hundred plates, demonstrating her versatility as a painter in handling stiff inflorescence, succulent sculptural leaves, and fragile translucent petals with equal conviction and technical deftness. Delicately nuanced, or brilliantly saturated, her colours assert their optical power against carefully drawn graphite-grey backgrounds. Rendering environments particular to the painted species, and utilizing format space to depict spacious landscapes, Batten positions her plants so that they proclaim their presence in foreground space but relate to receding terrain behind them.

Flowers of Southern Africa was not reviewed by art critics, probably because its title gave no indication that the contents might be interpreted as 'landscape', and it was received with reservations for its botanical content. Kerkham remarks that it 'is not necessarily a work of any great scientific value but the attractive plates are of a high standard'[22] but, as Killick observes, the paintings 'are of a very high standard and represent extreme scientific accuracy. Surely this contributes to its scientific value'.[23] Indeed it does, and to favour word over image is prejudice, not valid scientific criticism. Batten may have anticipated guarded reaction from the botany fraternity, for she observes in her preface, 'The accompanying text to each plate is merely intended to supply such details as may not be visible in the drawing, or other items of interest and as such, should be regarded as *supplementary to the plate,* rather than as a complete description of the plant in itself.'[24] With this statement of intent, Batten aligns herself with science through art.

The contradictory responses to Batten's work provide evidence that images occupying the grey

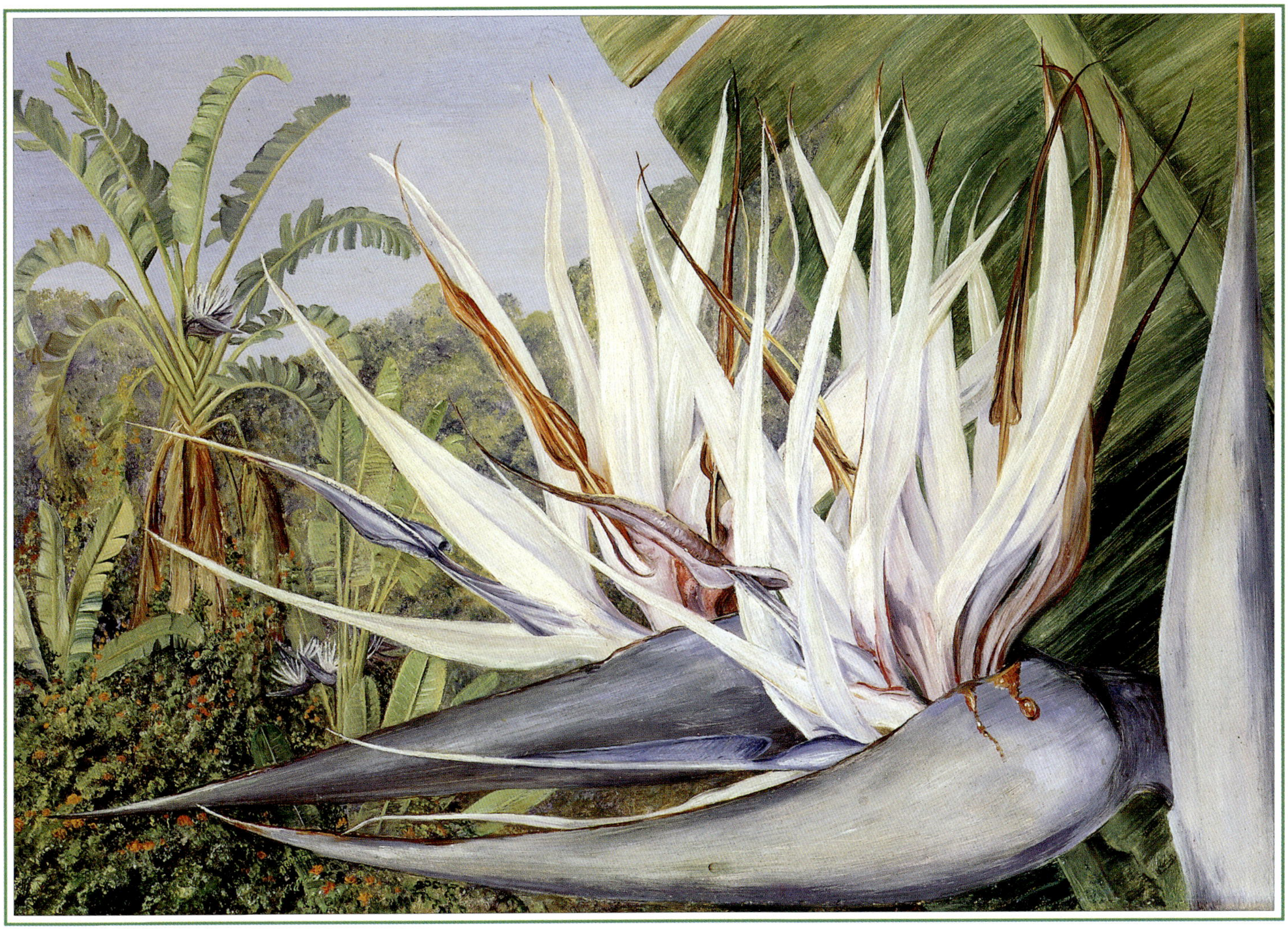

PLATE 106

Marianne North, *Strelitzia alba*, 1873, oil, 37 cm x 52 cm

In this painting North meets the primary requirements of botanists – she provides a recognizable likeness – but she also celebrates nature's lush growth, rhythms, forms and strong colours. By creating habitat studies she articulates a concept of interconnectedness, seeing and conveying her perceptions of natural space rather than designing conceptual arrangements of plant fragments. Her focus is more that of the ecologist than the taxonomist, and her eye is that of the artist not the specialist illustrator.

space of the floral landscape perplex viewers who have predetermined or purist attitudes to botanical art. This point is borne out by Kerkham's remarks about Thalia Lincoln's *Mimetes* illustrations which he considers 'do not match the brilliance of those of the eighteenth century artists, but nevertheless have a charm of their own'.[25] Both Batten and Lincoln take what they need from tradition but they make images for their own time, and make them in their own ways, revealing that artistic vision can accommodate scientific requirements and assert itself.

Style: the eye and the hand

Vision in art is expressed through 'style', a pivotal concept in art philosophy that has been subject to many vigorous debates about its historical implications. In essence, style – in the Fine Arts – is about visual identity; it is the distinctive 'look' of artworks or artefacts, and it manifests the personal and socio-historical visual responses of its makers to their internal and external worlds.

Through studying particular visual forms we identify the work of specific artists, periods within an artist's oeuvre, influences of other artists, historical epochs, and localities of production. In this way, style fulfils a classificatory function. Since style is 'the look' of the image, it is demonstrated through form, and the use and syntax of visual language. Although aestheticians argue about the distinction between form and content, ultimately form is the vehicle of content and, for painting, the shapes, colours, and tones that represent the external world express the meaning of depicted subjects and evoke aesthetic emotion.

Style is shaped partly by responses to methods and materials that are experienced in the process of realizing images, and partly by the intentional application of learned conventions. Stylistic analysis, which facilitates image classification, is undertaken when there is sufficient material to compare images. Through comparison, certain recurring characteristics become apparent, enabling one to determine the constituent elements of personal style (Auriol Batten's work), period style (Victorian representations), the style of a genre (botanical illustration), the style of a publication (*The Flowering Plants of Africa*) and so on. Similarly, just as recurring elements contribute to style formation, so difference is a key factor in distinguishing between artists, cultures, or historical periods.

Although plant portraits, good, bad and indifferent, may divulge sufficient botanical data to answer purely scientific needs, the expressive dimensions of style probably hold greater interest for viewers who look at plant portraits from the perspective of Art, than they do for botanists using images to verify scientific hypotheses. But, whatever the perspective of the viewer, the influences exerted by the hand and the eye make a significant contribution to style and its interpretation.

The hand: ways of doing

Over the centuries, the hand has found many ways of creating imagery and the processes producing representations of southern African plants have changed dramatically – from block printing to computer-generated imagery.

After the printing press was developed in about 1450 (Johann Gutenberg is traditionally credited with this invention in the West), book production escalated. The press offered the means to replicate drawings through woodcuts and, when Hans Weiditz startled his viewers with his 'living portraits of plants' in 1530, they did not see his lifelike watercolour studies, but linear black and white illustrations. Made by skilled craftsmen, these woodcuts were translations, not copies, of Weiditz's coloured paintings.

This translation process was common procedure for the reproduction of botanical pictures, varying only as new techniques and processes – metal engraving, etching, mezzotint, stipple engraving, and lithography – allowed a choice of reproduction methods. All these methods are accommodated in three basic printing methods. In relief printing (woodcuts) the image stands proud from the block to receive ink. With intaglio prints (engraving, etching, mezzotint) the image is bitten into the plate and ink fills the incisions. With the planographic method (lithograph), a greasy image drawn on the stone accepts ink which is repelled from damp areas of the stone.

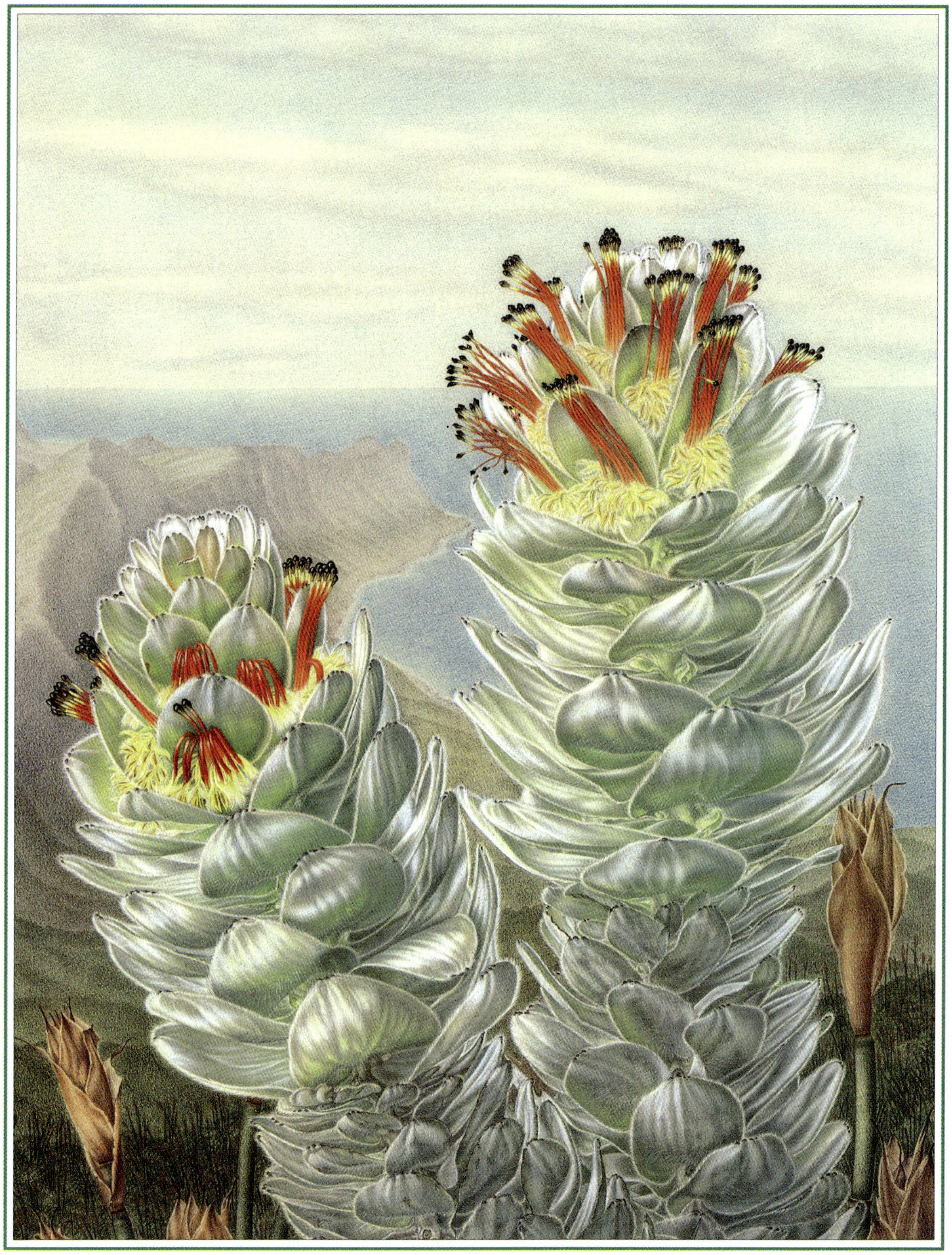

PLATE 107

Thalia Lincoln, *Mimetes hottentoticus*, 1975, coloured pencils on Bristol board

The proteaceous mimetes, endemic to the south-western and southern Cape, are renowned for their beauty, rarity, and inaccessibility. Lincoln's various studies of these sturdy, woody shrubs, are artworks of elegant precision.

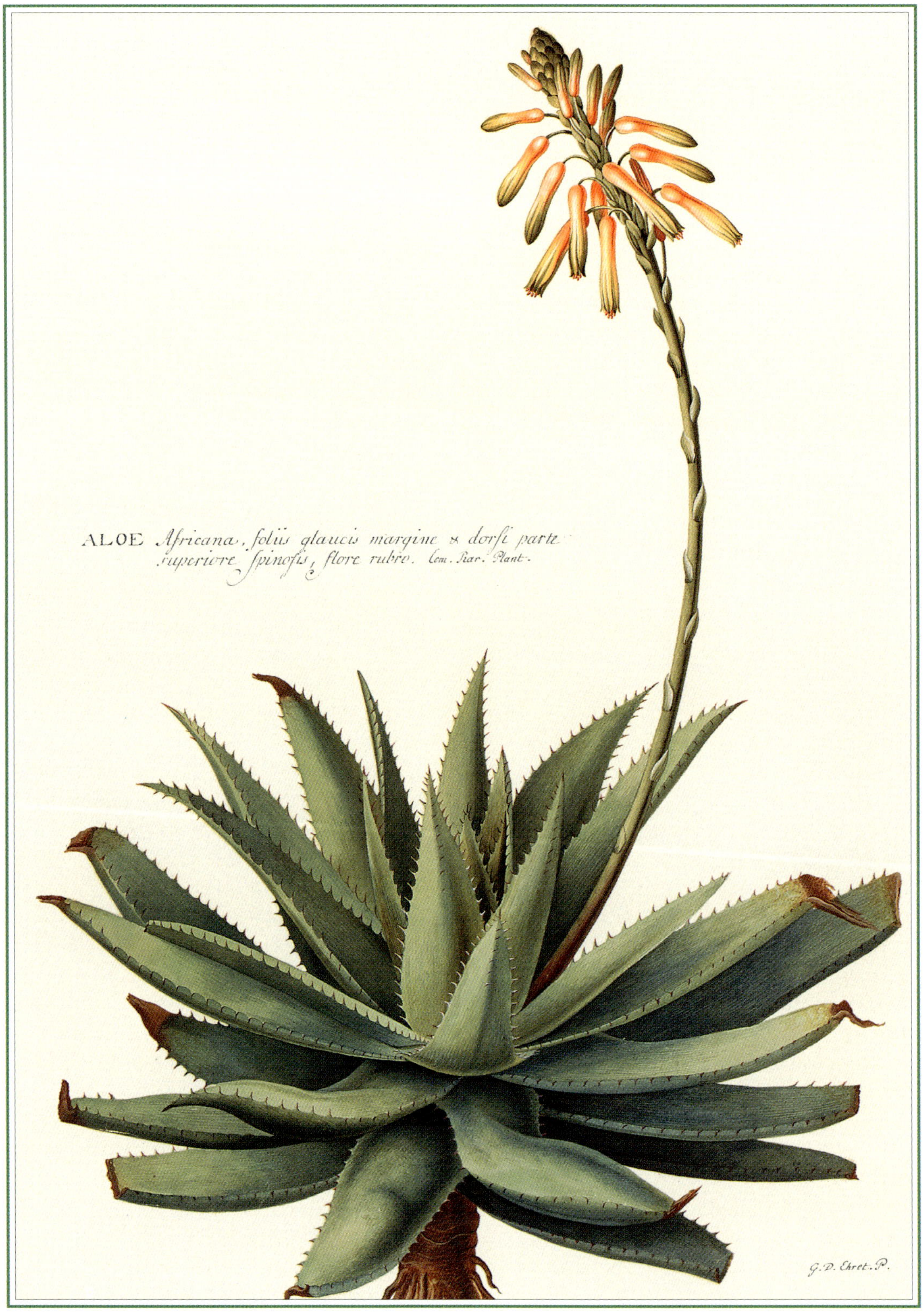

PLATE 108

George Ehret, *Aloe africana*, sixteenth/seventeenth century, body colour: 'Aloes, being 8 original drawings, 3 by G.D. Ehret, 2 by J. van Huysum, 3 unsigned, all painted between 1687–1770'

Since botanical art is primarily concerned with the accurate rendering of likeness and the distribution of illustrations through published texts, drawing (the delineation of shapes) takes precedence over other pictorial issues. Furthermore, the style of images has to be suitable for reproduction, and clarity is a primary objective: 'art' is qualified by the needs of botany. Hence, although Ehret is known for the stark elegance of his black and white illustrations (engraved not by him but by Jan Wandelaar), his dignified paintings (Plate 108) are less well known. In effect, Ehret's work, like that of most professional botanical artists, is familiar through reproductions, and the originals remain almost invisible.

The actual mechanics of printing were dramatically affected by the invention of the camera because cumbersome plates were largely superseded by photo-mechanical processes of reproduction. Images could be photographed, and film negatives used to make lithographic plates. This meant that a facsimile of the original painting of a plant became the final image without the intervention of a process that altered and transformed original statements into similar, but different images.

The term 'print' became ambivalent, and it now has two meanings. Original graphic prints (woodcuts, engravings, etchings, lithographs) are produced in limited, numbered editions controlled by the artist, and each image is an 'original' graphic print. Photo-mechanically produced prints, run off in large editions by commercial printers, replicate painted or drawn originals.

Traditionally, reproduction processes were controlled by the skilled eyes and hands of craftspeople – engravers making plates, or people employed to hand colour engravings (Plate 101) and lithographs (*The Flowering Plants of Africa* was hand coloured until 1945 and *Curtis's Botanical Magazine* until 1948). The old collaborative process explains the variations found in florilegia engravings, and accounts for the fact that some florilegium images exist in both black and white, and coloured versions.

The issue of colouring prints is an interesting one. When colour is added to pencil or pen and ink drawings, watercolour tends to be the preferred medium, probably because it can be used as a simple colour wash, or as a more sophisticated, painterly medium, often used in conjunction with body colour (now known as 'gouache'). Whether colour was used to create a watercolour painting on paper or vellum, or was added as washes on prints, it was subject to certain technical restrictions. Prior to the twentieth century, many pigments tended to be unstable; either they were fugitive and disappeared, or the hue altered as the painting was exposed to light. Some colours – cobalt blue and French ultramarine for instance – were invented only in the nineteenth century, and eighteenth-century artists had great difficulty in finding or mixing greens that would be durable or would simulate the greens of nature. With this problem in mind the *Botanical Magazine* announced when it was founded in 1784 that illustrations would be 'coloured as near to nature as the imperfection of colouring will admit.'[26]

To produce and publish book illustrations (as opposed to paintings for an exhibition) requires collaboration and, inevitably, compromise. Book design demands integration of text and images so that the reader/viewer is delighted and teased when turning pages. But visual surprise must be balanced with design unity – consistency in layout, balance between repetition and contrast, and harmonious interaction of quiet space and visually active areas. Text is both writing and image, and hence the size and style of font is as important as the number of double-page spreads and image sequence: book style has to be integrated with written and pictorial styles.

One of the most restrictive issues affecting book and journal illustrators is scale. Lincoln and Rourke's *Mimetes*, produced on the scale of a florilegium is impressive but unwieldy, and the format and size of *The Flowering Plants of Africa* (currently 25 cm x 19 cm) is very restrictive. While it permits information to be presented, it allows little scope for design and the result is that images are often cramped. Also, on a small format it is easier to illustrate small plants than large.

A generous format which does justice to good illustrations is seen in *Gladiolus in Southern*

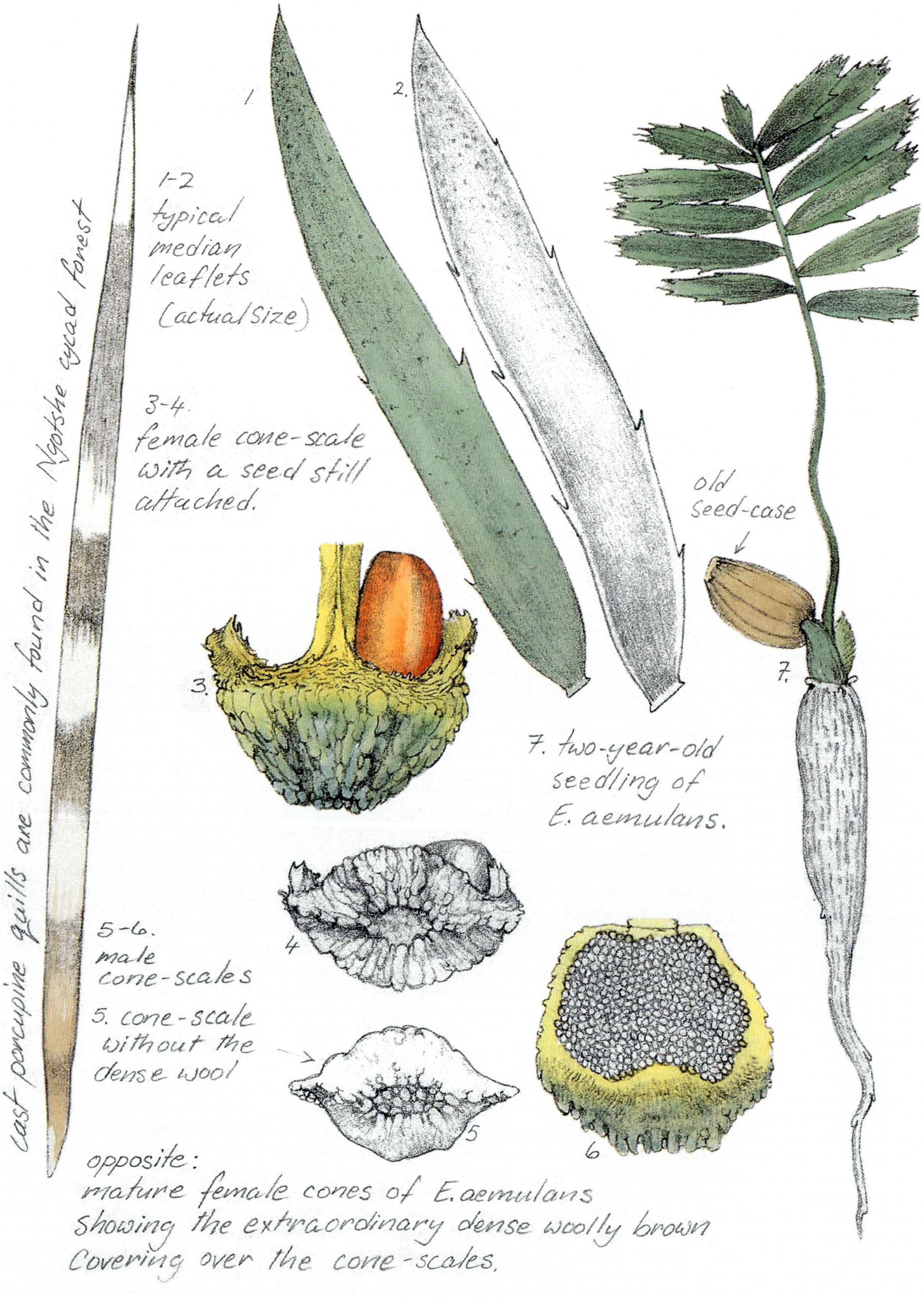

Plate 109 above and 110 right

Douglas Goode, *Encephalartos aemulans*, hand-coloured lithograph, double-page spread:
John Comrie-Greig, *The Cycad Collection*, illustrated by Douglas Goode (Balgowan, Natal, [1998])

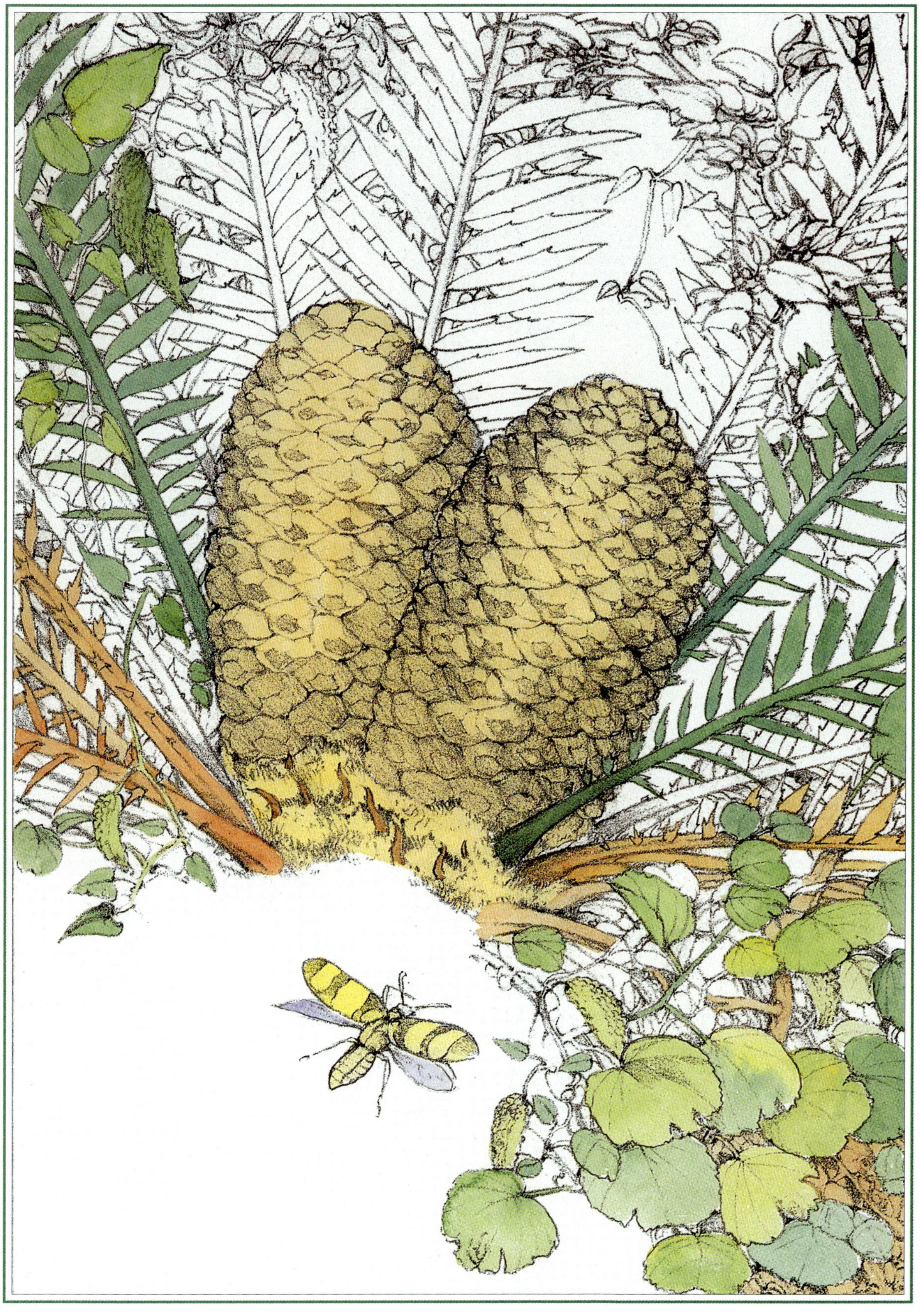

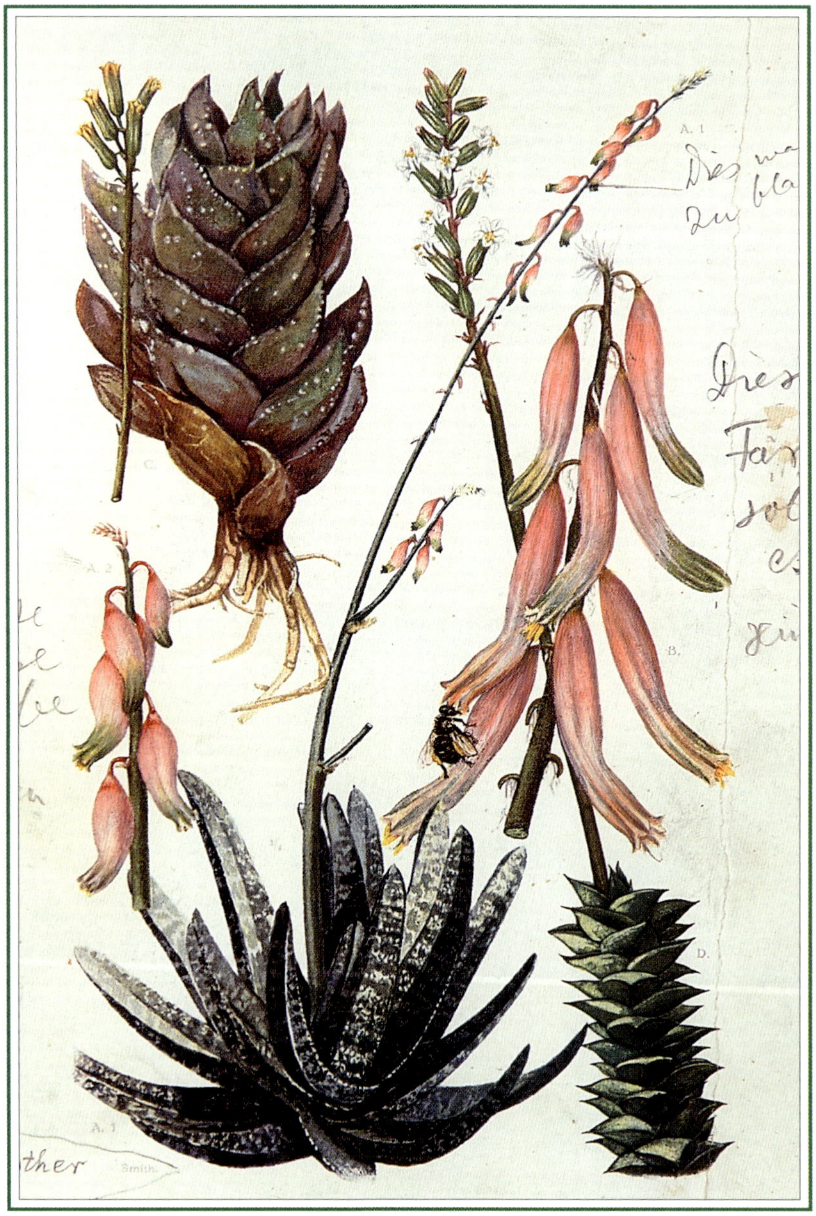

PLATE 111 above and 112 right

Esther Smith, Liliaceae: Aloineae, watercolour original for Rudolf Marloth's *The Flora of South Africa*, vol. 4 (Cape Town and London, 1915), and proof with annotations and corrections by Marloth

Africa.[27] This major revision of the species is the result of scientific research by botanists Goldblatt and Manning, with artwork by Anderson, Batten and Manning. The final visual product is an informative and aesthetic object. If the aim had been pure information, text would have sufficed. If the aim had been text with conceptual elaboration, the spare linearity of John Manning's fine line illustrations would have been an elegant solution. But the aim was a contemporary publication acknowledging the talents, vision and knowledge of artists and scientists, publisher (Pieter Struik), book designer (Joanne Simpson), and skilled technicians. The book is a synthesis of visual and verbal data – printed page, black and white line drawings, and richly coloured plates.

Beautiful books are a foil to the options

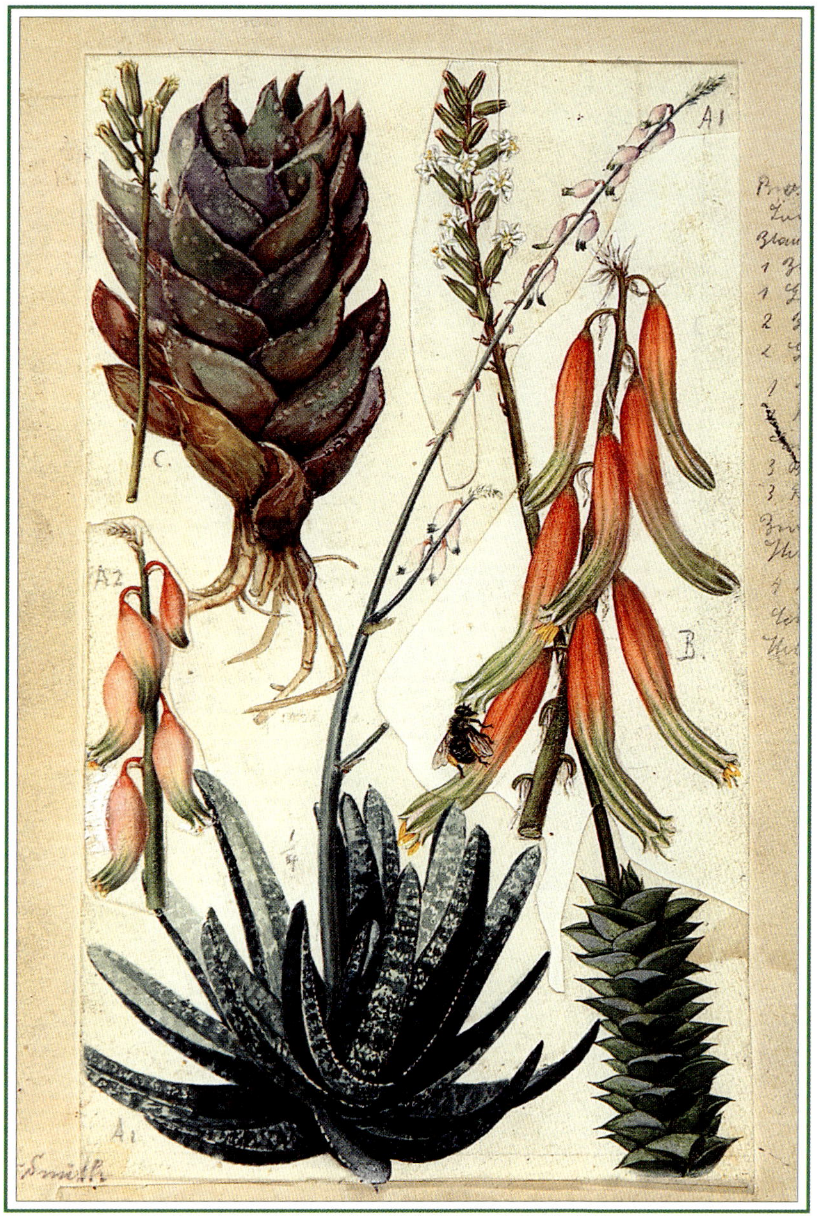

offered by desktop publishing, and the evidence of the subjective 'eye/I' is utterly different to impersonal software-generated images. Specialist books can now redefine themselves as aesthetic objects. Contemporary, limited edition, hand-printed botanical books are still rare in South Africa but The Caversham Press's production of *The Cycad Collection* by Douglas Goode (1944–) and master printer, Malcolm Christian (1950–), is one such book.[28] Handcrafted, this fine art book utilizes traditional ways of book-making. It invites lingering contemplation of paper surfaces, finely drawn and hand-coloured stone lithographs, relationships of writing and picturing, and ways of seeing (Plates 109, 110).

As techniques have become more sophisticated, so too has image manipulation. The loss of image quality that occurred with mechanical reproduction is largely a thing of the past. Today's contemporary computer technology makes very fine adjustments to effect the exact replication of the painted original. Colour can be digitally altered and images enhanced to heighten colour or eliminate errors, something particularly attractive to horticulturists wishing to advertise new, hybridized seeds and specimens. However, when Marloth was proofing the plates for *The Flora of South Africa*[29] he had to annotate proofs to inform his German printer when colours required adjustment or when spacing needed alteration. Looking at an original painting and proof (Plates 111, 112), the process of producing and printing book plates is made visible. In this process, prints were the end products and, at the end of this procedure, paintings were often considered redundant and discarded. In the case of Marloth's publication, paintings were literally discarded: the University of Stellenbosch librarian, Mrs Hanna Botha, retrieved many paintings from the floor of the botany department when it was clearing out before moving to new premises.

In considering botanical art, it is obvious that the requirements of botany control art, partly because the final destination of the image is the published text. A few of the original Weiditz watercolours were discovered by chance in 1930,[30] but the *Temple of Flora* paintings have largely disappeared. In the South African context, Auriol Batten donated her paintings for *Flowers of Southern Africa* to the National Botanical Institute (which also houses amongst its collection the original plates of *Flowering Plants of Africa)*, The Brenthurst Library acquired pelargonium and gasteria watercolours by Ellaphie Ward-Hilhorst (1920–1994), and Mesembryanthemaceae studies by Mary Page (1867–1925) are held by the Bolus Herbarium. Whilst these works have been preserved, they are not contextualized as 'art' but as products of science, and are stored largely unseen.

The eye: ways of seeing

When the philosopher, Ludwig Wittgenstein, observed memorably, 'We see the world the way we do, not because that is the way it is, but because we have these ways of seeing',[31] he implied that we learn to interpret the 'reality' we see. In effect, we see through the lens of knowledge and expectation and, for viewers of plant portraits, expectations are mediated by context.

One expects to see 'art' in art galleries, 'botanical art' in herbaria, and 'illustrations' in books. Viewers are shocked or amazed when ways of seeing are disrupted by context. Marianne North's work at Kew surprises art viewers and botanists alike. The paintings, located at a botanic garden, do not look 'botanical' and are installed like stamps in a giant album in a specially designed art gallery. Mary Page's work, consulted by scientists in the Bolus Herbarium rests on herbarium sheets in cupboards, but if it was mounted, framed and presented in an art gallery, South African art lovers would be astonished to discover an unknown watercolour painter.

Images seduce viewers into seeing 'reality', but in fact paintings look more like other artworks than like nature, and this is as true for plant portraiture as it is for narrative history painting. Paintings are products of conventions which, in the case of botanical art, have been derived from both art and science. They cater for specialist viewers whose primary search is for empirical information, not aesthetic experience. Horticultural literature, field guides, herbarium sheets, monographs and botanical journals all serve different needs and their imagery reflects different ways of seeing and ways of doing. Flower paintings on exhibitions, on the other hand, are contemplated as objects with aesthetic and material presence whose subjects are plants.

Ways of seeing plants were directed historically by the prevailing state of knowledge. The doctors of antiquity needed to differentiate between bulbs for their medical properties, and hence the bulbs and roots were included in ancient herbal illustrations. Linnaeus put the emphasis on the flower and so Ehret produced a new botanical art design that still exerts its influence over the format. In Ehret's schema, parts – flowers, fruit, leaves, stem, roots or bulb, and seeds, all isolated in neutral pictorial space – are in *conceptual relationships* to the whole. Features determining taxonomic classification are visible while reference to the ecological context is suppressed.

Depicting what they see is almost the easiest task confronting botanical artists. Like all

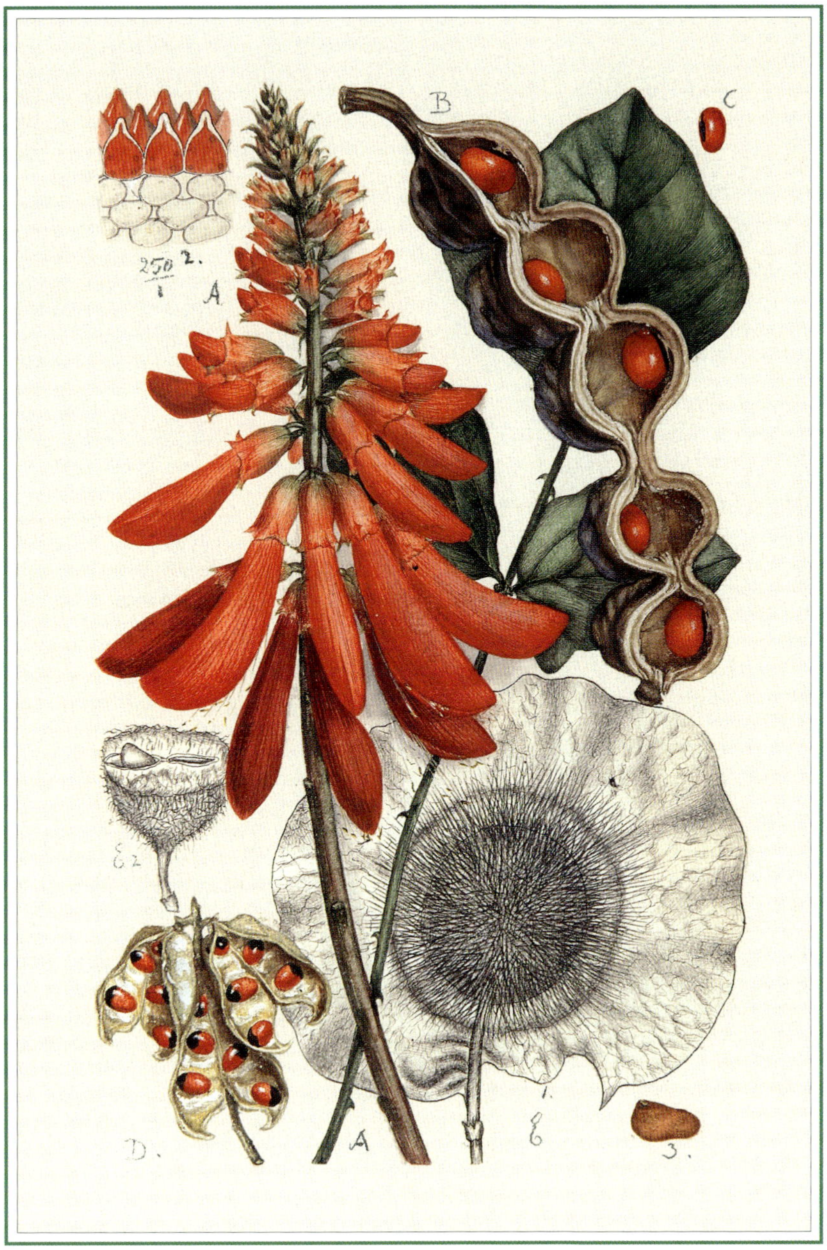

PLATE 113

Ethel Dixie, Leguminosae (Papilionatae), watercolour: reproduced in Rudolf Marloth, *The Flora of South Africa*, vol. 2 (Cape Town and London, 1925)

portrait artists, they first have to deal with the model. It may arrive damaged, ravaged by insects, or in bud. It might need to be immersed in water or supported by stakes. A succulent or strelitzia can be persuaded to sit patiently for days before drooping or dying, but a gladiolus could fold its graceful petals in exhaustion after a few hours. Artists must be resourceful and patient: out of living and dying forms, remnants and fragments, they have to resurrect and construct convincing images resembling typical living species, and collate information on flowers, fruit and seeds which do not occur simultaneously. To achieve these feats, they need good memories, knowledge of plant structure, and experience in the art of visual deception. They can then persuade botanists that they indeed serve their needs.

Botanists require accurate renderings of particular flowering plants, but writers of regional surveys or field guides usually refer readers to composite plates where several species are illustrated. Thus a plate in *The Flowering Plants of Africa*, where a single species is described and illustrated, looks different from another in *The Flora of South Africa* where a number of plants are crowded together on the format (Plate 113). And so the questions arise: given quite strict parameters, do botanical artists enjoy any freedom of expression and interpretation, and is personal style evident in plant representations?

In any creative enterprise it is virtually impossible to eliminate personal style because of the influence exerted by individual ways of seeing. Barbara Jeppe (Plate 114) and Auriol Batten (Plate 99) see differently although both relate plant details to environmental drawings. Plant portraits are replete with their maker's decisions concerning visual language, and even within the strict requirements of botanical art there is scope for personal style. This is evident in the plates of *The Flowering Plants of Africa* – see the contributions of Kathleen Lansdell (1888–1967), Stella Gower (1894–1991), Cythna Letty (1895–1985), M.E. Connell (1917–1997), Esmé Hennessy (1933–), Ellaphie Ward-Hilhorst and Gillian Condy (1952–).

Personal style is also apparent in artists employed by botanical institutions, such as Mary Page who worked as a botanical artist at the Bolus Herbarium, University of Cape Town, from 1915 to 1925. Her job – like that of Kathleen Lansdell, Cythna Letty and Gillian Condy at the National Botanical Institute – was to draw and paint plants on a daily basis. Her paintings supplement written information on types on herbarium sheets. Page's work also features prominently in *The Flowering Plants of Africa* and, nearly fifty years after her death, many of her illustrations were used in Herre's *Genera of Mesembryanthemaceae*,[32] which does both her and Beatrice Carter (1889–1939) disservice since the reproduction quality is very poor.

In her obituary on Page, Louisa Bolus (1877–1970) comments that when she met the artist in 1915 she was 'struck by the beauty of the flower paintings [Mary Page] showed me. She had not, up to this time, done any botanical drawing . . . For three months she lived with me, learning "to paint from the botanical point of view" '.[33] The redoubtable Mrs Bolus taught an already accomplished painter to look for the information that botanists want to see in plant representations, and she instructed Page about botanical art conventions, such as the inclusion of dissections and details. Much of Page's large oeuvre of botanical art (over a thousand works) is devoted to the Mesembryanthemaceae which were Louisa Bolus's speciality.

Page's paintings, created from gradated watercolour washes, possess convincing volumetric structures punctuated by meticulous observation of relevant detail (Plate 115). Characterized by precision and sensitivity, her work is concerned with perceiving and rendering living objects, and relating this to schematically drawn details (she drew in ink and her lines are still crisp and legible). Balancing the drawn and painted components of her image – part and whole, conceptual diagram and perceptual naturalism – Page painted intense, intimate portraits of flowering plants.

As a naturalistic artist favouring watercolour, Mary Page's personal style has much in common with that of Ellaphie Ward-Hilhorst, who, like Page and Batten favoured realism. Ward-Hilhorst created persuasive portraits of particular plant specimens. Through careful analysis of light, her paintings are suffused with a sense of the present moment. She considered form to be of primary significance: 'The decorative value is not as important as an absolutely accurate rendering so that form stands out from the paper.'[34] But it is a mark of her great artistic skill that her work possesses an impeccable feeling for decorative values, achieved through her strong design sense and years of work in the commercial art world. Seeing pictorial space with an artist's eye, she places bulbs in relation to flowers so that weight, rhythm, and spatial intervals are balanced (Plate 61). Combining the dominant forms with lucid diagrams and details, she establishes a pleasing interaction between conceptual and perceptual information.

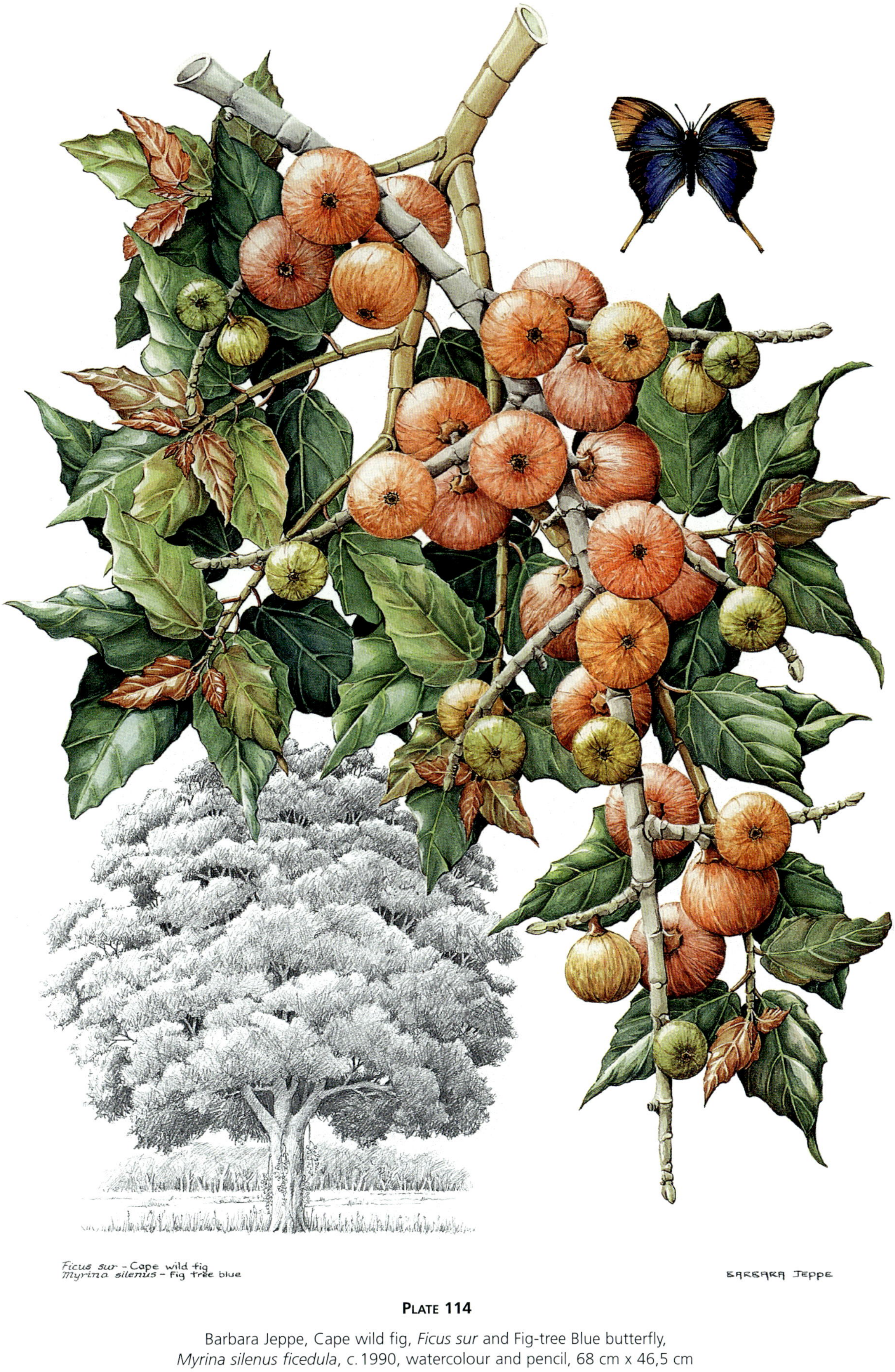

Plate 114

Barbara Jeppe, Cape wild fig, *Ficus sur* and Fig-tree Blue butterfly,
Myrina silenus ficedula, c. 1990, watercolour and pencil, 68 cm x 46,5 cm

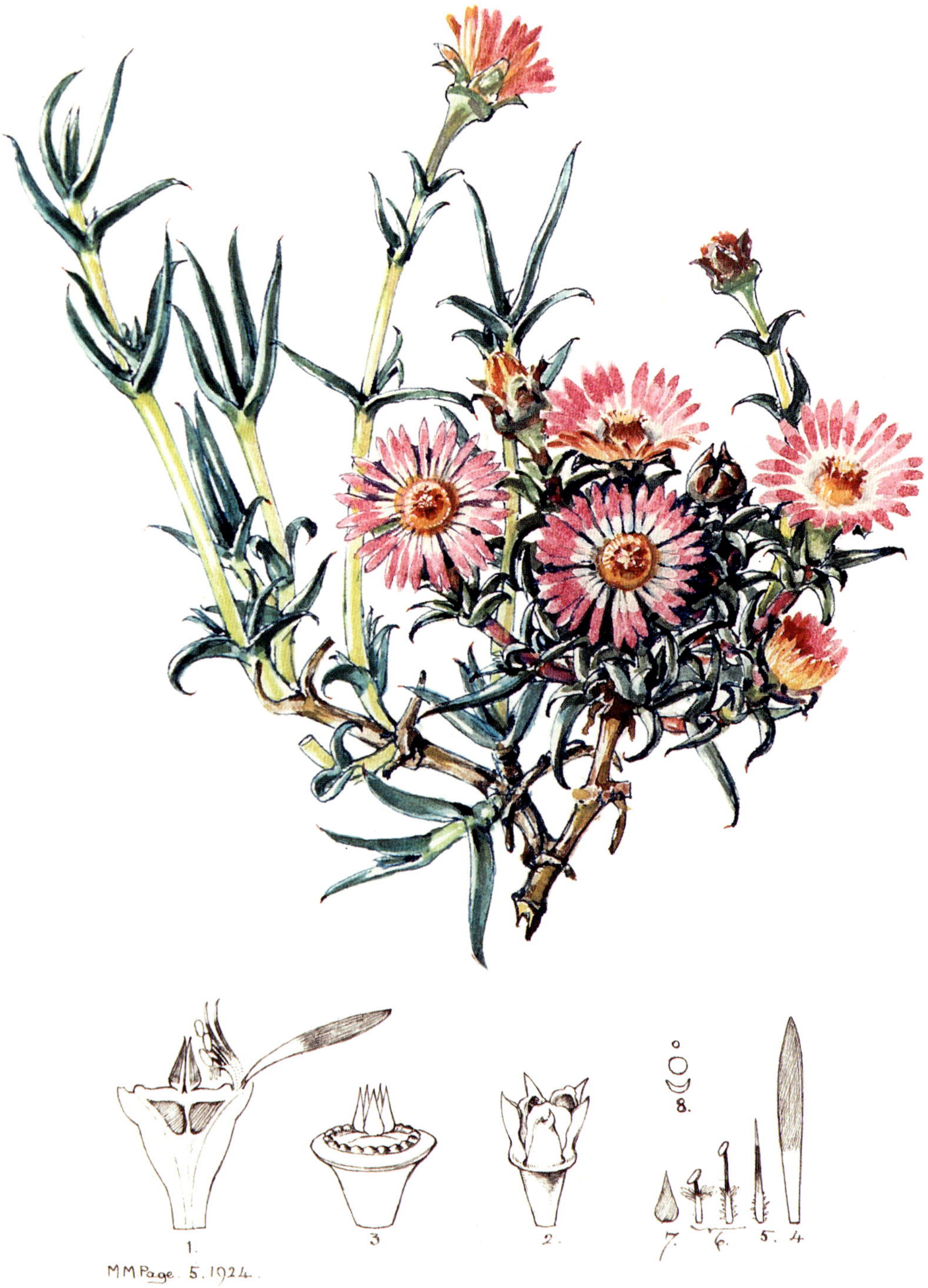

PLATE 115

Mary Page, *Lampranthus curvifolius*, 1924, watercolour and ink

Here and in Plate 46, the artist reveals an exceptional ability to imbue small succulent plants with character, playing off translucent, multi-petalled flowers with turgid, sculptural stems.

Ward-Hilhorst utilizes the fall of light on leaf edges to activate areas of foliage, to emphasize the significance of growth points and to convey surface undulations on leaves and petals (Plate 82). Her delight in the linearity and irregularities of bulbs is seen in *Haemanthus dasyphyllus* (Plate 61), and her enjoyment of agitated complex structures is evident in the confident rendering of the *Mimetes chrysanthus* inflorescence (Plate 63). An exceptionally versatile artist (frontispiece, Plate 1), she made a major contribution to the *Pelargonium* revision[35] where, in over three hundred watercolours and related pencil habitat sketches she produced a substantial family history, and gave each species its own personality (Plate 1).

The 'reality' of Ward-Hilhorst's work is persuasive but, like all convincing realism, it is achieved by the artist's focused looking, deep knowledge of form, structure and plant character, and manipulation of light, shadow and colour. Naturalistic painting is an elaborate game of skilled deception – of suppression, distortion, and exaggeration – but these strategies are disguised, and beguile the unsuspecting viewer into accepting 'reality'.

The devices of naturalism serve two portrait modes – realism and idealization. The concepts are characterized more by philosophical attitude, and particularly an attitude to time, than by technical visual language. Whereas some artists aim to portray particular plant specimens, and achieve this by emphasizing the fall of light and shadow created from a specific light source, others are more concerned with form existing under general light conditions in non-specific time. This approach yields a sense of the typical, essential – and sometimes idealized – nature of the plant. Although the language of description is utilized and the illusion of form is convincing, plants appear timeless. This approach distinguishes the work of two of the artists employed at the National Botanical Institute, Cythna Letty and Gillian Condy.

Letty, who made a major contribution to illustration through a very large oeuvre, worked relatively impressionistically, capturing the typical features of her subjects with an economical use of freely brushed colour (Plates 37, 40, 62, 71, 75, 91). Condy's paintings possess a classical restraint, being characterized by tonal harmony, balanced design, shape clarity and ordered, lucid structure (Plates 79, 116). An emphasis on the typical appearance is also the hallmark of Fay Anderson's work. The illustrator for *The Proteas of Southern Africa* (Plate 47)[36] and *The Moraeas of Southern Africa*[37] (Plate 117) and a contributor to many other publications, Anderson favours the drawn profile, enhanced by simple colour washes to represent identity.

Although personal style is certainly discernible in the work of experienced artists working for botanical publications, artists working in an art context, rather than a science context, have more scope for expression. But here one finds the parameters of art exert an influence that directs ways of seeing. In the late nineteenth century when it was expected that ladies would acquire the ability to sketch and paint in watercolours, the women artists of the Cape painted the indigenous flowers dutifully or, in the case of Emily Thwaits (*c.*1860–1906), with passion.

Thwaits, whose grandfather, Thomas Thwaits, was an art master in Cape Town, did not create paintings for a botanical institution or publication. She worked independently as an artist, producing flower paintings for exhibition, and was awarded a medal by the South African Fine Arts Association in 1880 for the best watercolour paintings shown that year. Her work is distinguished by a method of using very fine brush strokes to model volumes and create a sculptural sense of form. The presence of her plants is enhanced by cast shadows, a strategy that positions the flowers as still-life objects in a studio. *Carpobrotus sauerae* (Plate 119), constituted of a myriad of brilliant translucent petals and heavy swollen leaves, seems – like Thwaits's other paintings (Plates 24, 97) – to convey inner life force. In terms of interpretation and personal style, it is interesting to compare Thwaits's carpobrotus with that depicted by the unknown artist who illustrated the plant for Paterson (Plate 16).

The painstaking devotion to surface which characterizes Emily Thwaits's work also distinguishes Elise Bodley's paintings (Plate 118), and Thalia Lincoln's images. Lincoln, who works primarily for herself and not for botanists, is

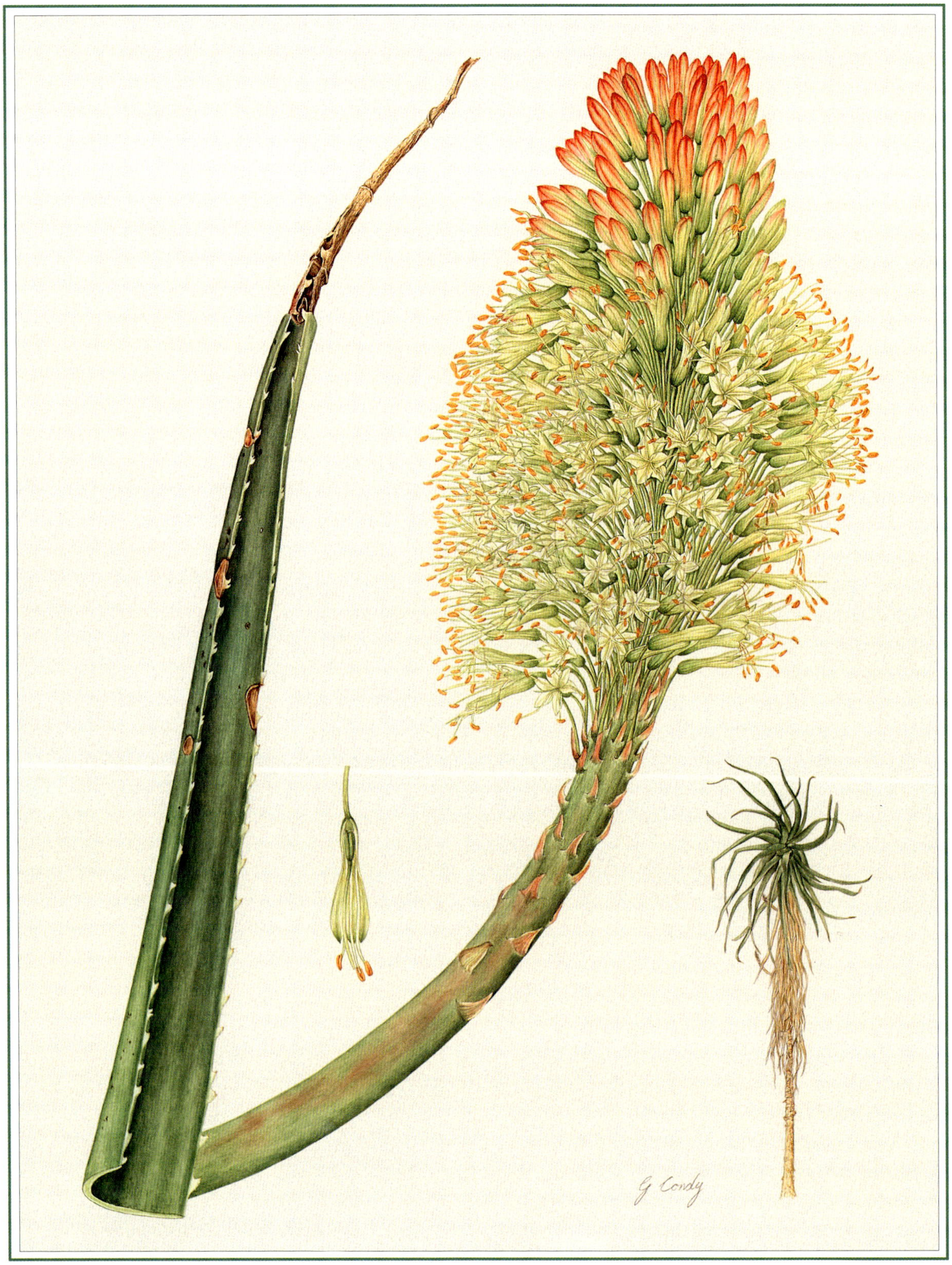

PLATE 116

Gillian Condy, *Aloe helenae*, 1986, watercolour: reproduced in *The Flowering Plants of Africa,* 1986

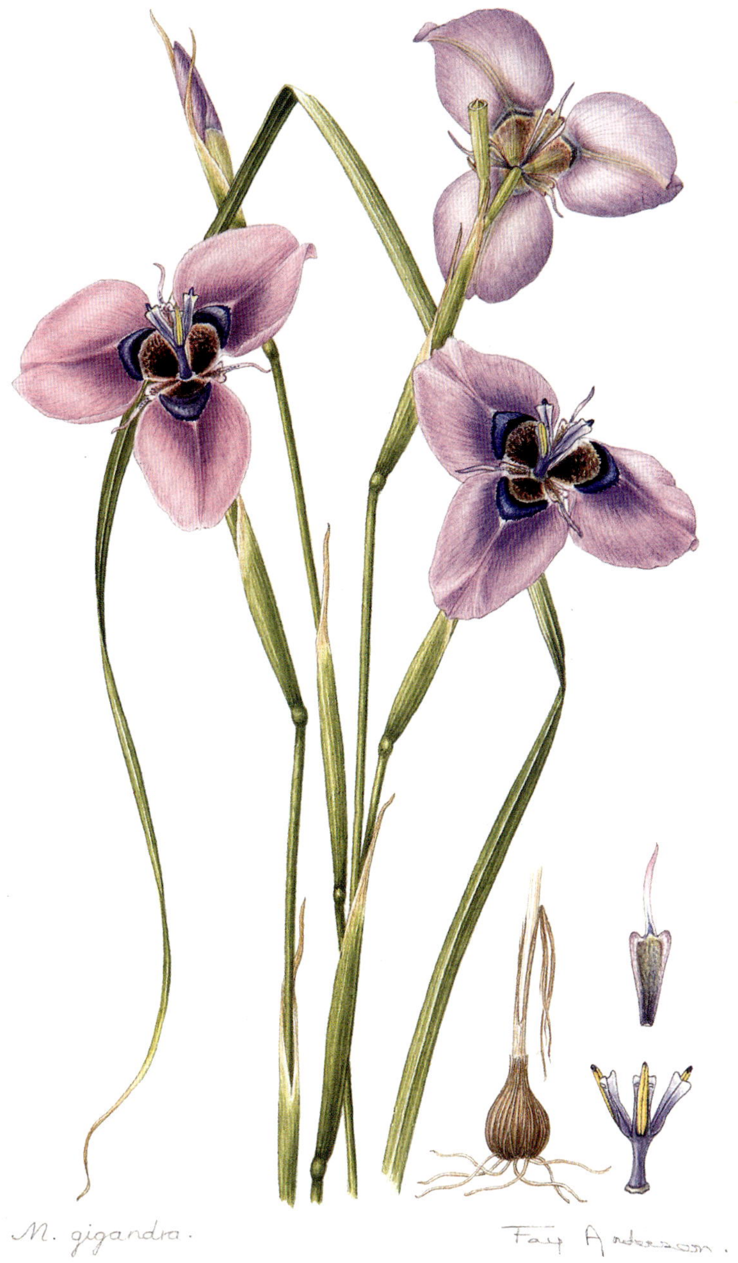

PLATE 117

Fay Anderson, *Moraea gigandra*, watercolour: reproduced in Peter Goldblatt, *The Moraeas of Southern Africa* (Cape Town, 1986)

motivated by aesthetic aims. Her primary objective is to communicate the beauty of plants and, whether depicting *Mimetes hottentoticus* in its habitat (Plate 107) or *Gardenia thunbergia* (Plate 43), her personal style which is founded on modulated layers of different hues, applied by coloured pencils is distinctive. This time-consuming technique is undertaken to express the very souls of plants.

Although personal style can be analysed, discussions on ways of seeing and doing are rendered redundant by eloquent images – which in turn demonstrate decisions concerning materials, methods, format, and scale. These factors

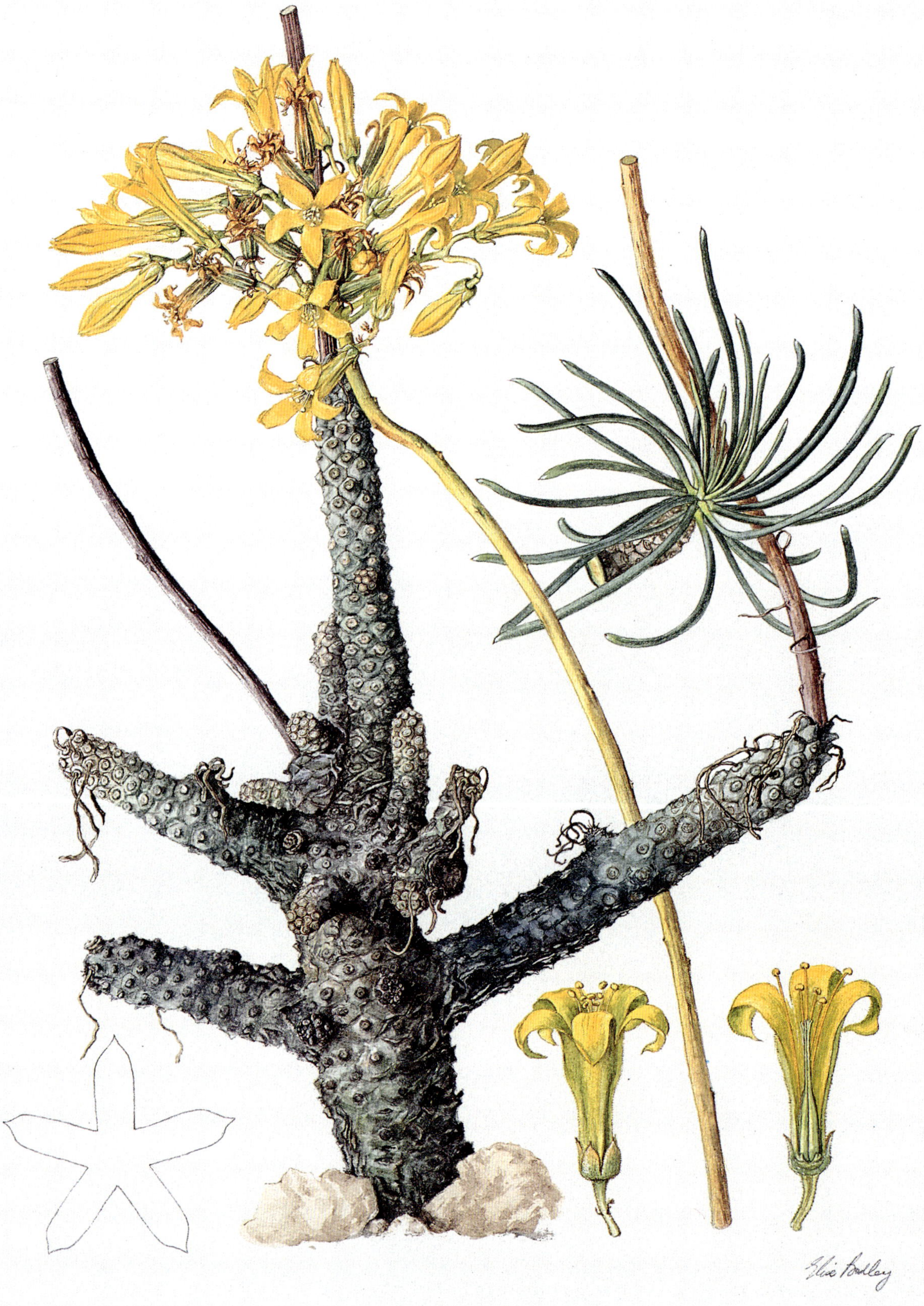

Plate 118

Elise Bodley van Wyk, *Tylecodon cacalioides*, c. 1985, watercolour and pencil: reproduced in Marion Arnold, *Art Meets Science: Flowers as Images* (Johannesburg, 1992)

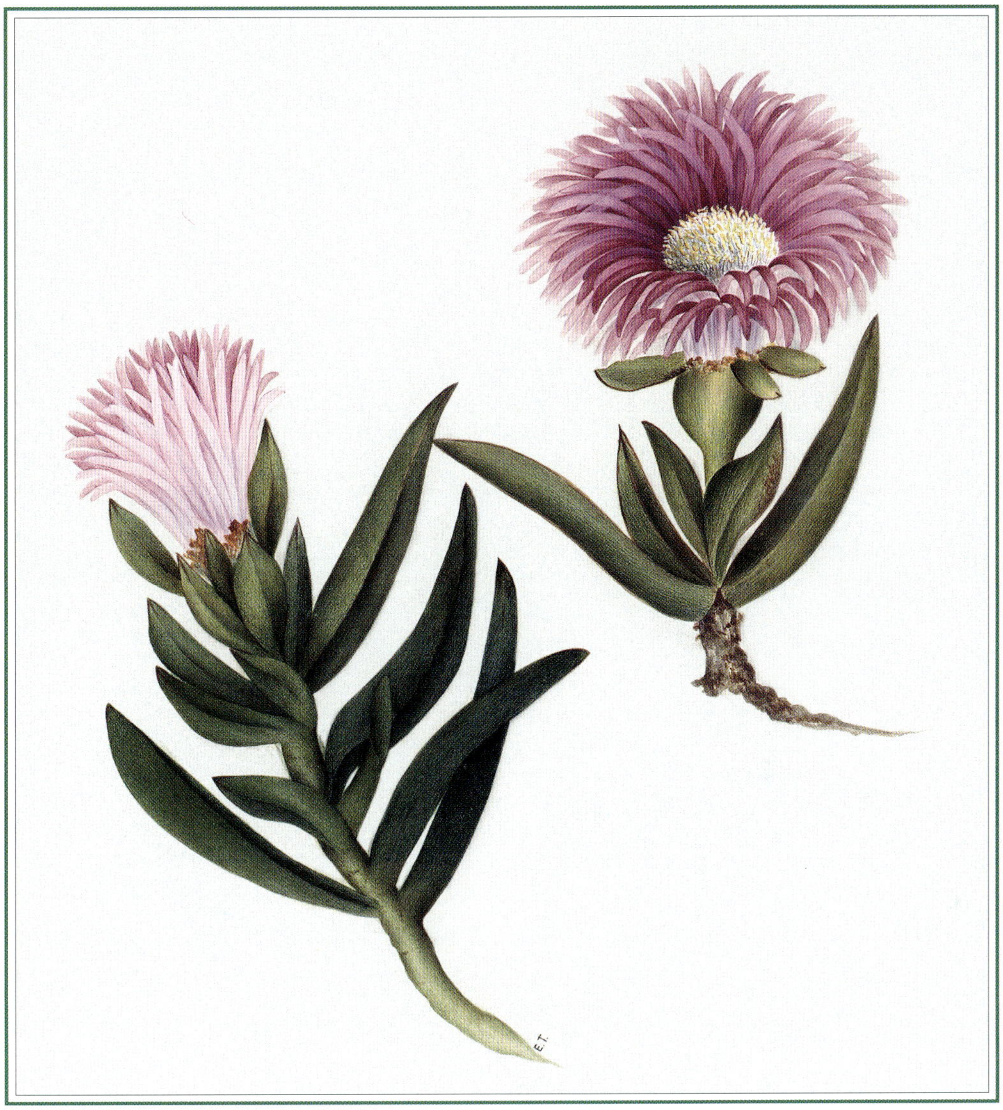

PLATE 119

Emily Thwaits, *Carpobrotus sauerae*, c. 1890, watercolour: reproduced in Marion Arnold, *Art Meets Science: Flowers as Images* (Johannesburg, 1992)

influence design which, if used insightfully, facilitates the process of creating plant portraits replete with signs of personal vision and style.

In *Cupid Inspiring the Plants with Love* (Plate 104), a clump of strelitzias articulates the foreground of this allegorical floral landscape. The plant has been the subject for many botanical artists and flower painters: Bauer, Curtis's artist, Redouté, Dixie (1876–1973), Lincoln, Batten and Starcke (1935–) show regina, the queen, in different regal moods which are as much a projection of the artist's sensibility, and their ways of seeing, as they are attributes of a striking flowering plant (Plates 120–126).

Like and like and like

In her novel *The Waves*, Virginia Woolf asks, ' "Like" and "like" and "like" – but what is the thing that lies beneath the semblance of the thing?'[38] Woolf poses a question that is central to biography and portraiture – both of which depend, as they do, on recording verifiable facts. Plant portraits, like all portraits, are purported to be 'like' the models. But can a piece of paper

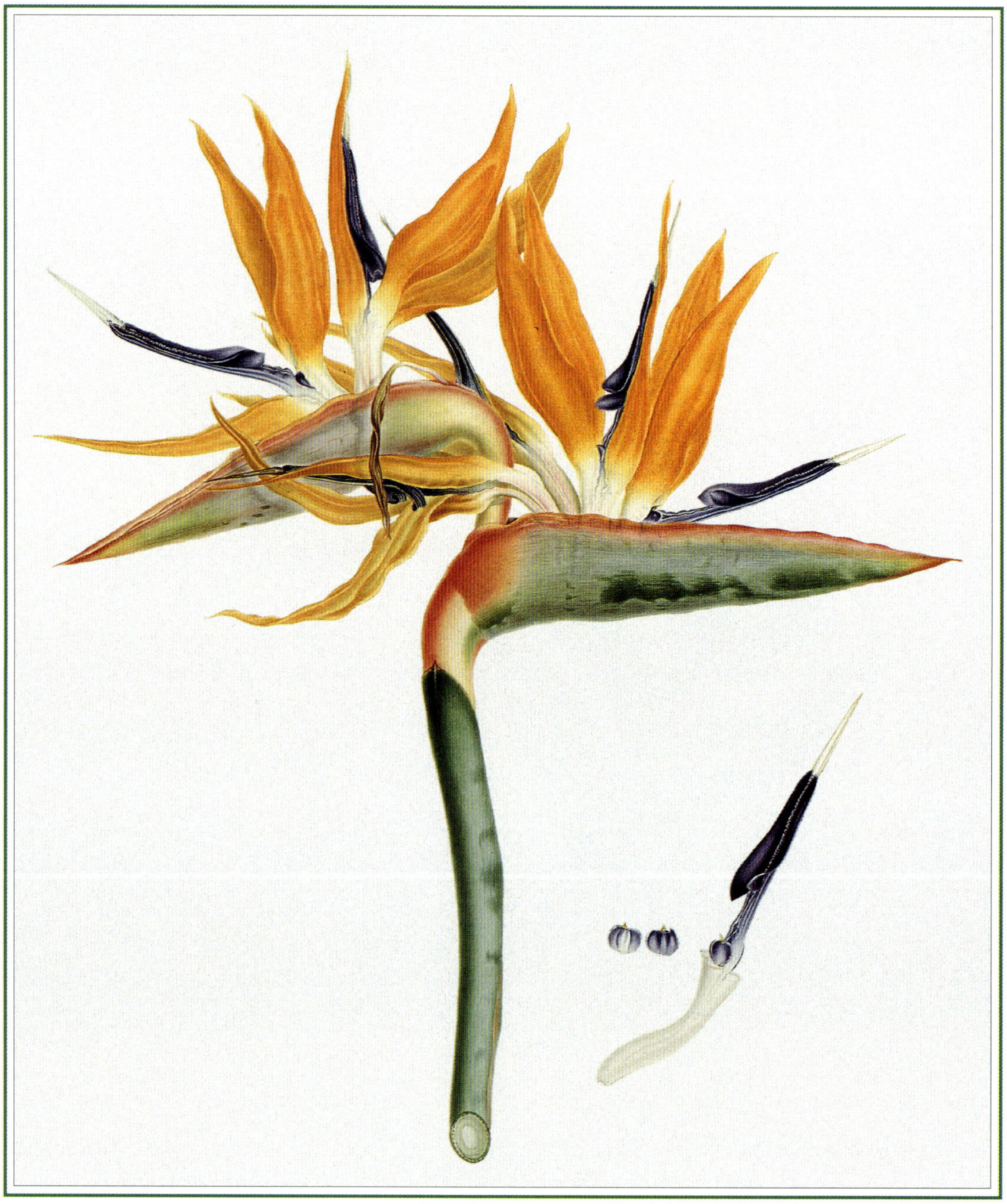

PLATE 120

Franz Bauer, *Strelitzia reginae*, c. 1818, hand-coloured plate: *Strelitzia Depicta or Coloured Figures of the Known Species of the Genus Strelitzia from the Drawings in the Banksian Library* (London, 1818)

with coloured marks on it truly resemble the structure and colours of a flowering plant? It cannot. Each possesses its own reality but, through the creative intervention of artists and scientists, both objects can enter a meaningful relationship.

Plant portraits are cunning acts of transformation. The most persuasive images serve as analytical partners in scientific research, but they also detain the eye with their power to evoke the life force of nature through their presence as silent and compelling aesthetic objects.

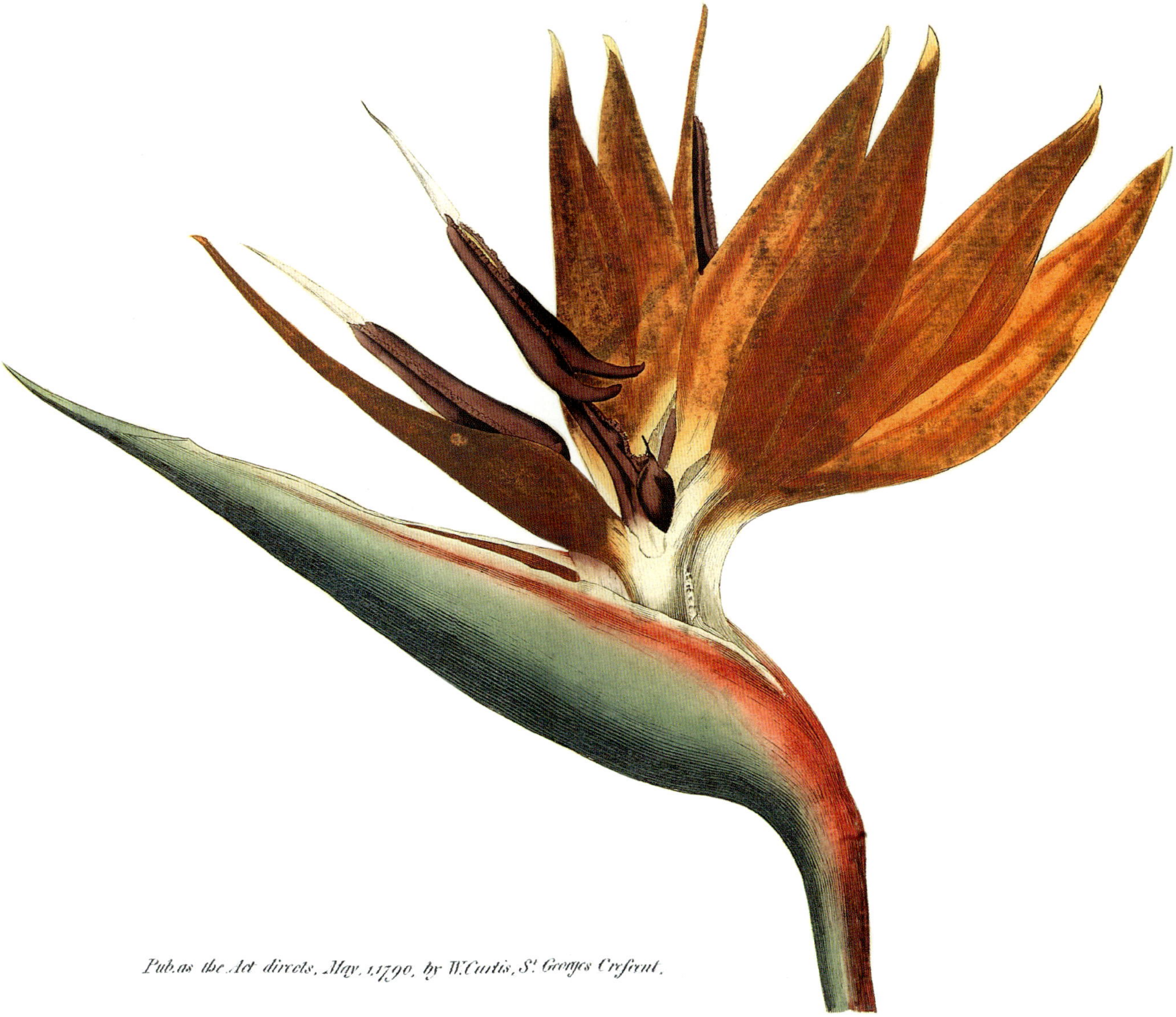

Plate 121

Strelitzia reginae, 1790, hand-coloured plate, by artist unknown, though possibly Sydenham Edwards or James Sowerby: *Curtis's Botanical Magazine*, 1790

Written description not only possesses limitations, it also says something completely different to visual description. The complex nature of interpretation is apparent when one considers, for example, the visual responses to *Strelitzia reginae* from a range of artists and the following 'General Description of the *STRELITZIA REGINAE*' in *Curtis's Botanical Magazine, 1790:* 'From a perennial stringy root shoot forth a considerable number of leaves, standing upright on long footstalks, from a sheath of some one of which, near its base, springs the flowering stem, arising somewhat higher than the leaves, and terminating in an almost horizontal long-pointed spatha, containing about six or eight flowers, which becoming vertical as they spring forth, form a kind of crest, which the glowing orange of the Corolla, and fine azure of the Nectary, renders truly superb.' *(unpaginated text for Plates 119, 120)*

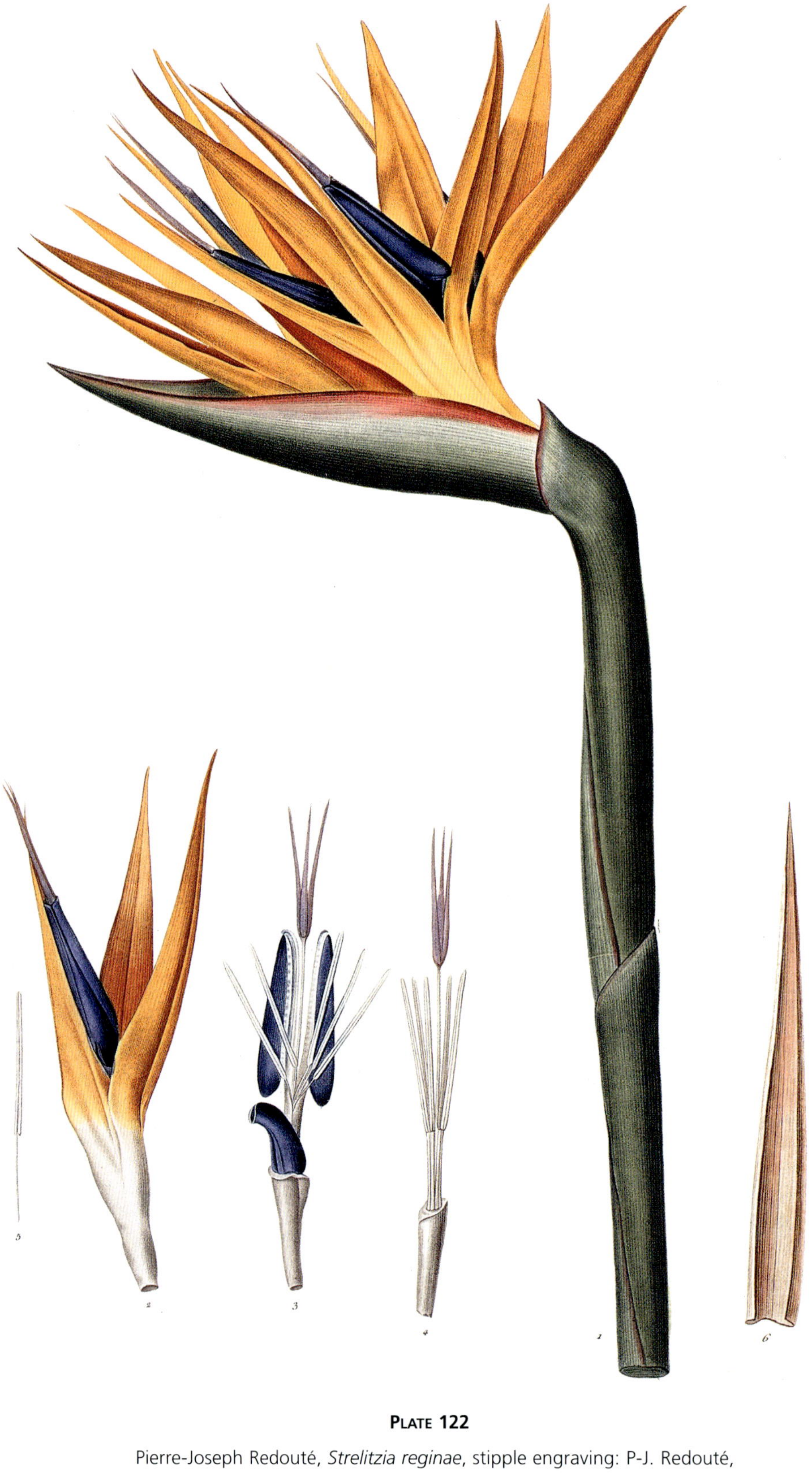

PLATE 122

Pierre-Joseph Redouté, *Strelitzia reginae*, stipple engraving: P-J. Redouté, *Les Liliacées*, vol. 2 (Paris, 1802–1816)

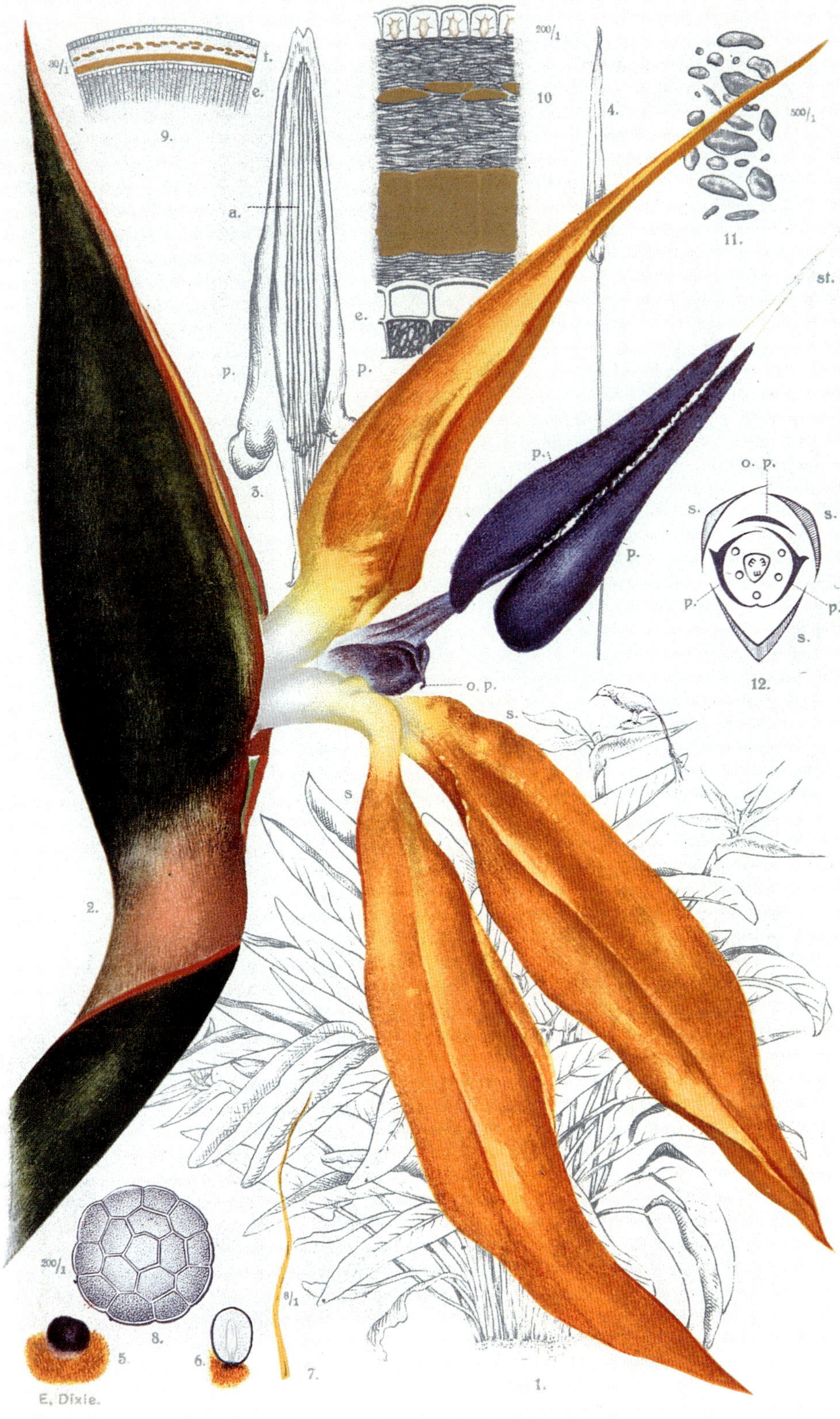

Plate 123

Ethel Dixie, *Strelitzia reginae*: Rudolf Marloth, *The Flora of South Africa*, vol. 4
(Cape Town and London, 1915)

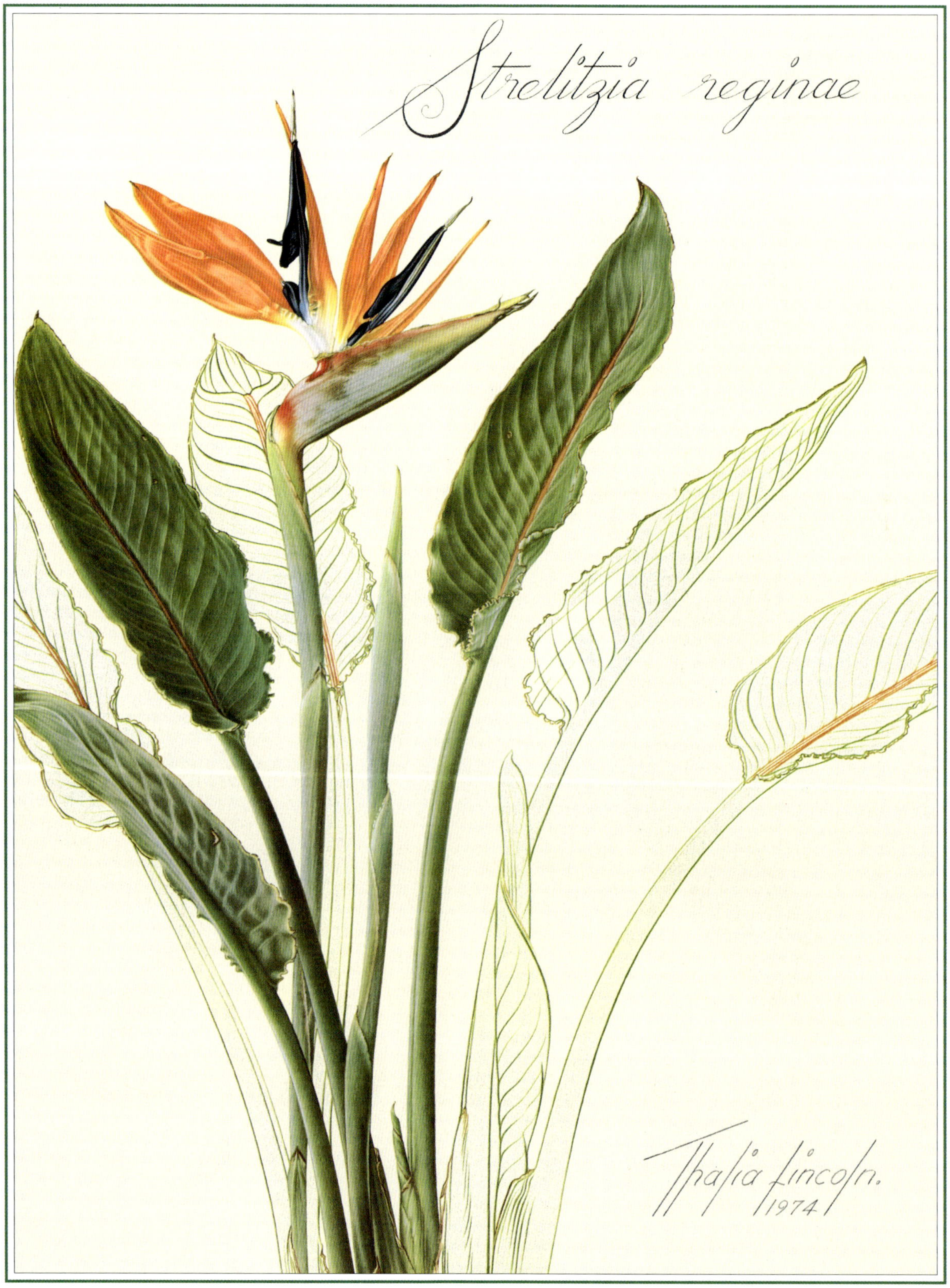

Plate 124

Thalia Lincoln, *Strelitzia reginae*, coloured pencils on Bristol board

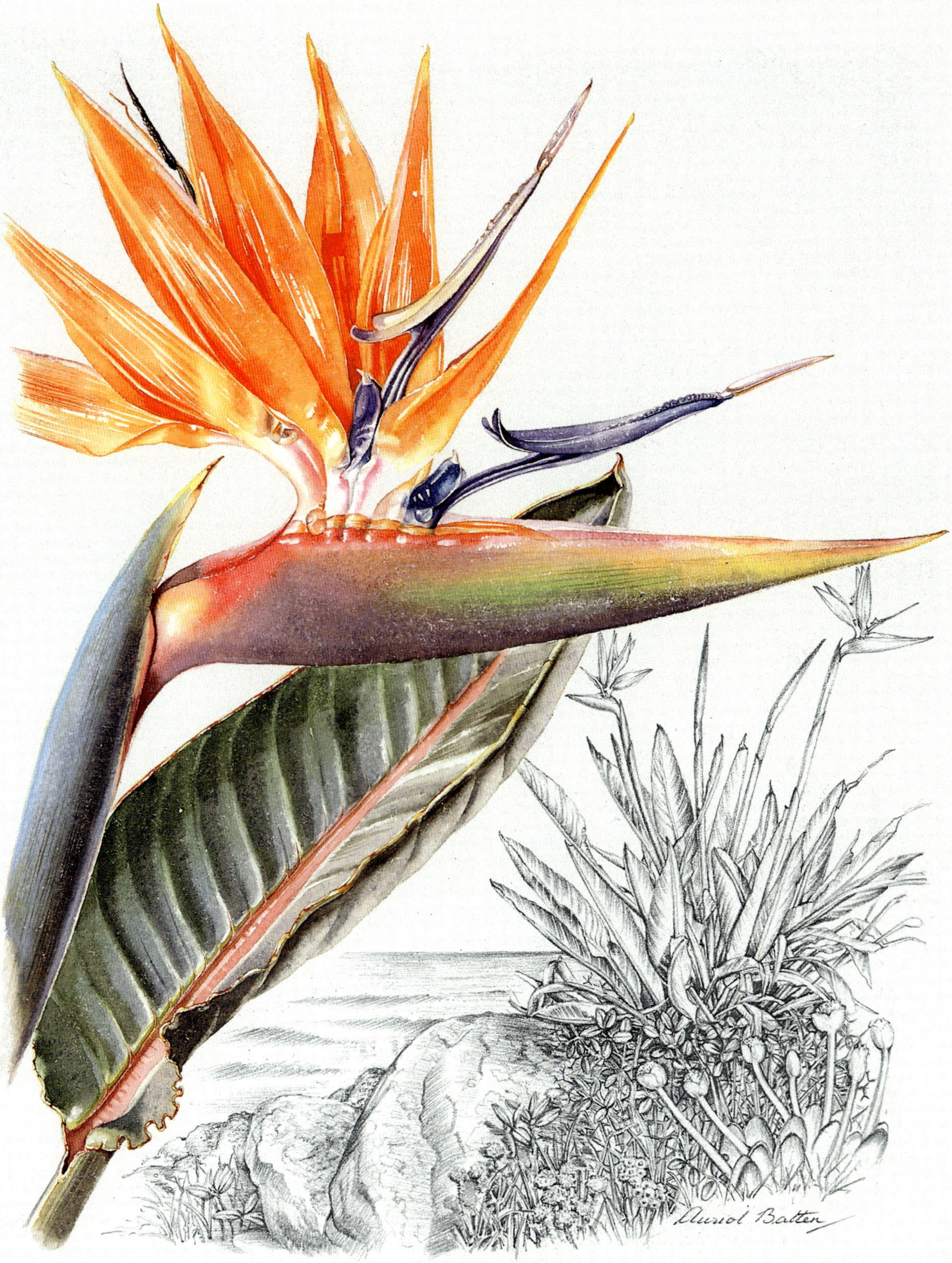

Plate 125

Auriol Batten, *Strelitzia reginae*, watercolour: reproduced in A. Batten, *Flowers of Southern Africa* (Sandton, 1986)

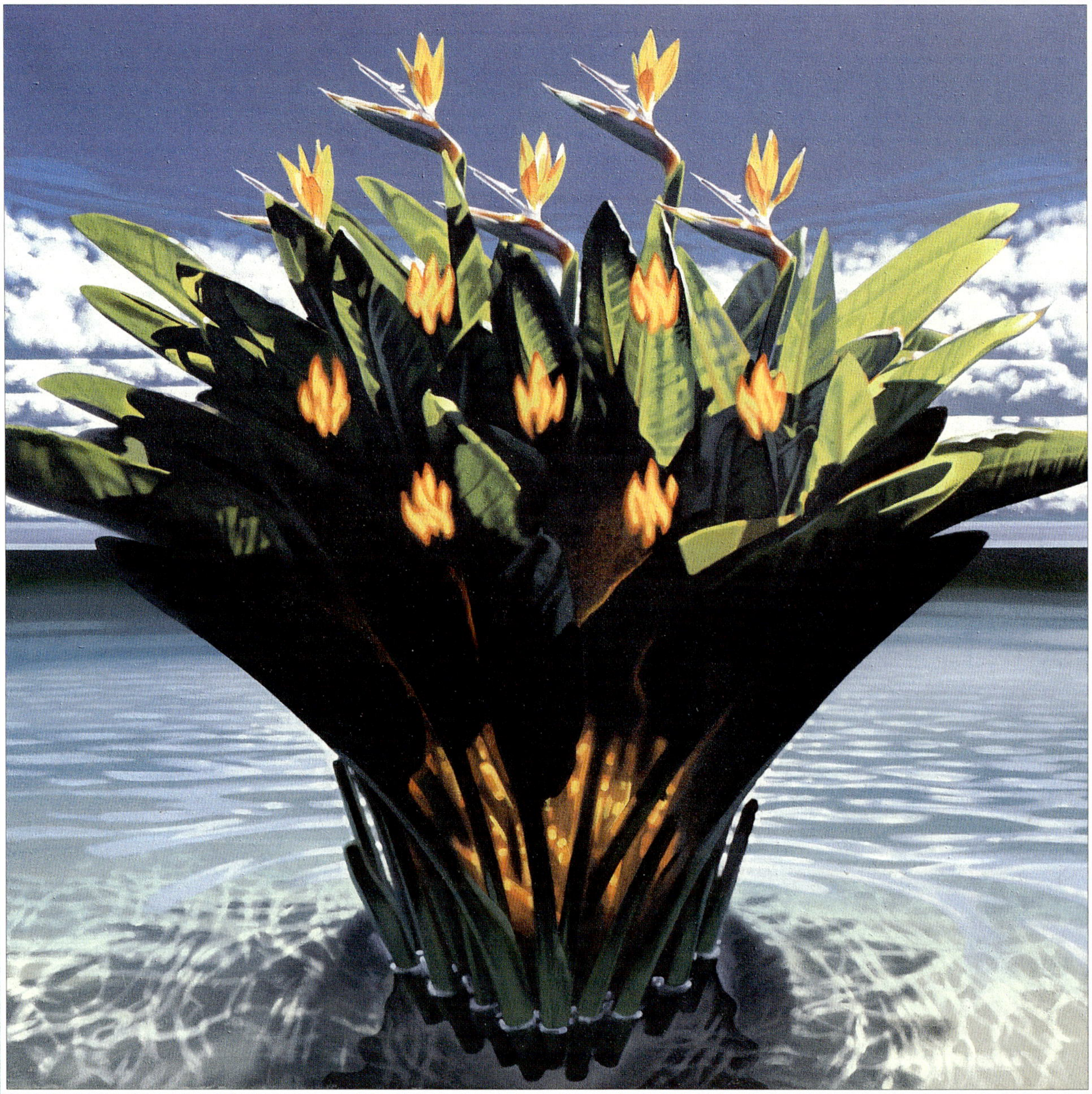

PLATE 126

Helmut Starcke, *Burning Bush,* 1983, acrylic on canvas,
152 cm x 152 cm

'The more I learn about man's tireless abuse of nature,' commented Starcke in the late 1980s, 'the less I am interested in "the human condition" as subject matter. [I feel] compelled to create images celebrating the life force, as well as order in nature, speaking of such things as fresh air, clear water, cool grass, and white clouds etc, but also of the mystery as well as the logic – all of which we perceive as beauty.' (*Lantern,* 38[2]1989, p. 43)

1 Lys de Bray, *The Art of Botanical Illustration: the Classic Illustrators and their Achievements from 1550 to 1900* (Secaucus, New Jersey, 1989), p. 20.
2 Frans A. Stafleu, *Linnaeus and the Linnaeans: the Spreading of their Ideas in Systematic Botany, 1735–1789* (Utrecht, The Netherlands, 1971), p. 37.
3 Carolus Linnaeus, *Hortus Cliffortianus* (Amsterdam, 1737).
4 Peter Goldblatt and John Manning, *Gladiolus in Southern Africa*, illustrated by Fay Anderson, Auriol Batten and John Manning (Cape Town, 1998).
5 Gill Saunders, *Picturing Plants: an Analytical History of Botanical Illustration* (London, 1995), p. 89.
6 (Leiden, 1798), pl. 6.
7 Wilfrid Blunt and William T. Stearn, *The Art of Botanical Illustration*, 2 ed. (Woodbridge, Suffolk, 1994), pp. 13–14 (Stearn's emphasis).
8 Ibid., p. 273.
9 See *Delineations of Exotick Plants Cultivated in the Royal Garden at Kew* (London, 1796–1803), illustrated by Francis Bauer.
10 De Bray, *Art of Botanical Illustration*, p. 40.
11 Marion Arnold, *Women and Art in South Africa* (Cape Town, 1996), pp. 63–76.
12 Charlotte Klonk, *Science and the Perception of Nature: British Landscape Art in the Late Eighteenth and Early Nineteenth Centuries* (New Haven and London, 1996), p. 39.
13 Donal P. McCracken, *Gardens of Empire: Botanical Institutions of the Victorian British Empire* (London and Washington, 1997), p. 3.
14 See Laura Ponsonby, *Marianne North at Kew Gardens* (Exeter, England, 1990); Marianne North, *A Vision of Eden: the Life and Work of Marianne North* (Exeter, England, 1980).
15 Arnold, *Women and Art*, pp. 69–71.
16 Blunt and Stearn, *Art of Botanical Illustration*, 2 ed., p. 277.
17 Ibid., p. 297.
18 John Rourke, *Mimetes: an Illustrated Account of* Mimetes Salisbury *and* Orothamnus Pappe, *Two Notable Cape Genera of the Proteaceae*, illustrated by Thalia Lincoln (Cape Town, 1982).
19 Lincoln to M. Arnold, personal communication, 1998.
20 Ibid., 1991.
21 Auriol Batten, *Flowers of Southern Africa* (Sandton, 1986), p. xiii.
22 A.S. Kerkham, comp., *Southern African Botanical Literature, 1600–1988* (Cape Town, 1988), p. 28.
23 D.J.B. Killick, 'Book review of Kerkham, A.S. (comp.), *Southern African Botanical Literature*', *South African Journal of Botany*, 56(5)1990, p. 590.
24 Batten, *Flowers of Southern Africa*, p. xiii (my emphasis).
25 Kerkham, comp., *Southern African Botanical Literature*, p. 28.
26 Alice M. Coats, *The Book of Flowers: Four Centuries of Flower Illustration* (London, 1974), p. 9.
27 Peter Goldblatt and John Manning (Cape Town, 1998).
28 John Comrie-Grieg, *The Cycad Collection*, illustrated by Douglas Goode (Balgowan, Natal, [1998]).
29 Rudolf Marloth, *The Flora of South Africa*, 4 vols in 6 (Cape Town and London, 1913–1932).
30 Blunt and Stearn, *Art of Botanical Illustration*, 2 ed., p. 63.
31 Bernard McGrane, *Beyond Anthropology: Society and the Other* (New York, 1989), p. ix.
32 A.G.J. Herre, *The Genera of the Mesembryanthemaceae* (Cape Town, 1971).
33 Louisa Bolus, 'Mary M. Page 1867–1925', *Annals of the Bolus Herbarium*, 4, 1925, p. 60.
34 Ward-Hilhorst to M. Arnold, personal communication, September 1991.
35 J.J.A. van der Walt and P.J. Vorster, *Pelargoniums of Southern Africa*, 3 vols, illustrated by Ellaphie Ward-Hilhorst (Cape Town, 1977–1988).
36 J.P. Rourke, *The Proteas of Southern Africa*, illustrated by Fay Anderson, Lura Ripley, *et al.* (Cape Town, 1980).
37 Peter Goldblatt, *The Moraeas of Southern Africa*, illustrated by Fay Anderson, Margo Branch and Janet Klein (Cape Town, 1986).
38 (London, 1977), p. 110 (1 ed., 1931).

FURTHER READING

Arnold, Marion, *Art Meets Science: Flowers as Images.* Johannesburg: Standard Bank, 1992. (Catalogue, Standard Bank National Arts Festival Exhibition, 1992)

Lang, Berel, ed., *The Concept of Style.* Ithaca and London: Cornell University Press, 1987, 2 ed.

POSTSCRIPT

MARION ARNOLD

Jean Cocteau called art 'science given flesh'. The trope is apt when botanical art is considered as the intersection of lean, factual science and sensuous, seductive painting. Botany does not need art to make itself visible. The very language of diagrams suffices. But botany got art. Art literally entered the picture when botany co-opted the skills of specialist artists to serve its own descriptive objects. Diagrams blossomed into plant portraits. Linear signs became beautiful facsimiles of reality.

Plant representations convey many meanings. But what they yield depends on what we ask of them, on our frames of reference, general knowledge, and the contexts in which images are studied. Consider Vivienne Koorland's collage, *South African Flora IV.* It is comprised of twenty protea images, immediately identifiable as flowering blooms. But what does the artwork mean?

Koorland found mass-produced reproductions of proteas, cut them out, and taped them to paper over a book cover. Her acts were significant: cutting is a brutal, dislocating process; taping imprisons things. The protea flowerheads, devoid of reference to their natural environment, are isolated on white paper and controlled by assertive masking tape. They resemble botanical specimens on herbarium sheets, faded remnants of their original splendour.

The artwork provokes speculation on the part of the viewer. Unlike much botanical art, the image is not goal-directed. The artist entrusts the spectator with the responsibility of interpreting her work, but she directs this visual scrutiny and intellectual inquiry by embedding clues in her construction.

South Africans at once identify the flowers as proteas. Botanists know them as *Protea grandiceps* (Red Sugarbush), *Protea cynaroides* (King Protea, South Africa's national flower), and *Leucadendron argenteum* (Silver Tree). Mere identification, however, does not resolve the matter of interpretation, and the viewer must go on to ask art many questions: Why did the artist choose these forms? Why were the blooms arranged in this manner? Why confer this title? We supply answers partly from information we are given, and partly from our own perceptual acuity and imaginative insights.

In this case, the artist was born in Cape Town and in her youth Koorland became familiar with local flora by accompanying her great-aunt (a horticulturist) and grandmother on field trips. Although she observed floral beauty, she grew up and studied in a repressive society where classification was a loaded term, controlling all socio-political and economic activity. In such a context, Koorland's later creative acts are not innocent. Her proteas, unmistakably flowering plants, can also be read metaphorically, as life forms, isolated, controlled, and classified.

South African Flora IV thus reminds us that systems are not in themselves threatening or benevolent. It is our application of systemic structures that destines them to be either damaging or beneficial. Texts, whether written or visual, are not eloquent monologues, but dialogues requiring the active participation of readers and viewers. Interpretation is a game of logic and speculation. The result is discovery – if we allow the images of art and science to saturate our senses and stimulate our minds.

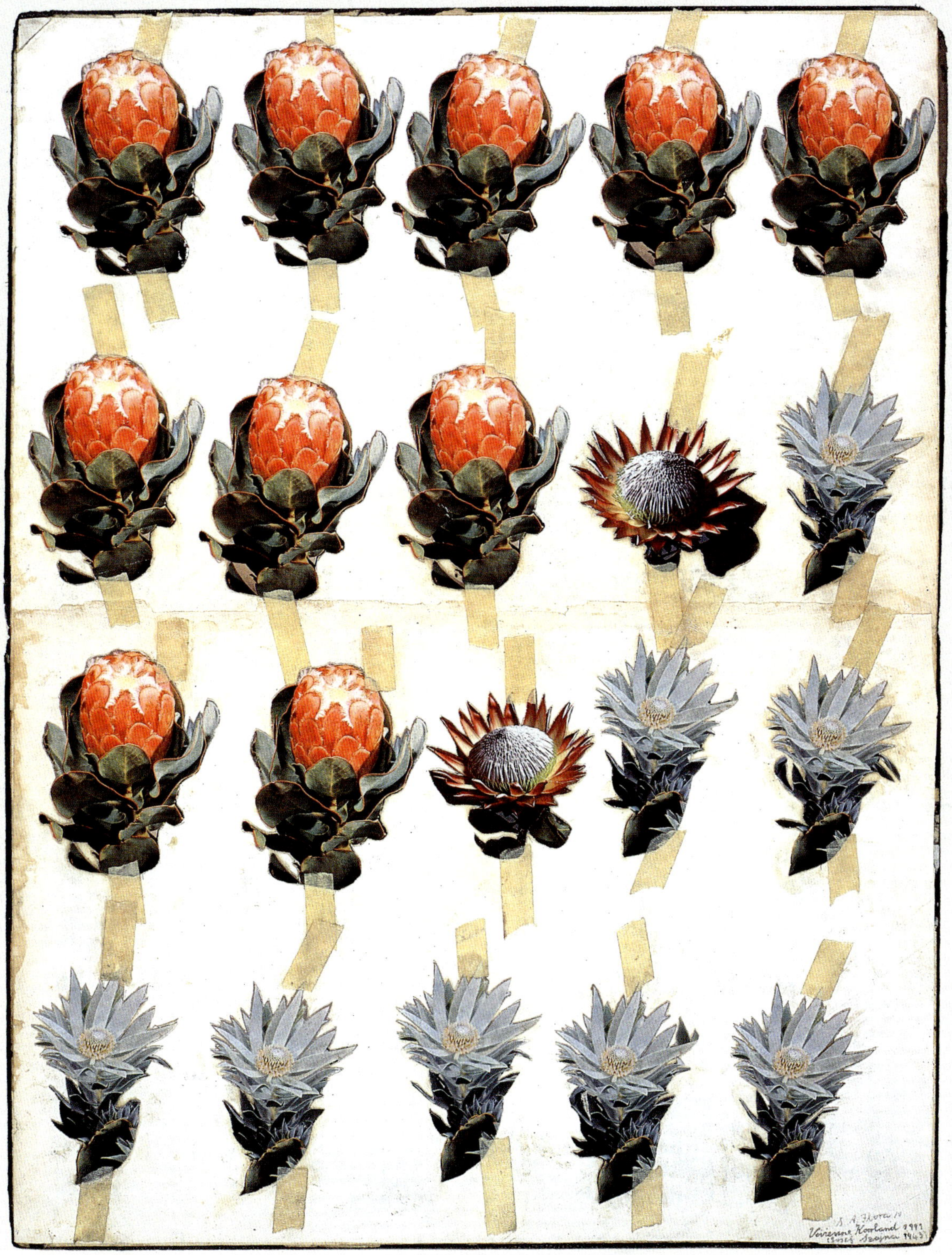

Plate 127

Vivienne Koorland, *South African Flora IV*, 1991–1993, card and tape on book cover, 54 cm x 40,6 cm

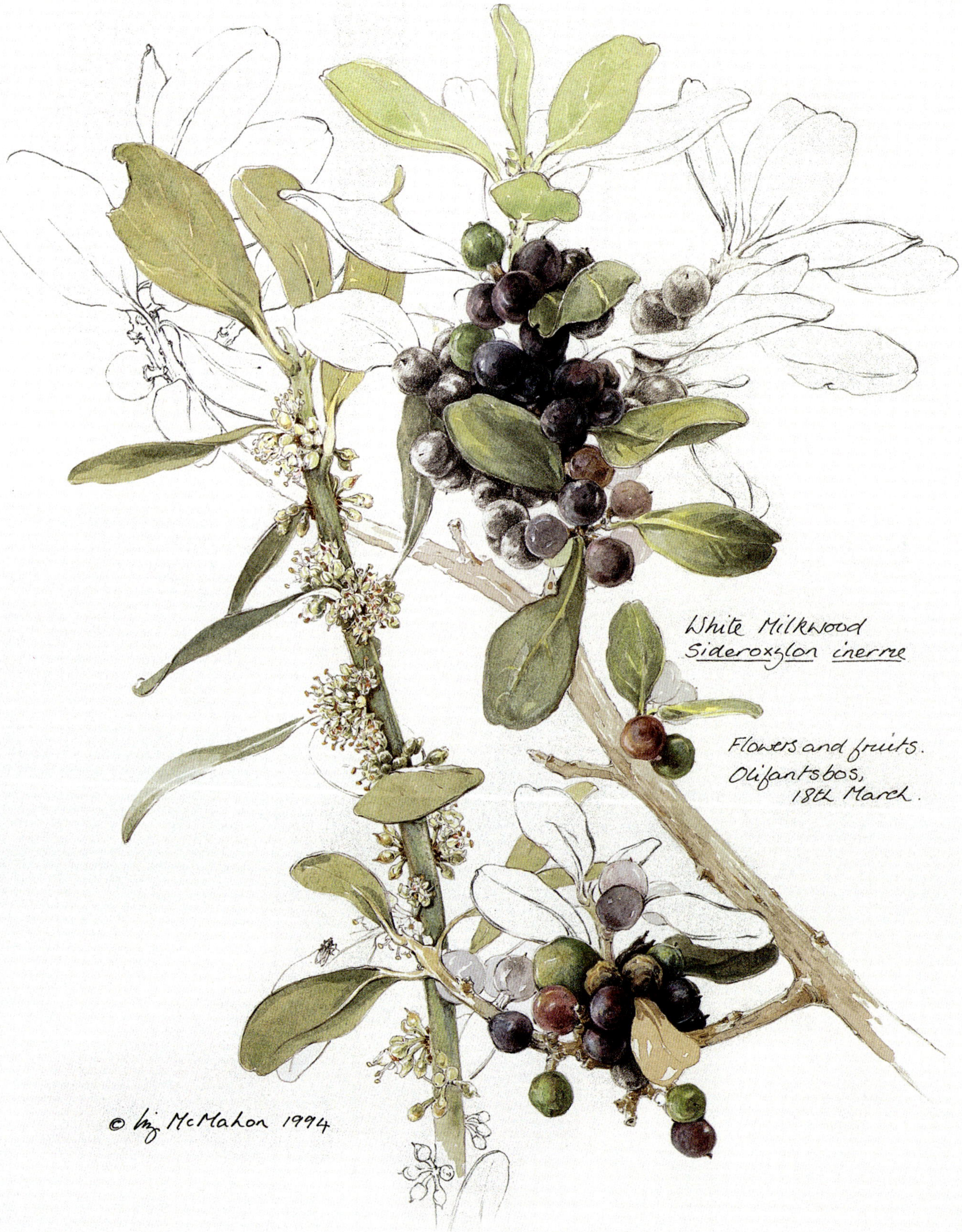

PLATE 128

Liz McMahon, *Sideroxylon inerme* (white milkwood), 1994, watercolour and pencil: reproduced in Michael Fraser, *Between Two Shores: Flora and Fauna of the Cape of Good Hope*, illustrated by Liz McMahon (Cape Town, 1994)

CONCISE DICTIONARY

OF SOUTH AFRICAN BOTANICAL ARTISTS

It is apparent from the essays in this book that 'botanical art' is a contested term. In defining this listing the term is used in both the traditional and revisionist senses, taking cognisance of terminological disputes within art theory, and striving at the same time to serve art as well as botany. As a result, most of the entries are comprised of traditional botanical artists, many of whom have contributed to botanical publications or have worked as illustrators for botanical institutions. In the minority are artists, more frequently located within the realm of 'fine art' and who may exhibit in galleries. Their inclusion here is based on a significant part of their oeuvre being devoted to plant portraiture, usually within the genres of still-life / flower painting and/or landscape imagery. In the latter instance, tree studies, which are often poorly represented in botanical art because of the issue of scale, constitute a significant subject. The imagery of the selected 'fine artists' fulfils a major requirement of botanical art in that, regardless of style, it is sufficiently descriptive to provide information that facilitates species identification.

The artists listed live and work in South Africa, or produced artworks in the region during the four-century period dating from the present to the 1620s. 'South Africa' is regarded as the subcontinent, taking in the present-day countries of South Africa, Namibia, Lesotho, and Swaziland. The yardstick used for selection has been the significance of an artist's contribution to the genre – in most cases through either formal exhibition or publication of artworks. No attempt has been made to evaluate the quality or artistic competence of images, and the dictionary makes no claim to be a definitive inventory.

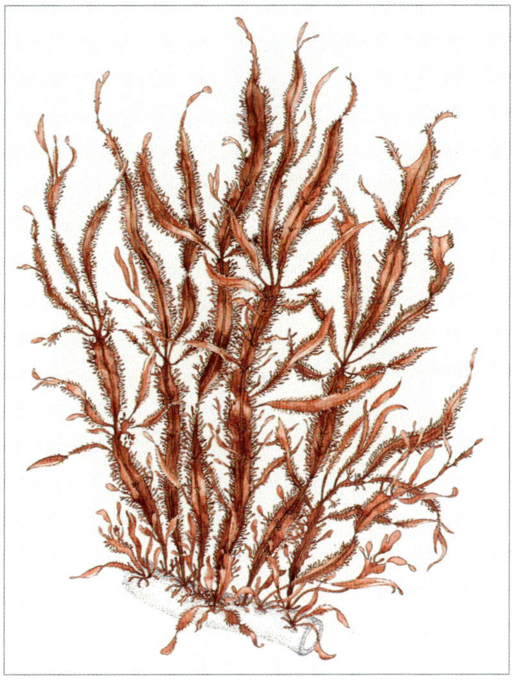

PLATE 129

Suhria vittata, a seaweed from Cape waters, c. 1990, watercolour and pencil, by Stanley Charles Seagrief

The Botanical Artists' Association of South Africa (BAASA)

A non-profit organization, formed in 1999, with a current membership of about 140. Periodic workshops, outings and exhibitions are arranged by its 3 branches – Cape Town, Pretoria and Durban.

address: BAASA, 1 Marcella Crescent, 7700 Newlands
e-mail: bulbin@iafrica.com
website: www.baasa.co.za

The Botanical Society of South Africa (BSSA)

A non-profit organization, established initially in 1913 to support Kirstenbosch National Botanical Garden, but today extended to include all 7 of the country's national botanical gardens. Membership exceeds 15 000, drawn from 45 nations around the globe. Fourteen branches run ongoing conservation programmes. Members of the society receive *Veld & Flora*, a quarterly journal which focuses on indigenous gardening, conservation and natural history.

address: Botanical Society of South Africa, Private Bag X10, 7735 Claremont
website: www.botanicalsociety.org.za

Bibliographic references

In keeping with the comprehensive nature of this dictionary, references are curtailed. Citation of a work does not define the extent of an artist's participation – some may have illustrated a work in total, others may have contributed only a single artwork. Oft-cited works and common abbreviations are:

A&A – Esmé Berman, *Art and Artists of South Africa*, 2 ed. (Cape Town, 1983)

Arnold – Marion Arnold, *Art Meets Science: Flowers as Images* – exhibition catalogue (Johannesburg, 1992)

BAASA – The Botanical Artists' Association of South Africa
Blunt & Stearn – Wilfrid Blunt and William T. Stearn, *The Art of Botanical Illustration*, 2 ed. (Woodbridge, Suffolk, and Kew, England, 1994)
BRI – Botanical Research Institute, Pretoria (previously Division of Botany; later National Botanical Institute)
BSSA – The Botanical Society of South Africa
CT – Cape Town
DSAB – *Dictionary of South African Biography*, 5 vols to date (Pretoria, 1968–1987)
DSAPS – G. Ogilvie, *The Dictionary of South African Painters and Sculptors* (Johannesburg, 1988)
DAG – Durban Art Gallery
Gunn & Codd – Mary Gunn and L.E. Codd, *Botanical Exploration of Southern Africa* (Cape Town, 1981)
Hunt Inst., USA – Hunt Institute for Botanical Documentation, Carnegie-Mellon University, Pittsburgh, USA
JAG – Johannesburg Art Gallery
Jhb – Johannesburg
Kew – Royal Botanic Gardens, Kew, England
KG, PE – King George VI Art Gallery, Port Elizabeth
MMus, Kimb. – McGregor Museum, Kimberley
NBG – National Botanical Gardens, Kirstenbosch, Cape Town
NBI, CT – National Botanical Institute (previously National Botanical Gardens), Kirstenbosch, Cape Town
NBI, Pret.– National Botanical Institute (previously Botanical Research Institute), Pretoria
PAM – Pretoria Art Museum
Pmbg – Pietermaritzburg
Pret. – Pretoria
RAU – Rand Afrikaans University, Johannesburg
RHS – Royal Horticultural Society, London
Rourke – John Rourke, *A Century of Cape Botanical Art* (Cape Town, 1988)
SA – South Africa / South African
SAAA – South African Association of Arts
SAAB – South African Association of Botanists
SANG – South African National Gallery, Cape Town
SESA – *Standard Encylopaedia of Southern Africa*, 12 vols (CT, 1970–1976)
Sherwood – Shirley Sherwood, *Contemporary Botanical Artists: The Shirley Sherwood Collection* (London, 1996)
SthA – southern Africa / southern African
UCT – University of Cape Town
UFS – University of the Free State
UND – University of Natal, Durban
UNP – University of Natal, Pietermaritzburg
UPE – University of Port Elizabeth
UStell. – University of Stellenbosch
UWC – University of the Western Cape
VOC – Vereenigde Oost-Indische Compagnie / Dutch East India Company
WH, Kimb. – William Humphrey Art Gallery, Kimberley
Women & Art – Marion Arnold, *Women & Art in South Africa* (Cape Town, 1996)
WU – University of the Witwatersrand, Johannesburg
Zim. – Zimbabwe

ACOCKS (also ACOCK), John Phillip Harrison (1911–1979)
Studies: BA, BSc (UCT, 1932); MSc (UCT, 1933)
Profile/career: pasture research section, Division of Plant Industry (1936–1945); botanical survey section, Division of Botany (later BRI, Pret.)(1945–1948); Grootfontein College of Agriculture, Middelburg, Cape (1948–1976)
Artwork published: *Veld Types of SA* (Pret., 1953) – also map artwork; D. Meredith, ed., *Grasses & Pastures of SA* (Jhb, 1954)
Artwork represented: NBI, Pret.
References: *Bothalia*, 13:1980; *DSAB;* Gunn & Codd

ADAMS, Jill (1932–)
Studies: matriculated St Andrew's School, Jhb; UCT school of architecture
Profile/career: mostly freelance; museum artist, MMus, Kimb. (10 years); several solo exhibitions
Artwork published: *Wild Flowers of the Northern Cape* (CT, 1976); R. Liversidge, *The Birds Around Us* (Jhb, 1991); *Reader's Digest Encyclopaedia of Gardening* (CT, 1984)
Artwork represented: Gant Family Trust; MMus, Kimb.; WH, Kimb.; numerous private collections
Reference: –

ANDERSON, Andrew Arthur (active 1850s–1897)
Studies: apparently self-taught as artist
Profile/career: born Scotland; settled as farmer near Pmbg (*c.*1850); travelled extensively in SthA, painting landscapes, botanical & zoological subjects; artworks in watercolour, often annotated; served in Anglo-Zulu War (1879), Transvaal War (1880–1881); also trader & cartographer; lived in England (1884–1888); returned to SA (1889)
Artwork published: *Twenty-Five Years in a Wagon*, 2 vols (London, 1887); J.M. Stuart, *The Ancient Gold Fields of Africa* (London, 1891)
Artwork represented: MuseumAfrica, Jhb
References: *DSAB; DSAPS*

ANDERSON, Olive Margaret (1938–)
Studies: self-taught as artist; commercial in-house training
Profile/career: National Herbarium, Harare, Zim. – produced over 1 000 b/w illustrations & numerous colour plates (1972–1980); botany dept, UNP – produced over 250 drawings & watercolours of seaweeds (1984–1988); draughtswoman, geology dept, UNP (1988–1995); cartographer, geography dept, UNP (1995–1999); freelance artist (1982–)
Artwork published: K.Coates Palgrave, *Trees of SthA* (CT, 1977); R.B. Drummond, *Common Trees of Central Watershed Woodlands of Zim.* (Salisbury, 1981); R.B. Drummond, *Arable Weeds of Zim.* (Salisbury, 1984); M. Gelfand, *et al.*, *The Traditional Medical Practitioner in Zim.* (Gweru, 1985)
Artwork represented: Harare Herbarium, Zim.; herbarium, WU; Hunt Inst., USA
References: *DSAPS;* NBI files

ANDERSON, Valerie Fay (Mrs R. Geary-Cooke) (1931–)
Studies: Diploma in Fine Art (Michaelis School of Fine Art, UCT, 1955)
Profile/career: illustrator, *Flowering Plants of Africa* & *Veld & Flora;* winner of several prestigious art awards (1967, 1988, 1998)
Artwork published: H.A. Baker & E.G.H. Oliver, *Ericas in SthA* (CT, 1967); G.J. Lewis, A.A. Obermeyer & T.T.

Barnard, *A Revision of the SA Species of Gladiolus* (CT, 1972); J.P. Rourke, *The Proteas of SthA* (CT, 1980; 2 ed., 1982); P. Goldblatt, *The Moraeas of SthA* (CT, 1986); P. Goldblatt, *The Woody Iridaceae* (Portland, Oregon, 1993); P. Goldblatt & J. Manning, *Gladiolus in SthA* (CT, 1998)
Artwork represented: Hunt Inst., USA; NBI, CT; NBI, Pret.; private collections SA & worldwide, including Shirley Sherwood, England; Robert Levitt, England
References: *A&A;* Blunt & Stearn; Gunn & Codd; Rourke; Sherwood

ARNOLD, Marion (1947–)
Studies: BA (UNP, 1968); BA Hons, Fine Art (UNP, 1969); BA Hons, History of Art (UNISA, 1978); MA, History of Art (UNISA, 1981); D Litt et Phil (UNISA, 1988)
Profile/career: born London; moved to Rhodesia (now Zim.); ran own art studio & taught art, Salisbury, Zim. (1971–1978); lectured, dept history of art, UNISA (1979–1991); lectured, dept fine arts, UStell. (1992–2000); moved to Norfolk, England (2000); author / co-author of several art catalogues, also monographs on artists Thomas Baines & Irma Stern; works as freelance art lecturer, writer & artist; artworks in watercolour, oil & pastel, often incorporating botanical images & themes; group & solo exhibitions, SA & overseas
Artwork published: 'Red Data AND etc.' (studies of landscapes, plants, objects & words, 1998)
Artwork represented: DAG; Gencor Collection, Jhb; JAG; National Gallery, Harare, Zim.; PAM; SANG; Sasol Collection, Jhb; Standard Bank Collection, Jhb; UFS; UNISA; UStell.; WU; private collections, SA & worldwide
Reference: *DSAPS*

ARRIDGE, Margaret Irene Chadwick ('Peggy') (Mrs Farquharson) (1921–)
Studies: Chelsea School of Art, London; private tuition from Bernard Adams & in miniature painting from Violet Butler
Profile/career: born Salisbury, England; moved to SA (1937); taught art privately; associate member, Royal Society of Miniature Painters, England (1949); associate member, National Society, London (1985); lived in England (1950–1954); artworks in watercolour, gouache, pastel & pencil include wide range of subjects, including botanical & landscape; exhibited group & solo, SA & overseas; third prize, International Artists in Watercolour Painting, Mall Galleries, London (1981)
Artwork published: unknown
Artwork represented: private collections, SA & worldwide
References: *Artlook*, Jan. 1971; *DSAPS; SAAPS*

BACKHOUSE, James (1794–1869)
Studies: art of gardening, Norwich, England (2 years)
Profile/career: with brother Thomas, ran established nursery in York, England (from 1816); as Quaker missionary left England (1831); in Australia & Tasmania (1831–1838); 3 months in Mauritius (1838); in Cape Colony, travelling extensively (mid 1838–mid 1840); on return to England (early 1841), resumed work at nursery
Artwork published: *A Narrative of a Visit to the Mauritius & SA* (London, 1844)
Artwork represented: –
References: *Bothalia*, 50:1950; *DSAB;* Gunn & Codd; *J. Bot. Lond.*, 7:1869; G.W. Reynolds, *The Aloes of SA* (Jhb, 1950)

BAINES, John Thomas (1820–1875)
Studies: apprenticed to William Carr, ornamental painter, King's Lynn, Norfolk, England (5 years)
Profile/career: left England for Cape (1842); marine & portrait painter, CT; unofficial war artist, attached to 74th Highlanders, Cape eastern frontier war (1850–1853); Royal Geographical Society expedition to Australia (1853–1855); artist & storekeeper, expedition of David Livingstone on Zambezi River (1858); travelled extensively in SthA, working as artist, & in contact with W. & J. Hooker, Kew (from 1860s until death)
Artwork published: *Illustrated London News* (war pictures); *Scenery & Events in SA* (London, 1852); *Explorations in South-West Africa* (London, 1864); *The Victoria Falls – Sketched on the Spot During the Journey of J. Chapman & T. Baines* (London, 1865); *Nature & Art*, 1:1866; M. & J. Diemont, *The Brenthurst Baines* (Jhb, 1975); E. Tabler, *et al.*, eds, *Baines on the Zambezi, 1858 to 1859* (Jhb, 1982); J. Carruthers, *Thomas Baines: Eastern Cape Sketches* (Jhb, 1990)
Artwork represented: Brenthurst Library, Jhb; MuseumAfrica, Jhb; National Archives, Zim., Harare; Natural History Museum, London; Royal Geographical Society, London; William Fehr Collection, CT; private collections
References: *A&A;* J. Carruthers & M. Arnold, *The Life & Work of Thomas Baines* (CT, 1995); *DSAB; DSAPS;* M. Stevenson, ed., *Thomas Baines: An Artist in the Service of Science in SthA* (London, 1999)

BARBER, Mary Elizabeth (née Bowker) (1818–1899)
Studies: apparently self-educated
Profile/career: brought up on Bowker family farm 'Tharfield' in Eastern Cape; collected plants & made observations on local flora; corresponded with eminent British scientists Charles Darwin, W. & J. Hooker, & W.H. Harvey; married (1845); moved to northern Cape diamond fields (1886) & then to Jhb
Artwork published: articles on plants & insects in *Jl Linn Soc. (Bot.)* & *Kew Bull.;* 'Wanderings in SA', *Q. Bull. SA Lib.*, 17(2,3&4)1962, 18(1&2)1963
Artwork represented: Albany Museum, Grahamstown – many destroyed in fire (1941); 1820 Settlers Memorial Museum, Grahamstown; Kew; Natural History Museum, London
References: Arnold; *DSAB; DSAPS;* Gunn & Codd; S. Schönland, *Records of the Albany Museum* (1904); *Women & Art*

BARCLAY, Dorothy (1892–1940)
Studies: apparently self-taught as artist
Profile/career: niece of artist Ethel May Dixie; worked closely with Louisa Bolus preparing colour plates of SA plants for educational projects
Artwork published: L. Bolus, *A Book of SA Flowers* (CT, 1925); L. Bolus, A *Second Book of SA Flowers* (CT, 1936); *SA Nature Notes; SA Gardening & Country Life*
Artwork represented: SA Library, CT; several private collections
References: *A&A; DSAPS; Veld & Flora* (1977)

BARKER, Winsome Fanny ('Buddy') (1907–1994)
Studies: BSc (Rhodes, 1928); self-taught as artist
Profile/career: Bentham Trustee, based at NBI, Kirstenbosch (1931–1933); botanical assistant, NBG (1933–1957); curator, Compton Herbarium, NBG (1957–1972)
Artwork published: *Flowering Plants of Africa* (1930–1938); 'The genus *Agapanthus*', *Jl of the SA Bot.*, suppl., 4:1965; H.A. Baker & E.G.H. Oliver, *Ericas in SthA* (CT, 1967)
Artwork represented: NBI, CT (107 unpublished plates of *Lachenalia); NBI*, Pret.; RHS
References: *Bothalia*, 25:1995; Gunn & Codd; Rourke

BARKLY, *Lady* Anne Maria (née Pratt) (1837–1932)
Studies: probably self-taught as artist
Profile/career: second wife of Cape governor, Sir Henry Barkly (1870–1877); made drawings of plants collected & also many grown by her husband
Artwork published: *Revised List of the Ferns of SA* (CT, 1875; repr. *Cape Monthly Magazine*, April 1875)
Artwork represented: Kew – folio of drawings of Stapelieae (Emily, daughter from Sir Henry's first marriage, assisted)
Reference: Gunn & Codd

BARROW, *Lady* Anna Maria (née Truter) (1777–1857)
Studies: self-taught as artist
Profile/career: only daughter of prominent Cape government official, Petrus Johannes Truter; married John Barrow, then Cape auditor-general (1799); couple left Cape for England (1803); possibly the earliest known SA-born woman to paint flowers; her husband became a founder of the Royal Geographical Society
Artwork published: not known
Artwork represented: MuseumAfrica, Jhb; NBI, Pret.
References: *Africana Notes & News* (1955); *DSAB;* Gunn & Codd; R.F. Kennedy, comp., *Catalogue of Pictures in the Africana Museum*, vol. 1 (Jhb, 1966)

BARTMAN, Charmaine (1956–)
Studies: National Diploma in Art & Design (Pret. Technikon); Higher National Diploma in Art & Design (Pret. Technikon); Diploma in Education – Secondary School (Pret. Technikon)
Profile/career: illustrator, NBI, Pret. (1981–1982); temporary & guest lecturer, Port Elizabeth Technikon (1992 & 1996); art teacher, Lawson Brown High School, Port Elizabeth (1997–2000); freelance botanical illustrator; exhibited, group & solo
Artwork published: G.E. Gibbs Russell, *et al., Grasses of SthA* (Pret., 1990); *Bothalia*, 14:1982

Artwork represented: consulting room foyer of St George's Hospital, Port Elizabeth; private collections
Reference: NBI files

BATHGATE, Jeanne (Mrs van Rensburg) (1950–)
Studies: National Diploma in Art & Design & Graphic Design (Pret. Technikon, 1970)
Profile/career: NBI, Pret. (1971)
Artwork published: *Flowering Plants of Africa*; illustrations for SA army veld survival handbook
Artwork represented: NBI, Pret.
Reference: NBI files

BATTEN, Auriol Ursula (1918–)
Studies: matriculated Pmbg Girls' High School; BSc (UNP, 1937); Higher Education Diploma (UNP, 1938); Durban Technical College
Profile/career: board of trustees, East London Museum (1983–); exhibited widely, group & solo, SA & overseas; winner of several prestigious art awards (1986, 1991, 1992, 1994, 2000)
Artwork published: *Wild Flowers of the Eastern Cape Province* (CT, 1966); *Flowers of SthA* (Sandton, 1986); O.M. Hilliard & B.L. Burtt, *Dierama* (Jhb, 1991); P. Goldblatt & J. Manning, *Gladiolus in SthA* (CT, 1998); *Flowering Plants of Africa*; *Curtis's Botanical Magazine*; designed 7 sets of stamps for Philatelic Services of the SA Post Office, Pret.
Artwork represented: NBI, CT; NBI, Pret.; Royal Botanic Garden, Edinburgh; numerous private collections & artist's personal collection
References: *A&A*; Arnold; Blunt & Stearn; *DSAPS*; Gunn & Codd; *SA Garden & Home* (1998); *SESA*; *Women & Art*

BEAUMONT, Angela Jane (1965–)
Studies: BSc Hons, applied biology (King's College, London University); MSc, botany (UNP, 1991); in preparation, PhD in systematic botany (UNP)
Profile/career: illustrator, Kew (1983, 1984, 1985); freelance artist for botanical & zoological scientific publications; private commissions; *Diascia* species (in preparation)
Artwork published: *A Field Guide to Wild Flowers: KwaZulu-Natal & the Eastern Region* (Durban, 1998); *Bothalia*, 29:1999
Artwork represented: NBI, CT
Reference: NBI files

BLAKE, Tamlin Gail (1974–)
Studies: BA Fine Art (UStell., 1995); BA Hons, Fine Art (Illustration) (UStell., 1999); MA Fine Art (UStell., 2001); top graduate student of year, fine art dept (UStell., 1995)
Profile/career: group & solo exhibitions, SA & overseas; Kirstenbosch Bronze Medal (2000); currently living & working in Ireland, with goal of future exhibition of paintings of SA plants; work on succulents in press
Artwork published: –
Artwork represented: private collections, SA & worldwide
Reference: NBI files

BODKIN, Ada Marie (née Shaw) (active 1880s–1890s)
Studies: probably self-taught as artist
Profile/career: wife of A.A. Bodkin, mathematics & science master at Diocesan College, CT; as friends of Harry Bolus, Bodkin couple accompanied the latter on outings & assisted with plant collection & recording; 2 albums of almost 400 watercolours of Cape plants completed
Artwork published: unknown
Artwork represented: NBI, CT
Reference: –

BODLEY VAN WYK, Elise (1921–1997)
Studies: Mowbray Teachers' Training College, CT, & Royal Academy of Fine Arts, Stockholm (1946–1948)
Profile/career: principal, Hugo Naudé Art Centre, Worcester; illustrator, BRI, Pret. (from 1980s); exhibited in SA, group & solo
Artwork published: N. du Plessis & G. Duncan, *Bulbous Plants of SthA* (CT, 1989); M. Arnold, *Art Meets Science: Flowers as Images* – exhibition catalogue (Jhb, 1992)
Artwork represented: Hunt Inst., USA; NBI, CT; NBI, Pret.; Smithsonian Institution, Washington, USA; UWC; private collections
References: Arnold; *Bothalia*, 29:1999; *DSAPS*; Rourke

BOHNEN, Pauline (née Kies) (1918–)
Studies: MSc (Huguenot College, Wellington, Cape)
Profile/career: Solly Scholar, Kirstenbosch (1940); junior botanist, NBI, Pret. (1941–1943); nurse, Groote Schuur Hospital, CT (1943–1945); botanist, NBI (1946–1948); NBI liaison officer, Kew (1948–1950); retired to Still Bay, Cape (1978); during post-retirement years, has painted local flora
Artwork published: *Flowering Plants of the Southern Cape* (Still Bay, Cape, 1986); *Bothalia*, 6:1951
Artwork represented: –
References: Gunn & Codd; NBI files

BOKELMANN, Hertha Ludovica (née Faekle) (1915–)
Studies: horticulture & botany, Technical High School, Ulm, Germany; diploma in art (Ulm-Donau, Germany)
Profile/career: left Germany for SA (June 1937); assistant, botanic gardens, UStell. (1 year); freelance artist
Artwork published: A. Batten, *Wild Flowers of the Eastern Cape* (CT, 1966); M. Courtenay-Latimer & G.G. Smith, *The Flowering Plants of the Tsitsikama Forest & Coastal National Park* (Pret., 1967)
Artwork represented: artist's personal collection
References: *A&A*; Gunn & Codd; NBI files

BOLUS, Harry (1834–1911)
Studies: Castle Gate School, Nottingham, England; self-taught as botanist & artist
Profile/career: left England for SA (1850); cofounder of Bolus Bros, earliest SA firm of stockbrokers; endowed chair of botany at UCT; bequeathed personal herbarium & library to UCT; prepared own colour plates of orchids to illustrate his publications; honorary DSc (UCT, 1902)
Artwork published: H.L. Bolus & A.M. Greene, eds, *The Orchids of the Cape Peninsula*, 2 ed. (CT, 1918); *Icones Orchidearum Austro-Africanarum Extra-Tropicarum*, 3 vols in 4 (CT & London, 1893–1913)
Artwork represented: Bolus Herbarium, UCT
References: *DSAB*; Gunn & Codd; Rourke

BORNMAN, Priscilla (née Postma) (1931–)
Studies: matriculated Klerksdorp High School (1948); self-taught as artist
Profile/career: mounting & filing clerk, NBI, Pret. (1949); moved to Pietersburg (*c.*1970)
Artwork published: *Veld & Flora*
Artwork represented: private collections
Reference: NBI files

BOWLER, Thomas (1812–1869)
Studies: initially self-taught as artist; later studied briefly under James Duffield Harding, England (1854)
Profile/career: arrived at Cape from England (1834); brief spell as lawyer's clerk; tutored children of commandant of prison settlement on Robben Island, CT (1836–1838); taught art at SA College School, CT & Diocesan College, CT (1842–1866); travelled extensively in SA (1840s, 1860s), Mauritius & England (1860s); oeuvre includes landscape studies
Artwork published: *Illustrated London News* (64 lithographic prints); *SA Sketches* (London, 1854); *The African Sketch Book* (London, 1855); *Pictorial Album of Cape Town* (CT, 1866; repr. CT, 1966)
Artwork represented: Albany Museum, Grahamstown; KG, PE; MuseumAfrica, Jhb; SANG; William Fehr Collection, CT; many private collections
References: *A&A*; E. Bradlow, *Thomas Bowler of the Cape of Good Hope* (CT, 1955); F. & E. Bradlow, *Thomas Bowler: His Life & Work* (CT, 1967); *DSAB*; *DSAPS*; *SESA*

BRANCH, Margo Lundie (1944–)
Studies: BSc, zoology, botany (UCT); BSc Hons, zoology (UCT); MSc, zoology (UCT); art courses, Cape Art Centre, CT; art courses, USA & Canada
Profile/career: research officer in entomology, Tobacco & Cotton Research Board, Zim. (1966–1968); junior lecturer, zoology dept, UCT (1968); research assistant, seaweed unit, Sea Fisheries Institute, CT (1969–1970); senior scientific officer, Marion Island sub-antarctic marine research programme, UCT (1984 & 1986–1994); freelance scientific author & illustrator (1971–1994); senior education officer, Two Oceans Environmental Education Trust, Two Oceans Aquarium, CT (1995–)
Artwork published: leaf drawings in K. Coates Palgrave, *Trees of SthA* (CT, 1977); *The Living Shores of SthA* (CT, 1981); P. Goldblatt, *The Moraeas of SthA* (CT, 1986); *Two Oceans: A Guide to Marine Life of SthA* (CT, 1995)
Artwork represented: NBI, Pret.; *Flowering Plants of Africa;* line drawings of *Moraea*, Iridaceae & *Watsonia* in numerous research papers
Reference: NBI files

BROWNING, Jane Beatrice Mary (née Ball) (1932–)
Studies: Eveline High School, Bulawayo, Zim.; BSc, botany, zoology (UCT, 1952); postgraduate teachers' certificate (1954); MSc (UNP, 1992); PhD (UNP, 1999)
Profile/career: high school biology teaching, Zim. (1954–1983); Wykeham School, Pmbg (1984–1986); sub-postmistress, Dishforth, Thirsk, Yorkshire (1964–1967); herbarium technician & researcher, UNP (July 1986–)
Artwork published: co-editor of J.S. Ball (late brother), *SthA Epiphytic Orchids* (Jhb, 1978); E. Pooley, *The Complete Field Guide to Trees of Natal, Zululand & Transkei* (Durban, 1993); K.D. Gordon-Gray, *Cyperaceae in Natal* (Pret., 1995); K. & W. Ellery, *Plants of the Okavango Delta: A Field Guide* (Durban, 1997); *SA Jl Bot.* (1989–1999)
Artwork represented: *Flowering Plants of Africa*; herbarium, UNP; NBI Pret.
Reference: NBI files

BRUYNS, Peter Vincent (1957–)
Studies: MA, DPhil, mathematics (Oxon); MSc, botany (UCT); LRSM, performer's piano
Profile/career: lecturer, mathematics dept, UCT; research associate, Bolus Herbarium, UCT
Artwork published: P.V. Bruyns & H.P. Linder, 'A revision of *Microloma*', *Bot. Jahrb. Syst.*, 112:1991; 'A revision of *Hoodia* & *Lavrania*', *Bot. Jahrb. Syst.*, 115:1993; 'The systematic position of *Quaqua* with a critical revision of the species', *Bot. Jahrb. Syst.* 121:1999
Artwork represented: used to illustrate own publications
Reference: –

BUITENDAG, Elise (née Van Rensburg) (1941–)
Studies: BSc Hons, botany (UPret.,1961); HOD (UPret., 1963); self-taught as botanical artist
Profile/career: assistant professional officer, BRI, Pret. (1961–1964); phenological research at Landboukollege, Grootfontein, Middelburg (1964–1965); botanical officer, Lowveld Botanical Gardens, Nelspruit (1969–1981); exhibited in SA, group & solo
Artwork published: D. Carr, *Combretaceae in SthA* (Jhb, 1988); brochures for Lowveld Botanical Gardens, Nelspruit (1971–1983); J.P. Rourke, *The Proteas of SthA* (CT, 1980; 2 ed., 1982)
Artwork represented: many private collections
References: Gunn & Codd; NBI files; *DSAPS; Veld & Flora* (1979)

BURCHELL, William John (1781–1863)
Studies: botanical apprentice, Kew, England; fellow of Linnean Society, then youngest ever elected (1803); also learnt craft of draughtsman
Profile/career: son of wealthy nurseryman; as schoolmaster taught drawing, St Helena – later became naturalist on island for English East India Company (1806–1810); arrived in CT (1810); undertook major journey into interior of SthA, during which several new plant & animal species discovered by him (1811–1815); explored in Brazil (1825–1830); returned to England
Artwork published: *Travels in the Interior of SthA*, 2 vols (London, 1822 & 1824); H.M. McKay, ed., *The SA Drawings of William J. Burchell*, 2 vols (Jhb, 1938 &1952)
Artwork represented: MuseumAfrica, Jhb; several private collections
References: *DSAB; DSAPS;* Gunn & Codd; J. Hutchinson, *A Botanist in SthA* (London, 1946); *Jl SAfr. Bot.* 7:1941; *SESA; Veld & Flora*

BURGES, Edith Kathleen (Mrs Metherell) (1919–)
Studies: Durban School of Arts & Crafts, Technical College (fine art, commercial art, pottery)
Profile/career: botanical artist, BRI, Pret. (1942–1944); participated in group exhibitions (after 1961)
Artwork published: *Flowering Plants of Africa* (1942–1966); designed Christmas stamps (1940 & 1941)
Artwork represented: NBI, Pret.
References: *DSAPS;* NBI files

BURROWS, Sandra Margaret (née Schultz) (1959–)
Studies: National Diploma, Horticulture (1977–1979); self-taught as botanical artist
Profile/career: Kirstenbosch scholarship (1980); freelance artist; in progress: 'Flora of Witwatersrand' (Moss Herbarium, WU); awaiting publication: 'Moraceae of South & Central Africa'
Artwork published: J.E. Burrows, *SthA Ferns & Fern Allies* (Sandton, 1990)
Artwork represented: Lowveld National Botanical Gardens, Nelspruit; NBI, Pret.; WU; private collections
Reference: NBI files

BUSH, Rowena Elizabeth (Mrs Calwick) (1917–)
Studies: West of England Academy of Art, Bristol (1933–1936); London School of Interior Design (1936–1937); sculpture, Natal Technical College, Durban (1969–1970)
Profile/career: lived in Ceylon (now Sri Lanka) (1917–1924); school teaching posts, Tanganyika (now Tanzania) & Kenya (1939–1943); illustrator for schoolbooks (1940–1946); settled in Durban (1962); taught art privately, Durban (1963–1971); several group & solo exhibitions, SA & overseas
Artwork published: *Veld & Flora*, Sept 1999
Artwork represented: Kew; NBI, CT
References: *A&A; DSAPS; Veld & Flora* (1999)

BUZZARD, Thelma (active 1970s–)
Studies: Natal Technical College, Durban (1974–1977); life drawing (London, 1977); UND (1978–1982)
Profile/career: associate of Watercolour Society of SA (1981); member, Natal Society of Arts (1981); painter of flowers & landscapes, in watercolour, ink & oil; exhibited, group & solo, SA & overseas
Artwork published: 4 portfolios: 2 of bougainvillaea (Jhb, 1984), 2 of azalea (Jhb, 1985)
Artwork represented: private collections, SA & worldwide
Reference: *DSAPS*

BYRNE, Estelle (1929–)
Studies: BA Fine Art (Rhodes); teachers' training postgraduate course (London University)
Profile/career: high school art teacher, Harare, Zim. (30 years); moved to Cape (1992); paintings of fynbos executed; exhibited widely, SA; work with Simon's Town Flora Conservation Group recording fynbos of the Cape Peninsula
Artwork published: *SA Country Life* (1997)
Artwork represented: private collections
References: NBI files; The Joan St Leger Lindberg Arts Foundation, Muizenberg

CARMICHAEL, Diana E. (1926–)
Studies: Hereford School of Art, England (interrupted by Second World War)
Profile/career: ambulance corps in England & Austria (Second World War); English civil service (until 1951); settled in Rhodesia (1952); worked as commercial artist, then taught fine art in schools; moved to Jhb (1985)
Artwork published: Fedsure calendar (1998)
Artwork represented: Fedsure; private collections
Reference: NBI files

CARR, John Denzil (1916–1997)
Studies: BSc Eng (WU, 1940)
Profile/career: bomber pilot, SAAF (Second World War); member, Sale Art Society, England (1946–1947); member, Manchester Graphic Club, England (1946–1947); chairman, Tree Society of SthA (1965–1977); produced landscape, still-life & botanical artworks worked in watercolour & pencil; also sculpture; several group exhibitions, England & SA; Marloth Medal from BSSA (1993)
Artwork published: *The SA Acacias* (Jhb, 1976); *Combretaceae in SthA* (Jhb, 1987); *The Propagation and Cultivation of Indigenous Trees & Shrubs on the Highveld* (Sandton, 1994)
Artwork represented: private collections, SA & worldwide
References: *DSAPS; Peltophorum*, 28:1997; *Trees in SA*, Dec. 1998

CARTER, Beatrice Orchard (1889–1939)
Studies: art school, CT
Profile/career: artist (ophthalmic drawings) for oculist; artist, Bolus Herbarium, UCT, mainly working on Mesembryanthemaceae under Dr Louisa Bolus (1926–1939)
Artwork published: *Flowering Plants of Africa*
Artwork represented: UCT
References: *A&A; Cape Times*, 9.11.1939; *DSAPS;* Gunn & Codd; H. Herre, *The Genera of the Mesembryanthemaceae* (CT, 1971); Rourke

CHESSELET, Pascale (1959–)
Studies: BSc Hons (UCT); MSc (UCT)
Profile/career: scientific illustrator, entomology dept, Transvaal Museum, Pret. (1981); scientific illustrator, zoology dept, UCT (1982); scientific illustrator (fungi) for Professor M. Wingfield, dept of agriculture (fungi); botanist, NBI, Pret.

(1993); botanist, NBI, CT (1997); illustrates own research papers on Mesembryanthemaceae
Artwork published: B.R. Davies & J. Day, *Vanishing Waters* (CT, 1998); 2 plates in *Smithers' Mammals of SthA*, 3 ed. (Jhb, 1996); variety of smaller publications
Artwork represented: –
Reference: –

CLARKE, Brenda (née Stiemens) (1917–)
Studies: MSc, botany (UPret., 1939)
Profile/career: mycologist, Low Temperature Research Station, CT (5 years); married (1944); lived in Zim., Cedara & Barberton; moved to Pret. (1958); research assistant in plant physiology, Margaretha Research Institute, UPret. (1958–1974, 1976–1977); met Eve Palmer & started illustrating (1980)
Artwork published: E. Palmer, *The SA Herbal* (CT, 1985); E. Palmer, *Under the Olive* (CT, 1989); E. Palmer, *Return to Camdeboo* (CT, 1992)
Artwork represented: NBI, CT
References: *DSAPS*; Gunn & Codd

CLAUDIUS, Heinrich (Hendrik) (*c.*1655–*c.*1697)
Studies: trained as apothecary (Germany)
Profile/career: sent to Batavia by fellow medical associate to collect plants as medical specimens; continued work at Cape, also making drawings of herbs & medicinal plants (1681); when employment lapsed, retained by Simon van der Stel, then governor of the Cape – joined two expeditions to explore western & north-western Cape (1683, 1685–1686); completed several albums of plant drawings; left Cape for The Netherlands (*c.*1693)
Artwork published: Guy Tachard, *Voyage de Siam* (Paris, 1686); G. Waterhouse, *Simon van der Stel's Journal of his Expeditions to Namaqualand, 1685–6* (London, 1932); *Jl SA Bot.*, 13:1947; A.H. Smith, *Claudius Watercolours in the Africana Museum* (Jhb, 1952)
Artwork represented: British Museum (Natural History), London; Jhb Central Library; SA Library, CT; SA Museum, CT; Trinity College, Dublin
References: *DSAB;* Gunn & Codd

COAPE-ARNOLD, Mary Genevieve (née Cummings) (1850–1936)
Studies: educated in England
Profile/career: lived briefly in SA (1911–1916); resident at 'Clifton', Rosebank, CT; visited Eastern Cape; painted landscapes & local flora
Artwork published: –
Artwork represented: several albums owned by granddaughter, Betty Dwight, Mowbray, CT
Reference: –

COLLETT, Rhona (née Brown) (1922–)
Studies: BA Fine Art (Natal University College, 1944); BA Hons, Fine Art (UNISA, 1960); National Art Teachers' Certificate (Bloemfontein, 1960)
Profile/career: Division of Botany (later BRI, Pret.) (1943–1947); matric art teacher (1955–1960); BRI, Pret. (1965–1969); junior school teacher (1962–1967); textile designer, David Whitehouse (1979–1983); exhibited in SA, group & solo
Artwork published: E. Palmer, *A Field Guide to the Trees of SthA* (London, 1977); *Flowering Plants of Africa* (1943–1981)
Artwork represented: NBI, Pret.; Roll of Honour, SA Air Force Base (Swartkops), Pret.; several murals in churches & clinics
References: *A&A; DSAPS;* Gunn & Codd; NBI files

CONDY, Gillian (1952–)
Studies: Scientific illustration (Middlesex Polytechnic, London); MA by Project – 'British Poisonous Plants' (Royal College of Art, London, 1976); 2 botanical art master-classes of Katie Lee & Jenny Phillips (CT)
Profile/career: freelance artist, London (1976–1978); international voluntary service, Phuthadikobo Museum, Mochudi, Botswana (1978–1980); primary curriculum, education dept, Lobatse, Botswana (1980–1982); resident botanical artist, NBI, Pret. (1983–); winner of several prestigious art awards (1990, 1992, 1975–1999, 2000); founder member of BAASA; exhibited SA & overseas
Artwork published: *Flowering Plants of Africa*; various NBI publications; *Curtis's Botanical Magazine;* 'Art Meets Science: Flowers as Images' (Balgowan, Natal, 1992: original Caversham Press prints first exhibited at Grahamstown Festival, 1992); C. Craib, *Geophytic Pelargoniums* (Pret., 2001); designed 13 sets stamps for Botswana
Artwork represented: The Blue Train; Kew; NBI, Pret.; Sherwood; numerous private collections, SA & worldwide
References: Arnold; C. Craib, *Geophytic Pelargoniums* (Pret., 2001); Sherwood; J. White, comp., *Catalogue: 7th International Exhibition of Botanical Art & Illustration* (Pittsburgh, USA, 1992)

CONNELL, Mary Elizabeth ('Betty') (Mrs Stutterheim) (1917–1997)
Studies: York College for Girls, England; Jhb High School for Girls; School of Art, Witwatersrand Technical College, Jhb (2 years)
Profile/career: Division of Botany (later BRI, Pret.) (1936–1941, 1958–1960); participated in group exhibitions
Artwork published: *Farming in SA; Flowering Plants of Africa*; D. Meredith, ed., *The Grasses & Pastures of SA* (Jhb, 1955); M. Henderson & J.G. Anderson, *Common Weeds in SA* (Pret., 1966)
Artwork represented: NBI, Pret.; private collections
References: *A&A; Bothalia*, 1998; *DSAPS;* Gunn & Codd

COPLANS, Peta Susan (1951–)
Studies: BA Fine Art – class medal for studiowork (UCT, 1972); postgraduate certificate in illustration (Central School of Art, London, 1974); botanical art master-classes of Katie Lee & Jenny Phillips (CT, 1998, 1999/2000)
Profile/career: many years in fine art, design & editorial work, London; bronze medal, 'Kirstenbosch Inaugural Exhibition of Botanical Art', CT (2000); participated in group exhibitions, SA; editor, BAASA's *Botanical Art News*, 4:2000
Artwork published: –
Artwork represented: Victoria & Albert Museum, London; private collections
Reference: *Catalogue: Kirstenbosch Inaugural Exhibition of Botanical Art* (CT, 2000)

CRAWLEY, Vicary Gibbs (1868–1909)
Studies: Greek & Latin literature – first class Hons (St Paul's College, Oxford University, 1889); modern history (St Paul's College, Oxford University, 1891); apparently self-taught as artist
Profile/career: worked in British civil service: Exchequer & Audit Dept, London (1892–1909); sent to SA to audit army accounts (1907–1909)
Artwork published: J. Burtt Davy & V.G. Crawley, 'The families, genera & species of Pteridophyta of the Transvaal, *SA Jl Sci.*, 6:1910
Artwork represented: –
References: Gunn & Codd; NBI files

CROSSMAN, Mrs F.G. (active *c.*1885)
Studies: probably self-taught as artist
Profile/career: unknown
Artwork published: –
Artwork represented: NBI, Pret. – 86 watercolour plates of SA flora in 2 albums, unsigned; Kew – 72 paintings, 8 pencil sketches in 3 albums
Reference: NBI files

DAVIS, Linda Susan (née Cowan) (1955–)
Studies: matriculated Pmbg Girls' High School; self-taught as artist, specializing in monochrome line drawings
Profile/career: scientific illustrator, NBI, CT (1976); scientific assistant & artist, UNP Herbarium (1977–1985); scientific assistant & artist, Mollusca dept, Natal Museum, Pmbg (1991–); paints natural history subjects, specializing in frogs & fish; mainly monochrome artwork, occasionally using watercolour
Artwork published: 'Studies on African Asclepiadaceae', *Kew Bull.*, 38(4)1984; O.M. Hilliard, *Trees & Shrubs of the Natal Drakensberg* (Pmbg, 1985); O.M. Hilliard, *Grasses, Sedges, Restiads & Rushes of the Natal Drakensberg* (Pmbg, 1987)
Artwork represented: Natal Museum, Pmbg; NBI, CT; UNP; private collections
Reference: –

DE VOS, Miriam Phoebe (1912–)
Studies: MSc, DSc (UStell., 1931–1940)
Profile/career: junior lecturer, botany (UCT, 1929); junior lecturer, botany (UStell., 1941); senior lecturer, botany (UStell., 1946); associate professor, botany (UStell., 1971–1977)
Artwork published: *The Genus* Romulea *in SA* (CT, 1972); *Flora of SthA*, vol. 7; *Flowering Plants of Africa*
Artwork represented: NBI, Pret.
Reference: *SA Jl Bot.*, 58(1)1992

DIXIE, Ethel May (1876–1973)
Studies: self-taught as artist (older sister received art tuition from Thomas Bowler)
Profile/career: freelance artist; principal artist for R. Marloth's *The Flora of SA*, 4 vols in 6 (CT & London, 1913–1932); lecturer, CT School of Art
Artwork published: R. Marloth, *The Flora of SA*, 4 vols in 6 (CT & London, 1913–1932) – many of these originals

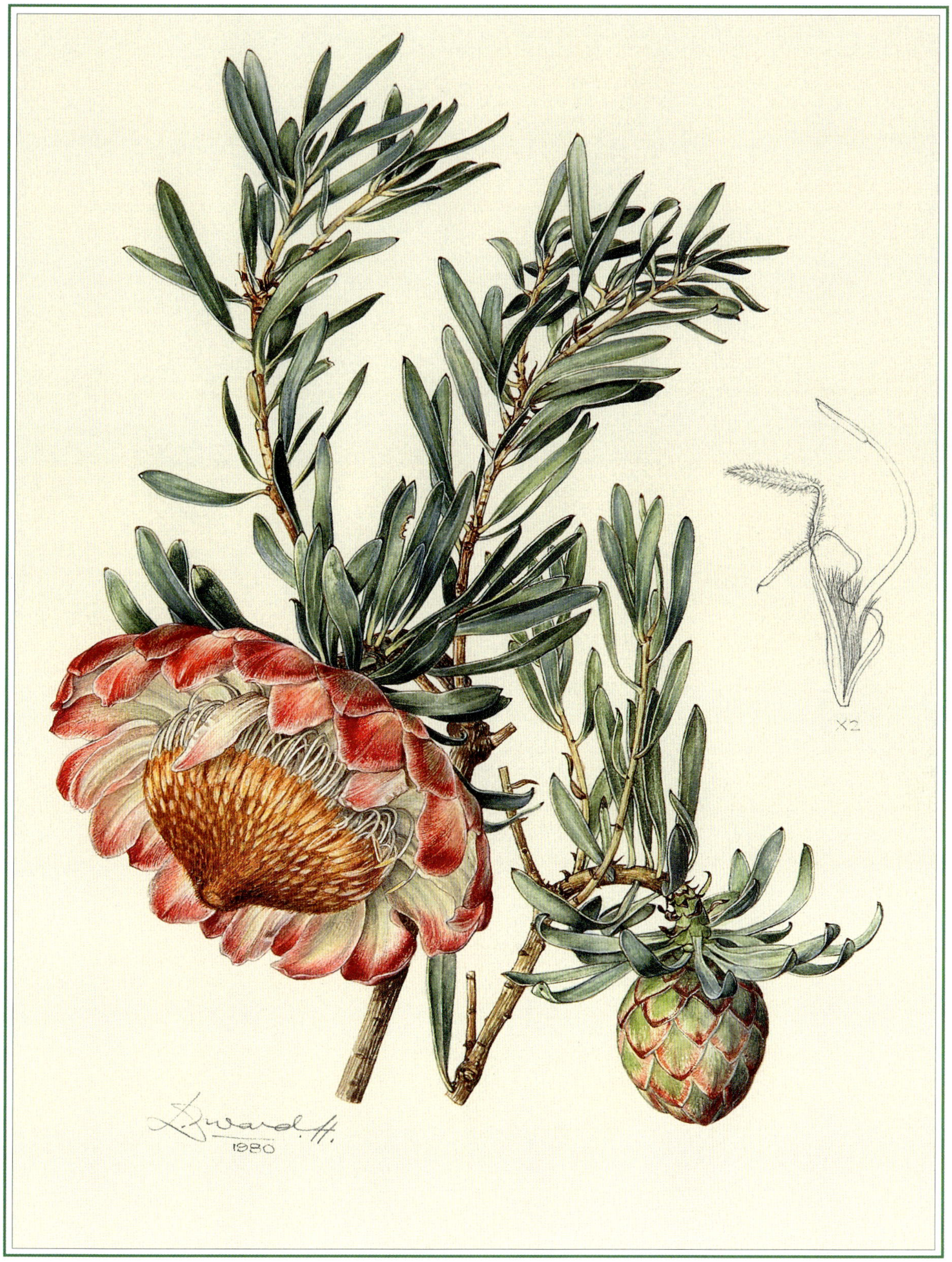

PLATE 130

Ellaphie Ward-Hilhorst, *Protea namaquana*, 1980, watercolour and pencil

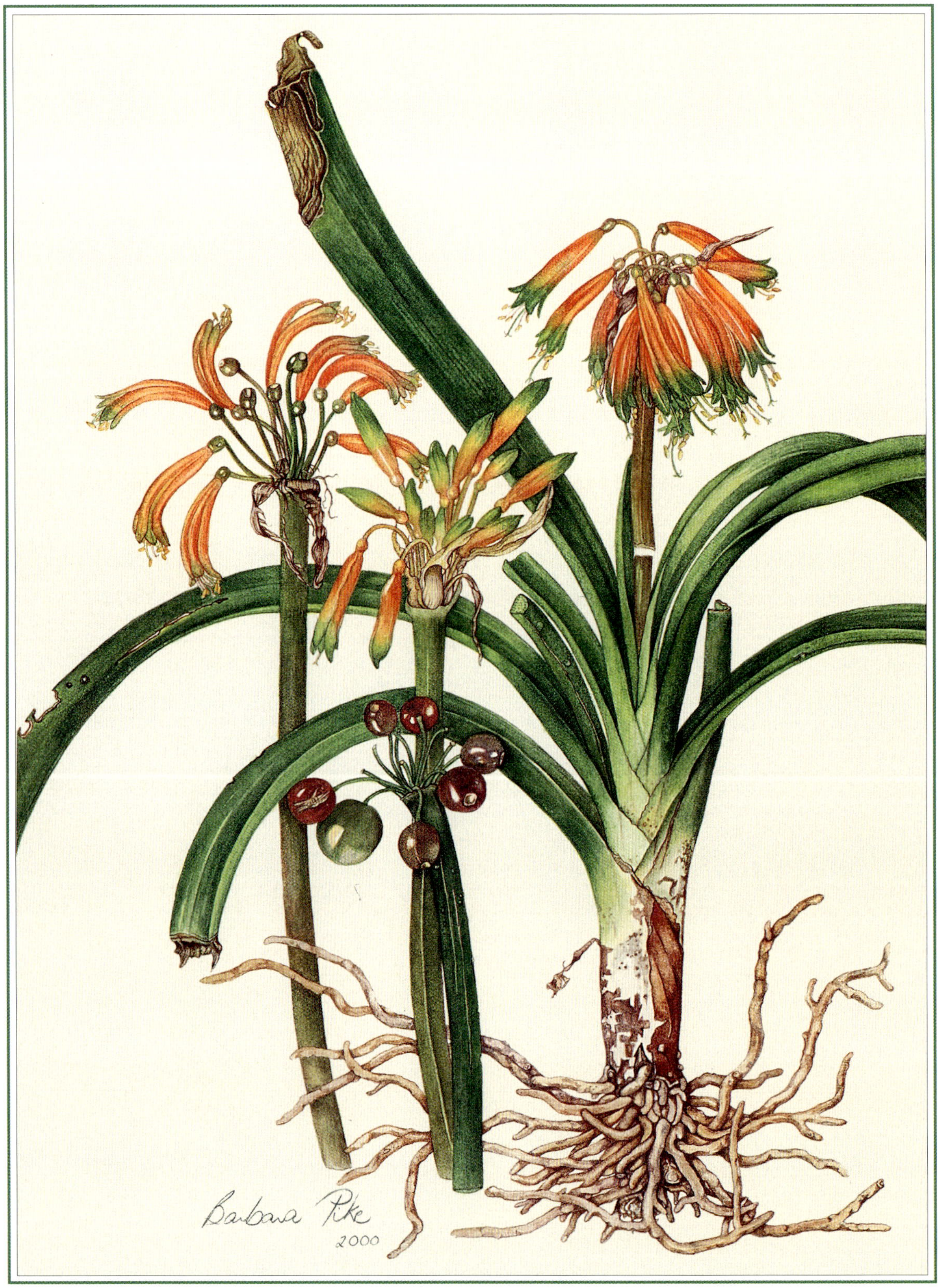

PLATE 131

Barbara Pike, *Clivia gardenii*, 2000, watercolour, 57 cm x 38 cm

destroyed in a fire at the publisher; R.H. Compton & E.M. Dixie, *Wild Flowers of the Cape of Good Hope* (CT, 1953); 2 portfolios of prints (published privately, 1990s)
Artwork represented: Brenthurst Library, Jhb; Carnegie Library archives, UStell.; MuseumAfrica, Jhb; NBI, CT; NBI, Pret.; SA embassies, London, Rome & New York; numerous private collections
References: *A&A;* Arnold; *Die Burger*, 10.5.1972; *Die Vaderland*, 4.6.1966; *DSAPS;* Gunn & Codd; NBI files; Rourke; *SESA*

DONALDSON, John (1959–)
Studies: PhD, zoology (UCT)
Profile/career: initially, painted insects (from 1984); first botanical art – cycads (1997)
Artwork published: in various scientific journals; also illustrates own research papers
Artwork represented: Montgomery Botanical Centre, Miami, USA; NBI, CT; personal collection
Reference: NBI files

DU TOIT, Hesta (née Wouda) (active 1970s–1990s)
Studies: unknown
Profile/career: technical assistant, NBI, Pret. (mid 1970s)
Artwork published: *Flora of SthA*, vol. 30; G.E. Gibbs Russell, *et al.*, *Grasses of SthA* (Pret., 1990)
Artwork represented: NBI, Pret.
Reference: NBI files

DWIGHT, Betty (née Bowker) (1924–)
Studies: commercial art (Podolinni's Art Studio, Dean Street, CT); self-taught as botanical artist
Profile/career: painted studies of wild flowers throughout lifetime for personal pleasure – mainly Western Cape area & especially Rondebosch Common, CT
Artwork published: 4 sets notelets depicting Rondebosch Common flora published by Metropolitan Life to support conservation of Rondebosch Common, CT
Artwork represented: several in private collections & personal collection
Reference: –

EKSTEEN, Helena (Mrs S. van Breda) (active 1880–1890)
Studies: unknown
Profile/career: lived at 'Zorgvliet' & 'The Woolsack', Rosebank, CT (1880s); apparently painted wild flowers for personal pleasure; most specimens painted were collected on farms belonging to Duckitt family at Darling in Western Cape
Artwork published: –
Artwork represented: NBI, CT
Reference: notes accompanying original artworks (NBI, CT)

ELIOVSON, Sima (1919–1990)
Studies: Witwatersrand Technical College, Jhb; private art tuition from Joyce Leonard & George Boys, Jhb, & Gofu Sano, Kyoto, Japan
Profile/career: painter of landscapes & botanical subjects; author of 11 horticultural books (1951–1985), some illustrated with own photographs; *Flowering Shrubs, Trees & Climbers* (CT, 1951); *SA Wild Flowers for the Garden* (CT, 1955); *The Complete Gardening Book* (CT, 1960); *Discovering Wild Flowers in SthA* (CT, 1962); *Proteas for Pleasure* (CT, 1965); *Namaqualand in Flower* (Jhb, 1972); *Garden Design for SthA* (Jhb, 1985); participated in several group exhibitions (from 1976); worked in oil, watercolour & Chinese ink
Artwork published: see above
Artwork represented: private collections
References: *A&A; DSAPS*

FABIAN, Anita Ann (née De la Harpe) (1934–)
Studies: Diploma in Fine Art (Michaelis School of Fine Art, UCT)
Profile/career: painter of landscape, still-life, & especially botanical artworks – in watercolour & gouache; bronze medal, 'Kirstenbosch Inaugural Exhibition of Botanical Art', CT (2000); exhibited widely, group & solo
Artwork published: Gerrit Germishuizen, *Transvaal Wild Flowers* (Jhb, 1982); David Hardy, *Succulents of the Transvaal* (Jhb, 1992); Gerrit Germishuizen, *Wild Flowers of Northern SA* (CT, 1997)
Artwork represented: Arts Association Namibia Collection
References: *DSAPS*; NBI files

FAWKES, Madeleine Charlotte (1880–1954)
Studies: Slade School of Art, London
Profile/career: exhibited in London; painted local flora while on visit to brother on 'Strathcona' farm, Ficksburg district, eastern Free State; Grenfell Medal of RHS for collection of paintings of Basutoland (Lesotho) flora
Artwork published: unknown
Artwork represented: NBI, CT
References: Gunn & Codd; *Veld & Flora*, 69:1983

FERN, Emily Isabel (1881–1953)
Studies: Port Elizabeth School of Art (1901); Jhb School of Art (1913); Royal Drawing Art Teachers' Certificate (Slade School of Art, London, 1920–1922); André Lhote Academy, Paris (1933, 1949)
Profile/career: born Victoria, Australia; painter of various subjects, including flowers & landscapes; founder member, Jhb Sketch Club & Friday Club (1916); member, SA Society of Artists (1918); taught art privately, Jhb (1926–1930); taught art, Witwatersrand Technical College, Jhb (1926–1930, 1934); lived in Europe (1931–1934); exhibited group & solo, SA & overseas
Artwork published: *A&A*
Artwork represented: Ann Bryant Gallery, East London; MuseumAfrica, Jhb; SANG; private collections
References: *A&A; DSAB; DSAPS*

FIELD, Natalie (Mrs Nash) (1898–1977)
Studies: Slade School of Art, London, & Westminster School of Art, London (1916–1922)
Profile/career: born London; at very early age moved with family to SA; life member, SAAA; founder member & honorary life member, Natal Society of Arts; exhibited widely, group & solo; oeuvre includes paintings of flowers; worked in pastel & oil
Artwork published: unknown
Artwork represented: MuseumAfrica, Jhb; SANG; SA National Museum of Military History, Jhb; WH, Kimb.
References: *A&A; DSAPS*

FINDLAY, Bronwen Eunice (1953–)
Studies: Natal Technical College, Durban (1973–1976); BA; Higher Diploma in Education (UNP, 1981); MA (UNP, 1994)
Profile/career: art teacher, various high schools, (1982–1993); lecturer, printmaking & painting, University of Durban-Westville (1993–2000); part-time lecturing, art depts, Natal Technikon, Durban & UNP; involved in several community art projects; exhibited widely, group & solo; paints variety of subjects, particularly flowers; works in oil
Artwork published: etching for portfolio (Durban, Playhouse,1986); 'Art Meets Science: Flowers as Images' (Balgowan, Natal, 1992: original Caversham Press prints first exhibited at Grahamstown Festival, 1992); M. Arnold, *Art Meets Science: Flowers as Images* – exhibition catalogue (Jhb, 1992); designed several dustjackets for books
Artwork represented: DAG; Gencor Collection, Jhb; Natal Technikon, Durban; Tatham Art Gallery, Pmbg; UCT; WU; private collections
References: Arnold; *DSAPS*

FINDLAY, Richard Innes (Dick) (1928–1996)
Studies: Pret. Boys' High School (art master: Walter Battiss); architecture (WU)
Profile/career: freelance natural history illustrator; scriptwriter of plays for radio & television
Artwork published: O.P.M. Prozesky, *Field Guide to the Birds of SthA* (London, 1970); R.H.N. Smithers, *The Mammals of the SthA Subregion* (Pret., 1983); designed over 250 stamps for Philatelic Services of the SA Post Office, Pret.
Artwork represented: NBI, Pret.; Philatelic Museum of SA, Pret.; UPret.; numerous private collections, SA & worldwide
References: *A&A; DSAPS; SA Panorama*, July/Aug:1991; *SA Stamp News*, 1:1996; *SESA; SETEMPE; Veld & Flora*, 63:1977

FLETCHER BROWN, Agnes (Mrs John Nobbs) (active 1870s–1880s)
Studies: probably self-taught as artist
Profile/career: apparently member of group of lady flower painters, active in CT; paintings signed with initials 'AFN'
Artwork published: –
Artwork represented: NBI, CT (collection first presented by her son Erik A. Nobbs to SA Museum, CT)
Reference: notes accompanying original artworks (NBI, CT)

FORD, George Henry (1809–1872)
Studies: unknown (father, J.E. Ford, painted miniatures)
Profile/career: English watercolour painter of landscapes & wildlife; lived in SA (1820–1837); accompanied Dr Andrew Smith on travels through interior of SthA (1834–1836); worked at British Museum,

London (after 1837)
Artwork published: A. Smith, *Illustrations of the Zoology of SthA*, 3 vols (London, 1849; repr. Jhb, 1973); R. Trimen, *SA Butterflies*, 3 vols (London, 1887–1889)
Artwork represented: WU
References: *DSAPS; SESA*

FORDYCE, Betsy Burns (née Moffat) (1907–1981)
Studies: Diploma in Art (Glasgow School of Art, 1930)
Profile/career: born Scotland; scenery designer & painter, Scottish National Theatre (1930–1938); art teacher, India (1942–1947); art teacher, Scotland (1947–1949); art teacher, Port Elizabeth Technical College & various schools; group & solo exhibitions, SA & overseas; painter of flowers, seascapes & landscapes, working in various media
Artwork published: unknown
Artwork represented: KG, PE
References: *A&A; DSAPS*

FRANKS, Millicent (Mrs Flanders) (1886–1961)
Studies: unknown; possibly self-taught as artist
Profile/career: junior assistant, Natal Herbarium, Pmbg (1901); painted & drew miscellaneous plants – some included in J. Medley Wood's *Natal Plants*; 3 months spent at Kew working on drawings for *Natal Plants* (1914)
Artwork published: J. Medley Wood, *Natal Plants*, 6 vols (Durban, 1899–1912); R. Marloth, *The Flora of SA*, 4 vols in 6 (CT & London, 1913–1932)
Artwork represented: Natal Herbarium, Pmbg; NBI, CT
Reference: Gunn & Codd

FREAN, Millicent Louie (1927–)
Studies: PhD (WU); FLS (Linnean Society of London); untrained as artist, but attended several workshops on botanical art
Profile/career: retired as senior lecturer in botany, WU (1993); founder member, BAASA; exhibited widely, SA, mainly group
Artwork published: numerous scientific papers
Artwork represented: several private collections
Reference: NBI files

FROOME, Bertha (1880–1978)
Studies: South Kensington School of Art, London (1894–1898); Durban Art School (late 1890s)
Profile/career: born England; moved to SA (late 1890s); art teacher, Coast High School, Durban; president, Natal Society of Arts; painted flowers, working mainly in watercolour; participant in group exhibitions
Artwork published: –
Artwork represented: DAG; private collections
References: *A&A; DSAPS*

FURNESS, Dorothy (née Taylor) (active 1960–1970s)
Studies: self-taught as artist
Profile/career: specialized in artworks of grasses & Restionaceae
Artwork published: –
Artwork represented: private collections
Reference: *Veld & Flora*, old series, 2:1972

GARRETT RICE, Elsie (1869–1959)
Studies: Slade School of Art, London
Profile/career: art teacher, Bedales School, England; emigrated to SA (1933); commissioned to paint artworks of Cape flora; worked mainly in watercolour
Artwork published: R.H. Compton & E. Garrett Rice, *Wild Flowers of the Cape of Good Hope* (CT, 1950); H. Hall, *Common Succulents* (London, 1955); designed Kirstenbosch Botanical Gardens commemorative set of stamps for Philatelic Services of the SA Post Office, Pret. (1963)
Artwork represented: NBI, CT – most originals & some unpublished plates for Compton & Rice's *Wild Flowers of the Cape of Good Hope*; NBI, Pret.; private collections
References: *A&A; DSAPS;* Rourke; *Veld & Flora*, 63:1977

GIBSON, Janet Mary (née Sands) (1908–)
Studies: Diocesan School for Girls, Pret.; commercial art (Natal Technical College Art School)
Profile/career: left England to live in Natal (*c.*1912); freelance artist, painting in watercolour
Artwork published: *Wild Flowers of Natal: Coastal Region* (Durban, 1975); *Wild Flowers of Natal: Inland Region* (Durban, 1978)
Artwork represented: NBI, Pret.; private collections
References: *DSAPS;* NBI files

GIESS, Heinrich Johann Wilhelm (1910–2000)
Studies: Adlerflucht Realschule, Frankfurt (1925); Neudamm Agricultural College, near Windhoek (1928–1929); taxonomy – during war internment (1940–1946)
Profile/career: left Germany for SWA (later Namibia) (1926); internship during Second World War in SA at Andalusia (Jan Kempdorp) (1940–1946); curator, SWA Herbarium (1957–1980)
Artwork published: many scientific papers; *Bestimmungsschlüssel für südwestafrikanische Grasgattungen* – engravings of grasses using tomato-box wood & toothpaste tubes to make type (privately printed, 1944)
Artwork represented: Swakopmund Museum, Namibia; National Herbarium, Windhoek, Namibia
References: *Bothalia*, 30:2001; Gunn & Codd

GINN, Michael G. van der R. (active 1960s–1980s)
Studies: BA Fine Arts (Rhodes)
Profile/career: schoolteacher, Rhodesia (later Zim.); housemaster, St Andrews School, Grahamstown
Artwork published: N.K. Hobson, *Veld Plants of SthA* (Jhb, 1975); *Veld & Flora*, 67:1981
Artwork represented: –
Reference: –

GLEDHILL, Eily Edith Agnes (née Archibald) (1914–)
Studies: St Dominic's Priory, England; BSc, botany, mathematics (Rhodes); MSc, botany (Rhodes); PhD (University College, London)
Profile/career: research assistant, Rhodes University; president, Historical Society, Grahamstown; chairman, St Johns Ambulance, Grahamstown; Commander Sister of the Order of St John; president, Soroptimist International, Grahamstown; director, first Settlers National Monument Council; chairman, Grahamstown Guild of Tour Guides
Artwork published: *Flowering Plants of Africa*; *Eastern Cape Veld Flowers* (CT, 1971; 2 ed., 1981)
Artwork represented: NBI, Pret.
Reference: NBI files

GOLDIN, Alice (1925–)
Studies: Pretoria Technical College (1955–1960); Rome (1965); Camden Institute for Adult Education, London; screenprinting
Profile/career: born Vienna, Austria; moved to SA (1948); member, SAAA (1960); member, Artists' Guild, CT; lived in England (1938–1948); lived in Pret. (1948–1971); moved to Cape (1971); visited Rome (1964–1965); visited London (1978); visited Jerusalem (1984); Cité Internationale des Arts, Paris (1986); painter & graphic artist, producing variety of subjects including flowers & trees; numerous group & solo exhibitions, SA & overseas
Artwork published: *A&A;* P. du Plessis, *Gebed vir Wilhelm* (Jhb, 1984)
Artwork represented: Hester Rupert Art Museum, Graaff-Reinet; KG, PE; PAM; Potchefstroom Museum; Potchefstroom University for CHE; RAU; SANG; UCT; UStell.; Willem Annandale Art Gallery, Lichtenburg; WH, Kimb.; WU; private collections
References: *A&A; DSAPS*

GOODE, Douglas Leslie (1944–)
Studies: art school, Pret.
Profile/career: Transvaal Museum, Pret.; artist, display dept, Durban Natural History Museum
Artwork published: *Cycads of Africa* (CT, 1989); 'Art Meets Science: Flowers as Images' (Balgowan, Natal, 1992: original Caversham Press prints first exhibited at Grahamstown Festival, 1992); M. Arnold, *Art Meets Science: Flowers as Images* – exhibition catalogue (Jhb, 1992); J. Comrie-Grieg, *The Cycad Collection* (Balgowan, Natal, 1998); 3 posters: cycads of Natal, Transvaal & Cape
Artwork represented: Durban Natural History Museum; private collections
References: Arnold; NBI files

GOODMAN, Robert ('Gwelo') (1871–1939)
Studies: part-time art tuition from J.S. Morland, CT (*c.*1890s); Académie Julian, Paris (1895–1897); special medal (Académie Julian, Paris, 1897)
Profile/career: born England; moved to SA (1886); began using name 'Gwelo' (1900); member, SA Society of Artists & honorary member, Jhb Sketch Club; participated in restoration of old buildings, Cape, Natal & Jhb; lived in CT, France, Jhb; travelled to SWA, Europe & India; painter of landscapes (especially goldmines) & general

scenes, including flowers, using oil, pastel & watercolour; winner of several prestigious art awards; numerous group & solo exhibitions, SA & overseas
Artwork published: *A&A; DSAPS;* D. Fairbridge, *Historic Houses of SA* (London, 1922); E. Roworth, *Art of the British Empire* (London, 1917)
Artwork represented: Albany Museum, Grahamstown; DAG; galleries in Birmingham, Liverpool, Manchester & Oldham, UK; JAG; Julius Gordon Africana Centre, Riversdale, Cape; KG, PE; MuseumAfrica, Jhb; National Gallery, Canada; National Museum, Bloemfontein; PAM; SA House, London; SANG; Tatham Art Gallery, Pmbg; Toronto Gallery, Canada; UCT; UNatal; WH, Kimb.; numerous private collections
References: *A&A; DSAB; DSAPS;* J. Newton-Thompson, *Gwelo Goodman: South African Artist* (CT, 1951); *SESA*

GORDON, Robert Jacob (1743–1795)
Studies: languages & natural sciences, University of Harderwijk, The Netherlands (1759–1760s); military training, Scottish Brigade, The Netherlands
Profile/career: cadet, Scottish Brigade, Netherlands forces (1758); brief visit to Cape – travelled with botanists Carl Thunberg & Francis Masson (1773); returned to Cape & travelled more extensively (1777–1786); meticulous scientific notes & observations made on travels – aided by artists William Paterson & Johannes Schumacher (& possibly others); difficult to assign authorship of artworks as few are signed; as colonel, commanded Dutch garrison at Cape (1780); married Swiss-born S.M. Nicolet, CT (1780); committed suicide, due to ill health
Artwork published: P. Raper & M. Boucher, eds, *Robert Jacob Gordon: Cape Travels, 1777 to 1786* (Jhb, 1988)
Artwork represented: Brenthurst Library, Jhb; Rijksprentenkabinet, Amsterdam
References: P. Cullinan, *Robert Jacob Gordon, 1743–1795* (CT, 1992); *DSAB*; P. Raper & M. Boucher, eds, *Robert Jacob Gordon: Cape Travels, 1777 to 1786* (Jhb, 1988); *SESA*

GOWER, Stella Irene (Mrs Louw) (1894–1991)
Studies: teaching diploma (Grahamstown Training College, 1910); art school, under Professor F.W. Armstrong (Grahamstown, 3 years)
Profile/career: taught school art & needlework; artist, Division of Botany (later BRI, Pret.) (1920–1924); worked briefly for Professor S. Schönland in Grahamstown, then for Professor J.W. Bews, Natal University College, Pmbg; after marriage worked as freelance artist
Artwork published: *Flowering Plants of Africa*; S. Schönland, *SthA Cyperaceae* (Pret., 1922); designed protea emblem on SA monetary coins (3d & 6d)
Artwork represented: Albany Museum, Grahamstown; NBI, Pret.
References: *A&A;* Arnold; *DSAPS;* NBI files

GREENING, Mary Elspeth (1921–)
Studies: architecture (WU, 1940–1945); oil painting, with Pascal Chandler & Daniele Malherbe (Durban, *c.*1980–1985)
Profile/career: member, BAASA (2000); 2 commissions for KwaZulu-Natal Nature Conservation Service, Giant's Castle Camp; group & solo exhibitions, SA
Artwork published: –
Artwork represented: unknown
Reference: –

GROSE, Jean (Mrs Robinson) (1926–)
Studies: Rochester School of Art, Kent, England; Royal College of Art, London; Associate of the Royal College of Art, London (Fine Art); British School, Rome
Profile/career: taught in art schools, London, & schools, SA; New English Art Club, London; Women's International Art Club, London; exhibited SA & overseas
Artwork published: –
Artwork represented: private collections, SA & worldwide
References: *Catalogue: Kirstenbosch Inaugural Exhibition of Botanical Art* (CT, 2000); NBI files

GURNELL, Ethel (active 1940s)
Studies: unknown
Profile/career: born in England; exhibited work as fundraising for Second World War; participated in group exhibitions, London; presented 6 paintings to Queen Elizabeth, England; watercolour artist
Artwork published: R.A.H. Flugge-de-Smidt, *Flowers by the Roadside* (Jhb, 1947)
Artwork represented: East London Museum
Reference: *East London Daily Dispatch*, 6.11.1964

GUTHRIE, Isobel Louise Sophie (1879–1966)
Studies: self-taught as artist
Profile/career: worked at Bolus Herbarium, UCT (1918–1927); moved to Caledon, Cape; set herself task of painting all species of SA Proteaceae – 264 watercolours later presented to Bolus Herbarium, UCT
Artwork published: some line drawings in scientific papers
Artwork represented: archives, UCT; SA House, London; private collections including artist's descendants
References: Rourke; *Veld & Flora*, 87:2001

HAHN, Charles Theophilus (HEADLY after 1919) (1870–1930)
Studies: BA (Pembroke College, Oxford, 1892); MA (1895)
Profile/career: Anglican clergyman; missionary in Zululand (1908–1913); painted 235 life-size watercolours of Natal flora; also landscapes
Artwork published: –
Artwork represented: private collections
Reference: *Bothalia*, 22:1992

HANDFORTH, Catherine Jane (Mrs Steiner) (1964–)
Studies: BA Fine Art (UCT); Postgraduate Diploma in Fine Art (UCT)
Profile/career: freelance at NBI, CT – pen-&-ink algae illustrations for Dr R. Norris, plant illustrations for Kirstenbosch Botanic Gardens horticultural notes, & a few colour plates for Dr K. Steiner; graphics dept, SA Museum, CT; moved to USA (1999)
Artwork published: R.E. Norris, *The Marine Red Algae of Natal, SA* (CT, 1992); K.E. Steiner, 'A new *Nemesia* from the interior of the southern Cape, SA', *SA Jl Bot.*, 60:1994; Kirstenbosch Gardens, *Horticultural Notes* (CT, 1989–1994); G. Duncan, *Growing SA Bulbous Plants* (CT, 1996)
Artwork represented: NBI, CT
Reference: –

HARRIS-DEPPE, Ann Judith (1947–)
Studies: Freie Kunst Schule, Stuttgart; self-taught as porcelain painter; botanical art master – classes of Jenny Phillips (CT & Jhb)
Profile/career: Swiss-born; moved to SA (1967); started porcelain painting, often depicting botanical themes (1981); winner of prestigious art awards (1998, 1999); exhibited widely, SA & overseas
Artwork published: –
Artwork represented: several private collections
Reference: NBI files

HASELGROVE, Anne Vipya (1933–)
Studies: no formal art training
Profile/career: worked as architectural draughtswoman; began painting (in watercolour) images of flowers (*c.*1999); participant in group exhibitions
Artwork published: A. Lambert, *The Palm Collection of the Durban Botanic Gardens* (Durban, 1994)
Artwork represented: Three Tree Hill Lodge, Spioenkop, KwaZulu-Natal; KwaZulu-Natal Nature Conservation Service, Giant's Castle Camp
Reference: –

HAYGARTH, Walter (1862–1950)
Studies: unknown
Profile/career: son of Williamson Haygarth whose sister married John Medley Wood – Walter later adopted by Medley Wood
Artwork published: J. Medley Wood, *Natal Plants*, 6 vols (Durban, 1899–1912)
Artwork represented: unknown
Reference: –

HAZELL, Beatrice (1864–1946)
Studies: various art schools, England (1833–1885); CT School of Art (1885–1893)
Profile/career: left England for SA (1885); taught drawing, CT; member, SA Society of Artists; travelled extensively, SA & Europe; working mainly in oil & watercolour, painted still-life & landscape, though early works included floral & fruit studies; exhibited in SA & overseas, group & solo
Artwork published: *A&A*
Artwork represented: MuseumAfrica, Jhb
References: *A&A; DSAPS*

HEMING, Patricia Joan (1930–)
Studies: taught by George Boys (1978/1979, 1982/1983)
Profile/career: freelance artist; commissions in SA include Steenberg Estate (Manor House), CT, & Cellars Hohenhort Hotel, CT; also commissioned overseas; group & solo exhibitions
Artwork published: –
Artwork represented: Steenberg Estate (Manor House); Cellars Hohenhort Hotel, CT; several private collections, SA & worldwide
Reference: –

HENNESSY, Esmé Frances (née Franklin) (1933–)
Studies: Durban Girls' College; BSc, botany, zoology (UNP, 1956); BSc Hons, botany (UNP, 1958); MSc, botany (UNP, 1956); PhD; self-taught as botanical artist
Profile/career: schoolteacher, Durban Girls' College (1956); research assistant, Amoebiasis Unit (CSIR) (1958–1960); lecturer, University College, Durban (1961–1972); associate professor, University of Durban-Westville (1972–); exhibited SA & overseas, group & solo
Artwork published: *SA Erythrinas* (Durban, 1972); E.F. Hennessy & J. Stewart, *Orchids of Africa* (Jhb, 1981); *The Slipper Orchids* (Jhb, 1989); *Flowering Plants of Africa*
Artwork represented: NBI, Pret.; many private collections
References: *A&A; Catalogue, 1977*, Hunt Inst., USA; *DSAPS; SA Garden & Home*, April 1989; 'Varsity Voice', University of Durban-Westville, March 1988

HERSCHEL, John William Frederick (1792–1871)
Studies: MA (Cantab); honorary DCL (Oxon)
Profile/career: self-taught astronomer; baronetcy conferred (1838); stayed at Cape (1834–1838) with family & wife, **Margaret Brodie Herschel (1810–1884)**; set up Cape's first major telescope, 'Feldhausen' estate, CT, making important astronomical observations; used camera lucida to provide accurate outlines of landscapes & botanical studies; produced few colour plates, most botanical studies being outlines provided by him & then painted in by his wife Margaret
Artwork published: B. Warner & J. Rourke, *Flora Herscheliana* (Jhb, 1998)
Artwork represented: SA Library, CT; Brenthurst Library, Jhb; private collections, including John Herschel, Norfolk, England; Robert Levitt, England
References: *DSAB*; B. Warner & J. Rourke, *Flora Herscheliana* (Jhb, 1998)

HERZOG, Heidi (1905–1967)
Studies: several periods of tuition in Europe, notably under Marc Chagall in France
Profile/career: born Switzerland; moved to SA (1951); artworks in oil & pastel include flower studies; group & solo exhibitions, SA & overseas, including retrospectives
Artwork published: unknown
Artwork represented: Kunstmuseum, Basle, Switzerland; RAU; WH, Kimb.; private collections, SA & worldwide
References: *Artlook*, Apr. 1968; *DSAPS*

HEURNIUS, Justus (1587– c.1653)
Studies: MD (Leiden, 1611); studied divinity (Groningen, The Netherlands)
Profile/career: urged VOC to promote Christianity in East Indies; worked as missionary in Batavia (1624–1638); botanized at Cape (1624) en route to Batavia – made simple sketches of 10 Cape Peninsula plants
Artwork published: J.B. van Stapel, *Theophrasti Eresii; De Historia Plantarum, Libri Decem* (Amsterdam, 1644)
Artwork represented: –
References: *DSAB*; Gunn & Codd

HOBBS, Philippa Anne (1955–)
Studies: Diploma in Fine Art (Jhb College of Art, 1975); Philadelphia College of Art, USA (1977–1978); Higher Diploma in Printmaking (Witwatersrand Technikon, 1986)
Profile/career: lecturer, Witwatersrand Technikon, Jhb (1979–1984); head, printmaking dept, Witwatersrand Technikon, Jhb (1984); visited Italy (1973 & 1987); winner of several prestigious art awards; printmaker, working in woodcuts, silkscreens & etchings; subject matter sometimes botanical; exhibited widely, SA & overseas, group & solo
Artwork published: G.K. O'Neill, *Philippa Hobbs* (Jhb, [1993]); P. Hobbs & E. Rankin, *Printmaking in a Transforming SA* (CT, 1997)
Artwork represented: DAG; Durban Technikon; Gencor Collection, Jhb; JAG; MuseumAfrica, Jhb; PAM; RAU University; Reserve Bank, Pret.; Roodepoort Town Collection; SABS Collection; Sandton Town Collection; SANG; Sasol Collection, Jhb; Standard Bank Collection, Jhb; Tatham Art Gallery, Pmbg; UFS; UNISA; WH, Kimb., private collections
References: Arnold; *DSAPS*; G.K. O'Neill, *Philippa Hobbs* (Jhb, [1993]); *Women & Art*

HOLCROFT, Rosemary (née Roos) (1942–2000)
Studies: Dipl. in Dance Teaching (UCT Ballet School, 1963); studied fine art (UNISA, 1974–1978)
Profile/career: taught ballet, Pret. & CT (7 years); resident artist, NBI, Pret. (1975–1985); freelance artist – worked for Pret. Engineers' Dept, Pret. Post Office, & *Optima*; hostel supervisor for SA Council for the Blind (1988–1994)
Artwork published: many NBI in-house publications, including *Flowering Plants of Africa*
Artwork represented: NBI, Pret.; several private collections
Reference: NBI files

HULME, Jillian Frances Mary (Mrs Gilliland) (1943–)
Studies: Pmbg Girls' High School; BA Fine Art & English Lit. (UNatal); postgraduate UED
Profile/career: art teacher, various Natal high schools; botanical artist, UNatal (1968–1970); botanical artist, NBI, Pret.; emigrated to Canada
Artwork published: Phyllis Savory, *Fireside Tales from the North* (CT, 1966); *Flowering Plants of Africa*; J.H. Ross, *A Conspectus of the African* Acacia *species* (Pret., 1979)
Artwork represented: NBI, Pret.
Reference: NBI files

HULME, Mairn M. (active 1940s–1950s)
Studies: horticulture (Reading University, England); apparently self-taught as artist
Profile/career: painted 346 species of Natal flora, concentrating on midland & south coast areas
Artwork published: *Wild Flowers of Natal* (Pmbg, 1954)
Artwork represented: private collections
References: *A&A; DSAPS*

JEPPE, Barbara Joan (née Brereton) (1921–1999)
Studies: Parktown Girls' High School, Jhb; art (Witwatersrand Technical College, Jhb); private art tuition; self-taught as botanical artist
Profile/career: worked initially as draughtswoman, then as professional freelance botanical artist; married Dr Carl Jeppe; nominee, *The Star* Woman of the Year (1977); winner of several prestigious botanical art awards; *Strumaria barbarae* named after her (1981); several exhibitions, group & solo
Artwork published: *A&A; Trees & Shrubs of the Witwatersrand* (Jhb, 1964); *SA Aloes* (CT, 1969); *Aloes* (CT, 1974); C. Giddy, *Cycads of SA* (CT, 1974); K. Grabandt, *Effective Weed Control in Maize & Grain Sorghum* (Jhb, 1975); *Natal Wild Flowers* (CT, 1975); L. Davidson, *Acacias* (Jhb, 1981); *SA is My Garden* (Jhb, 1984); K. Grabandt, comp., *Weeds of Crops & Gardens in SthA* (Jhb, 1985); *Spring & Winter Flowering Bulbs of the Cape* (CT, 1989); *Irises* (Pret., 1999)
Artwork represented: numerous private collections, SA & worldwide
References: *A&A; Bothalia*, 30:2000; *DSAPS; SESA; SA Panorama*, Sept. 1979; *Sunday Times*, 18.7.1990

JOHNSTONE, Patricia (1933–)
Studies: self-taught as artist
Profile/career: participated in group botanical art exhibitions, Durban (1999, 2000)
Artwork published: calendar, Durban Botanic Gardens (2000)
Artwork represented: KwaZulu-Natal Nature Conservation Service, Giant's Castle Camp (2001)
Reference: –

JOUBERT, Elizabeth Beatrice ('Elbe') (1951–)
Studies: Pret. Art Centre; botanical art master-classes of Katie Lee & Jenny Phillips (CT)
Profile/career: art teacher, Rustenburg High School; pottery & fine art lecturer, Technical College, Rustenburg; established own art school, Standerton; winner of several prestigious art awards (2000); group exhibitions in SA
Artwork published: book illustrations for faculty of natural science, UStell.; Western Cape flora & marine plants for UWC; environmental calendars (1998, 1999)
Artwork represented: conference centre, Spier Wine Estate, Cape; natural science dept, UStell.; UWC; several private collections
Reference: NBI files

JUDD, Eric (1935–)
Studies: Springs Boys' High School; self-taught as artist
Profile/career: administrative/sales position, SA Pulp & Paper Industries (SAPPI); freelanced as professional artist (1973–); exhibited widely in SA, group & solo
Artwork published: *What Aloe is That?* (CT, 1967); *Veld & Flora*, 75:1989; *Aloe* magazine; *Cactus & Succulent Journal* (USA); various calendars
Artwork represented: Vanderbijlpark Town Council (1970); Hida Aloe Institute of Japan; private collections
Reference: *A&A; DSAPS*; NBI files

KADAK, Tiiu (Mrs Kaul-Andries) (active 1960s)
Studies: BA Fine Art (Tartu University, Estonia, 1935)
Profile/career: born St Petersburg, Russia; worked in variety of positions: journalist & editor; schoolteacher; inspector of homecrafts, Chamber of Domestic Sciences, Tallinn, Estonia; director, Arts & Crafts Museum, Tallinn, Estonia; author of several books on Estonian folk art; arrived in SA (1948); began painting SA flora (1961); exhibited widely, group & solo, SA & overseas
Artwork published: *A Bouquet of Wild Flowers* (Jhb, 1964)
Artwork represented: BRI, Pret.; Hunt Inst., USA; MuseumAfrica, Jhb; Queenstown Art Gallery; private collections
References: *A&A; DSAPS*

KIMPTON, Jill (Mrs Baker) (1950–)
Studies: Jhb College of Art (Diploma in Graphic Art, 1971)
Profile/career: botanical artist, NBI, Pret. (1972–1974); lived in Somerset East, E. Cape (few years); layout artist, agricultural technical services dept, Pret. (1977–1979); freelance artist (1988– 1990) – worked for NBI, Pret; laboratory technician, CSIR (1990)
Artwork published: *Flowering Plants of Africa; Flora of SA;* designed botany medal for SAAB
Artwork represented: NBI, Pret.
References: Arnold; NBI files

KOORLAND, Vivienne Asya (1957–)
Studies: BA Fine Art (UCT, 1978); Meisterschulerin (Hochschule der Kunste, Berlin, 1980); Ecole Nationale Superieure des Beaux-Arts, Paris (1982); Skowhegan School of Painting & Sculpture, Maine, USA (1983); MA Fine Art (Colombia University, New York, 1984)
Profile/career: left CT to study & live overseas (1980s); exhibited widely, group & solo, SA & overseas; winner of several prestigious art awards, SA & USA
Artwork published: E. Lipton, 'Vivienne Koorland', *Arts Magazine*, March 1989; K. Levin, 'Voice choices', *Village Voice*, 29 Sept 1992; M. de Bord, 'Apparent signatures', *NKA Journal for Contemporary African Art*, Spring/ Summer 1995; E. Bedford, ed., *Contemporary SA Art 1985–1995 from the SANG Permanent Collection* (CT, 1997)
Artwork represented: Jewish Museum, New York; Mint Museum of Art, Charlotte, North Carolina, USA; SANG; UCT; private collections, SA & overseas
Reference: E. Bedford, ed., *Contemporary SA Art 1885–1995 from the SANG Permanent Collection* (CT, 1997)

KRÄMER, Evelyn (active 1940s)
Studies: unknown
Profile/career: painted 46 colour plates of *Lithops*, reproduced in G.C. Nel, *Lithops* (Stellenbosch, [1946])
Artwork published: G.C. Nel, *Lithops* (Stellenbosch, [1946])
Artwork represented: not known; possibly held by descendants
Reference: –

KRIGE, François (1913–1995)
Studies: Michaelis School of Fine Art, UCT; Koninklijke Academie voor Schone Kunsten en de Hogere Instituten, Antwerp, Belgium; Spain (1934–1937)
Profile/career: member, SAAA; official SA war artist, Egypt & Italy (1941–1945); medal of honour, SA Akademie vir Wetenskap & Kuns (1949); painter of variety of subjects, including flowers, working in oil, watercolour, ink, pencil & charcoal; exhibited widely, group & solo, SA & overseas
Artwork published: *A&A;* Eitemal, *Jaffie* (CT, 1953); Uys Krige, *Sol y Sombra* (Pret., 1956); V. Pohl, *Their Secret Ways* (CT, 1960); W.A. de Klerk, *Witwyne van Suid-Afrika* (CT, 1967)
Artwork represented: Hester Rupert Art Museum, Graaff-Reinet; JAG; Julius Gordon Africana Centre, Riversdale, Cape; KG, PE; National Museum, Bloemfontein; Potchefstroom Museum; PAM; SA Akademie vir Wetenskap & Kuns; SANG; UFS; UPret.; UStell.; WH, Kimb.; private collections
References: *A&A; DSAPS;* J. Fox, *The Life & Art of François Krige* (CT, 2000); M.J. Strydom, *François Krige* (CT, 1993)

LANSDELL, Kathleen Annie (1888–1967)
Studies: Durban Girls' High School; Government Art School (Natal Technical College), Durban
Profile/career: junior herbarium assistant (under J. Medley Wood), Botanic Station, Durban (1914); lay assistant, Division of Botany (later BRI, Pret.) (1917); pathological & horticultural section, Division of Botany (later BRI, Pret.) (1927); artist & modeller, Deciduous Fruit Exchange, CT (1934); seconded to defence dept, Imperial Institute, London (1940–1941); group & solo exhibitions
Artwork published: J. Medley Wood, *Natal Plants,* 6 vols (Durban, 1899–1912) – 100 pencil drawings for vol. 7; *Flowering Plants of Africa*
Artwork represented: Killie Campbell Library, UND; Natal Herbarium, Pmbg; NBI, Pret.
References: *A&A; DSAPS;* NBI files; *PlantLife,* 14:1996; *The Transvaal Gardener,* 24:1976

LAUBSER, Maria Magdalena ('Maggie') (1886–1973)
Studies: tutored by Edward Roworth, CT (1903); further art studies, Laren, The Netherlands (1913–1914); Slade School of Art, London (1914–1919); night classes, Berlin (1922)
Profile/career: teacher, Transvaal (1912); travelled & studied, Europe (1913–1921); lived in SA (1921–1922); returned to Europe (1922–1924); lived in Transvaal & Cape (1924–1973); member, SA Society of Artists; member, New Group (1938); painted in oil & gouache; known for brightly coloured landscapes, portraits, flower studies & still-life; influenced by German expressionism (1922); winner of several prestigious art awards; exhibited widely, group & solo, SA & overseas
Artwork published: *A&A;* J. van Rooyen, *Maggie Laubser* (CT, 1974)
Artwork represented: DAG; Hester Rupert Art Museum, Graaff-Reinet; JAG; Julius Gordon Africana Centre, Riversdale, Cape: KG, PE; National Museum, Bloemfontein; PAM; Potchefstroom Museum; RAU; SA Akademie vir Wetenskap & Kuns; Sandton Town Collection; SANG; UFS; UNISA; UPret.; UStell.; WU; WH, Kimb.; many private collections, SA & worldwide
References: *A&A;* E. Delmont, *Maggie Laubser: Early Works from the Silberberg Collection* (CT, 1987); *DSAB; DSAPS;* J. Meintjes, *Maggie Laubser* (CT, 1944); J. van Rooyen, *Maggie Laubser* (CT, 1974)

LAURENCE, Gertrude Elizabeth (1916–1990)
Studies: La Rochelle Girls' School, Paarl; correspondence course in art
Profile/career: anatomical artist, Onderstepoort Veterinary Institute, Pret. (1939–1942); ambulance driver, Europe (Second World War); Division of Botany (later BRI, Pret.) (after 1946); in latter position illustrated grasses for D. Meredith, ed., *The Grasses & Pastures of SA* (Jhb, 1955)
Artwork published: D. Meredith, ed., *The Grasses & Pastures of SA* (Jhb, 1955) – 12 of 26 colour plates & 340 of 420 b/w plates; *Flowering Plants of Africa*
Artwork represented: Arnold Theiler Museum of African Diseases, Onderstepoort Veterinary Institute, Pret. (wax models of poisonous plants); NBI, Pret.
Reference: *Veld & Flora,* 87:2001

LAUTH, Frieda (Mrs Floyd) (1879–1949)
Studies: unknown
Profile/career: born in England; family moved from Cape to Natal (1882); assistant at Natal Herbarium, Pmbg – responsible for majority of illustrations for first three volumes of J. Medley Wood's *Natal Plants* (until 1903); married (1903); later lectured in botany for some years at Durban Technical College
Artwork published: J. Medley Wood, *Natal Plants,* 6 vols (Durban, 1899–1912)
Artwork represented: –
Reference: Gunn & Codd

LE VAILLANT, François (1753–1824)
Studies: French ornithologist, born & educated in Surinam
Profile/career: travelled extensively in SthA (1781–1784); amassed collection of 165 topographical, zoological, ethnographic & botanical watercolours (47 botanical); some artworks executed by Le Vaillant himself, others are equivalent copies of plates from another collection (that of R.J. Gordon)
Artwork published: J.C. Quinton & A.M. Lewin Robinson, eds, *François le Vaillant: Traveller in SA,* 2 vols (CT, 1973); F. le Vaillant, *Voyage . . . dans l'Intérieur de l'Afrique,* 2 vols (Liège, 1790)
Artwork represented: Library of Parliament, CT
References: *DSAB; DSAPS;* V.S. Forbes, *Pioneer Travellers in SA* (CT, 1965); *SESA*

LEACH, Leslie Charles (Larry) (1909–1996)
Studies: British Army Technical School (electrical technology); self-taught as pen-&-ink artist

Profile/career: army technician, electrical engineering; army post in Southern Rhodesia (now Zim.) (1938); established own business, Salisbury (1944); developed interest in succulents (*c.*1950); later concentrated on systematic collection of *Stapelia, Euphorbia* & *Aloe* species (1981); on staff of NBI, Pret. (1982–1989); on staff of National Botanical Gardens, Worcester (1989–1996); honorary research fellow, botany dept, University of the North; numerous awards including Golden Cactus award of the International Organisation for Succulent Plant Study in Zurich (1990)
Artwork published: illustrates own publications – over 100 published in various international journals
Artwork represented: private collections
Reference: *Bothalia*, 27:1997

LETTY, Cythna Lindenberg (Mrs Forssman) (1895–1985)
Studies: finished schooling at Pretoria Girls' High; taught art of drawing by mother, Josina; honorary LLD (WU, 1975)
Profile/career: in early years worked in variety of capacities: governess, nurse & railways stores clerk; technical draughtswoman, Onderstepoort Veterinary Research, Pret. (1925–1927); botanical artist, Division of Botany (later BRI, Pret.) (1927–1968); with a few breaks, career of botanical artist at National Herbarium spanned four decades; several group & solo exhibitions, SA & overseas
Artwork published: *A&A; Flowering Plants of Africa;* D. Meredith, ed., *The Grasses & Pastures of SA* (Jhb, 1955); R.A. Dyer, I.C. Verdoorn & L.E. Codd, *Wild Flowers of the Transvaal* ([Pret.], 1962); *Trees of SA* (CT, 1975); *Children of the Hours* (Jhb, 1981); floral designs on SA monetary coins (10c, 20c, 50c)
Artwork represented: NBI, Pret., Onderstepoort Veterinary Research, Pret.; numerous private collections
References: *A&A; Bothalia,* 16:1986; *DSAPS;* Gunn & Codd; *Lantern,* Sept. 1968; *SESA*

LETTY, Josina (née Lindenberg) (1861–1938)
Studies: interest & ability in languages; self-taught as artist
Profile/career: while on travels through Transvaal, completed 2 volumes flower paintings: 1 at Standerton (1897), 1 near Heidelberg (1914); earned income painting illustrated genealogies or memorials (often bartered for supplies)
Artwork published: portfolio of 6 plates
Artwork represented: NBI, Pret.
References: *A&A; DSAPS; SA Garden & Home,* Sept. 1996

LEWIS, Gwendoline Joyce (1909–1967)
Studies: BA (UCT); PhD (UCT); FRSSAf
Profile/career: botanist in charge, Herbarium, SA Museum, CT (1938–1956); senior research officer, Compton Herbarium, National Botanic Gardens, Kirstenbosch (1956–1967); illustrated own scientific papers with own line drawings & colour plates
Artwork published: *The Genus* Babiana (CT, 1959); G.J. Lewis & A.A. Obermeyer, *Gladiolus: A Revision of the SA Species* (CT, 1972); *Flowering Plants of Africa.*
Artwork represented: NBI, Pret.
References: Gunn & Codd; Rourke

LINCOLN, Thalia Nina (1924–)
Studies: fine arts (UCT, 1 year); commercial art (UCT, 6 months)
Profile/career: commercial artist, CT (1944–1949); began botanical drawing (1961); developed unique technique of layered colour-pencil illustration; produced *Mimetes* illustrations (1975–1983); exhibited widely, group & solo, SA & overseas; nominee, *The Star* Woman of the Year (1982) for *Mimetes* book
Artwork published: J.P. Rourke, *Mimetes* (CT, 1982); variety of calendars & portfolios (1984–1995); Endangered Wildlife Trust, *Vision of Wildlife* (Jhb, 1994, 1996); *A&A; Flowers of SthA* (Jhb, 1995); M. Arnold, *Women & Art in SA* (CT, 1996); Sherwood
Artwork represented: First National Bank Collection, Jhb; Standard Bank Collection, Jhb; private collections SA & worldwide, including Shirley Sherwood, England
References: *A&A; DSAPS;* Rourke; Sherwood; *Women & Art*

LINDER SMITH, Claire (1954–)
Studies: BA (UNP); botanical art master-classes of Katie Lee & Jenny Phillips (CT)
Profile/career: artist, NBI, Pret. (1980–1981); Kew (1982–1985); exhibited widely, SA & overseas; winner of several prestigious art medals (1985, 1999); founding member, BAASA
Artwork published: H.P. Linder & H. Kurzweil, *Orchids of SthA* (Amsterdam, 1999); *Curtis's Botanical Magazine; Flowering Plants of Africa; Veld & Flora*
Artwork represented: The Blue Train; Brenthurst Library, Jhb; Hunt Inst., USA; private collections, SA & worldwide
References: NBI files; Rourke

LINSMAYER, Eva-Maria (1924–)
Studies: 3 months' school art tuition – terminated by Second World War (Munich, Germany)
Profile/career: variety of occupations during & shortly after Second World War – postal apprentice, farm labourer, library assistant; married (1949); returned to SWA (now Namibia); started botanical drawing (grasses); moved to Omaruru; freelance artist
Artwork published: designed 2 sets stamps for SWA (1984, 1990) for Philatelic Services of the SA Post Office, Pret.; 1 aerogramme designed (1995)
Artwork represented: State of Namibia – about 40 paintings, formerly displayed in State House, Windhoek, & official summerhouse, Swakopmund
Reference: NBI files

LOEDOLFF, Jeanette (1937–)
Studies: BSc (UPret.); self-taught as artist
Profile/career: scientific illustrator, NBI (1982–); self-initiated project – posters featuring plants & animals (Environmental Education, 1989–2001)
Artwork published: P.L. Perry, 'A revision of the genus *Eriospermum* (Eriospermaceae)', *Contributions to the Bolus Herbarium,* 17:1994; G. Duncan, *Growing SA Bulbous Plants* (CT, 1996); P.L. Perry, '*Bulbinella* in SA', *Strelitzia,* 8:1999; 11 full-colour posters for Environmental Education (1989–2001)
Artwork represented: NBI, CT; private collections
Reference: –

LÜCKHOFF, Carl August (1914–1960)
Studies: MB ChB (UCT); MD (UCT); self-taught as botanical artist
Profile/career: medical doctor & keen amateur botanist specializing in asclepiads; trustee & one-time chairman of NBG of SA; author of books on Table Mountain & CT's Malay quarter; several solo exhibitions, Schweikerdt Gallery, Pret.
Artwork published: *The Stapelieae of SthA* (CT, 1952)
Artwork represented: unknown
References: *A&A;* H.M. Bolus & J. Lückhoff, *Jl Bot. Soc. SA,* 1960; *DSAB; DSAPS*

MACLEAN, Anne (1969–)
Studies: BSc, botany, ethno/economic botany (UNatal)
Profile/career: scientific illustrator, mainly SA flora & fauna (since 1990)
Artwork published: D.A. Barraclough, 'The SthA species of Neminidae (Diptera: Schizophora)', *Annals of the Natal Museum,* 40:1999
Artwork represented: KwaZulu-Natal Nature Conservation Service, Giant's Castle Camp; several private collections
Reference: NBI files

MACLEAR, Augusta Emily (1840–1929)
Studies: unknown; probably educated at home & possibly self-taught as artist
Profile/career: daughter of Cape astronomer Sir Thomas Maclear & Mary Pearse; sister of Mary Susanna Maclear, also a painter of flowers; produced collection of about 200 watercolours of Cape wild flowers; also painted landscapes; usually signed work 'AEM'
Artwork published: –
Artwork represented: in private collection
Reference: Maclear family records

MACLEAR, Mary Susanna (1826–1907)
Studies: unknown; probably educated at home & possibly self-taught as artist
Profile/career: daughter of Cape astronomer Sir Thomas Maclear & Mary Pearse; sister of Augusta Emily Maclear, also a painter of flowers; originally lived at Royal Observatory, CT; as a child, influenced in technique of flower painting by Margaret Herschel; painting wild flowers in minute detail a hobby; became friend of artist Marianne North when she visited CT (1882); in later life lived at 'Wayland', Kenilworth, CT, with sister Augusta Emily (from 1896)
Artwork published: –
Artwork represented: in private collection
Reference: Maclear family records

MANNING, John Charles (1962–)
Studies: BSc Hons (UNP, 1983); PhD, botany (UNP, 1987); self-taught as artist
Profile/career: postdoctorate research, NBI, CT (1988); thereafter taxonomist,

Compton Herbarium, NBI, CT; authority on African Iridaceae & Hyacinthaceae & pollination biology; specializes in pen-&-ink artwork; also watercolours of indigenous butterflies
Artwork published: extensively in scientific journals, often self-illustrated; also several popular books, articles & field guides; *Flowering Plants of Africa*; 'Art Meets Science: Flowers as Images' (Balgowan, Natal, 1992: original Caversham Press prints first exhibited at Grahamstown Festival, 1992); P. Goldblatt & J. Manning, *Gladiolus in SthA* (CT, 1999)
Artwork represented: NBI, CT
References: Arnold; NBI files; Rourke

MARAIS, Christine Anne (1935–)
Studies: BA Fine Arts (WU, 1954–1957)
Profile/career: senior art teacher, Jeppe Girls' High School, Jhb (1958); married (1958); private art teaching, adults & children (1965–1989); moved to SWA (later Namibia) (1970); freelance artist (1986–); Herbert Evans prize for painting (1957); group & solo exhibitions, SA & Namibia
Artwork published: P. Craven, *Namib Flora* ([Windhoek], 1986); P. Craven, *Waterberg Flora* ([Windhoek], 1989); P. Craven, *Damaraland Flora* ([Windhoek], 1992)
Artwork represented: several corporate collections in Namibia, including Commercial Bank of Namibia, Land Bank in Namibia, Land Bank in Windhoek; official residence of state president, Namibian embassy, London
References: *DSAPS;* NBI files

MARCH, Herrat Lydia Erika (1935–)
Studies: private tuition to matric, Angola; certificate in photography (Bavarian State College of Photography); postgraduate degree in design (Hochschule für Kunst & Design, Köln, 1957–1961)
Profile/career: senior botanical artist, NBI, Pret. (1970–1974); exhibited group & solo, SA & overseas
Artwork published: E.A. Schelpe & N.C. Anthony, *Flora of SthA: Pteridophyta* (Pret., 1986); *Flowering Plants of Africa*; *Bothalia*
Artwork represented: Hunt Inst., USA; private collections
References: *A&A*; Arnold; *DSAPS*; NBI files

MARTIN, Lois (Mrs Povall) (1905–)
Studies: Penhrôs College, Colwyn Bay, Wales; Liverpool Art College; textile design (Regent Street Polytechnic Art School, London); self-taught as botanical artist
Profile/career: freelance botanical artist (1938–1943); instructor of Royal Drawing Society's syllabus, private schools, Devon, England; moved to SA (1951); group & solo exhibitions
Artwork published: E. Rosenthal, comp., *Encyclopaedia of SthA* (London, 1961); *Flowering Plants of Africa; Veld & Flora*, old series, 2:1972
Artwork represented: Hunt Inst., USA; Killie Campbell Library, UNP; private collections
References: *A&A*; *DSAPS*; Hunt Inst., USA; NBI files

MARX, Gerhard (1956–)
Studies: BA Fine Arts (UPret., 1976–1979)
Profile/career: freelance artist – mainly children's books (1980–1983); military training (1983–1984); exhibition artist, Albany Museum, Grahamstown (1985); emigrated to USA (1999)
Artwork published: *Art Portfolio of Four Rare SA Aloes* (Pret., 1983–1984); numerous cover designs, *Cactus & Succulent Society Journal* (USA); designed set of aloe stamps for Venda: Philatelic Services of the SA Post Office, Pret. (1990); 'Art Meets Science: Flowers as Images' (Balgowan, Natal, 1992: original Caversham Press prints first exhibited at Grahamstown Festival, 1992)
Artwork represented: several private collections
References: Arnold; NBI files

MASON, Hilda (Mrs Eric Axelson) (1909–1986)
Studies: associate in design – ARCA (Royal College of Arts, London, 1936)
Profile/career: born in Chile; educated in England; moved to SA (1936); lecturer, Port Elizabeth Technical College (1937–1938); lecturer, Witwatersrand Technical College, Jhb (1938–1942); lecturer, Pretoria Technical College (1947); settled in CT (1962); began painting local flora; exhibited in SA, group & solo
Artwork published: *Flowering Plants of Africa*; *Western Cape Sandveld Flowers* (CT, 1972); J.P. Rourke, *The Proteas of SthA* (CT, 1980; 2 ed., 1982); designed SA monetary coins – 5 shilling (1959) & 1cent (1960); also designed several medals
Artwork represented: NBI, CT; SANG; Axelson family & other private collections
References: *A&A; DSAPS;* Rourke; *Veld & Flora*, 72:1986

MASON, Marianne Harriet (1845–1932)
Studies: unknown; probably self-taught as artist
Profile/career: employed in English Civil Service; in SA (1910); travelled extensively through Eastern Cape, making watercolour drawings; 3 plants named after her
Artwork published: 'Some flowers of Eastern & Central Africa', *Jl of the RHS*, Aug. 1913; *Kew Bulletin*, 1925, 1932
Artwork represented: Kew – 110 large paintings & 5 albums plant studies
Reference: NBI files

MASSON, Francis (1741–1805)
Studies: trained as gardener (Kew)
Profile/career: visited Cape to collect plant material for cultivation (Kew, 1772–1775, 1786–1795); plants illustrated in *Stapeliae Novae* (London, 1796–1797) apparently cultivated in own garden, CT
Artwork published: *Stapeliae Novae* (London, 1796–1797) – 41 of 42 illustrations by Masson
Artwork represented: British Museum (Natural History), London – plates in a collection attributed to Masson, but clearly by several different hands (not clear which are by Masson)
References: *DSAB*; Gunn & Codd; *Jl SA Bot.*, 27:1961

MAYER, Ernst Karl Erich (1876–1960)
Studies: architecture, Charlottenburg Technische Hochschule, Berlin (1894–1896); drawing & painting, Karlsruhe (1907), Stuttgart (1909–1911), Florence (1935–1936)
Profile/career: left Germany to live in SA (1898); prisoner-of-war, St Helena (1900–1902); assistant land-surveyor, Vrede, OFS; repatriated to Germany (1902); lived in SWA (now Namibia) (1903–1906); lived in Germany (1907–1911); returned to SA (1911); war internment, Pmbg (1914–1916); thereafter worked as architect, mainly Pretoria, travelling & painting extensively in SA & SWA, in oil & watercolour; medal of honour, SA Akademie vir Wetenskap & Kuns (1943); exhibited widely, group & solo, SA & Zim., including retrospective
Artwork published: *A&A;* L. Maré, *Die Nuwejaarsfees op Palmietfontein* (Pret., 1898); A. Schiel, *23 Jahre Sturm und Sonnenschein in Südafrika* (Leipzig, 1903); commemorative issue of SA postage stamps featuring work of artist (1976)
Artwork represented: Ann Bryant Gallery, East London; DAG; JAG; Julius Gordon Africana Centre, Riversdale, Cape; KG, PE; MuseumAfrica, Jhb; National Museum, Bloemfontein; PAM; Potchefstroom Museum; SA Akademie vir Wetenskap & Kuns; SA Cultural History Museum, CT; UCT; UFS; UNISA; UPret.; UStell.; WH, Kimb.; Windhoek Museum; Deutsche Afrika Bank, Windhoek; many private collections, SA & worldwide
References: *A&A; DSAB; DSAPS*; M. Schoonraad, *Erich Mayer* (Pret., 1990); *SESA;* H.M. van der Westhuysen, *Erich Mayer Album* (CT, 1953)

MAYTHAM KIDD, Mary Catherine (1914–2001)
Studies: Roedean School, Jhb; Université de Paris (Sorbonne); Royal Academy School, London
Profile/career: painted native flora of Cape Peninsula & fungi of Western Cape
Artwork published: *Flowering Plants of Africa; Wild Flowers of the Cape Peninsula* (CT, 1950); E.L. Stephens, *Some SA Edible Fungi* (CT, 1953); E.L. Stephens, *Some SA Poisonous & Inedible Fungi* (CT, 1953); *Protected Wild Flowers of the Cape Province*, 3 vols (CT, 1951–1956); *Cape Peninsula: SA Wild Flower Guide 3*, 3 ed. (CT, 1983)
Artwork represented: Bolus Herbarium, UCT; NBI, CT
References: *A&A; DSAPS;* Rourke; *Veld & Flora*, 69:1983

McMAHON, Liz (1959–)
Studies: St Aiden's High School, Wishaw, Scotland; BA Hons (Duncan of Jordanstone College of Art, Dundee, Scotland, 1983)
Profile/career: moved to SA (1984); commercial artist, CT (1984–1986); fieldwork & research for book on fynbos (1987); design & marketing of 'The Fynbos Collection' (1989–1994); fieldwork & research for book on Cape peninsula (1994); winner of Marloth Medal of BSSA
Artwork published: M. Fraser, *A Fynbos Year* (CT, 1988); M. Fraser, *Between Two Shores* (CT, 1994); illustrated numerous publications by husband M. Fraser
Artwork represented: several private collections
Reference: NBI files

MCMANUS, Peter (active 1920–1930)
Studies: unknown
Profile/career: unknown
Artwork published: R. Marloth, *The Flora of SA*, vol. 1 (CT & London, 1913)
Artwork represented: UStell.
Reference: Rourke

MILLER, Ralph Rillman (1915–)
Studies: sculpture, under Alexander Archipenko, University of Washington (1935); part-time, Art Students League, New York (1964); drawing, Florence (1971–1972)
Profile/career: born USA; lived in New York (1947–1970); director, Museum of the City of New York (1960–1970); lived in Florence (1970–1973); moved to SA (1974); part-time art lecturer, UCT (1973–1974); member, SA Society of Artists; exhibited widely, group & solo, SA & overseas; paints in oil: variety of subjects including flowers
Artwork published: dustjacket portrait: David Bloomberg, *Meet the People* (CT, 1975)
Artwork represented: Pietersburg Collection; UCT; private collections, SA & worldwide
Reference: *DSAPS*

MORIARTY, Audrey (1918–)
Studies: Edinburgh College of Domestic Science; self-taught as artist
Profile/career: specialist messing officer, British Army (Second World War); moved to SA (1946); moved to Rhodesia (now Zim.) (1958); lived in Malawi, painting wild flowers – designed 2 sets Malawi postage stamps; later moved to George, Cape; painted southern Cape flora
Artwork published: *Wild Flowers of Malawi* (CT, 1975); *Outeniqua, Tsitsikamma & Eastern Little Karoo* (CT, 1982; 2 ed., 1997)
Artwork represented: artist's personal collection
Reference: –

MURRAY, Eliza (active 1830s)
Studies: unknown
Profile/career: accompanied husband to Cape: wife of Dr John Murray, surgeon to British forces at Cape & physician to household of Cape governors (1820s–1830s); friends of Sir John & Lady Herschel while at Cape (1834–1838); painted over 100 varieties of *Erica*
Artwork published: unknown
Artwork represented: fate of Eliza Murray's collection unknown
References: Q. *Bull. SA Lib.*, 23(3)1969

NICHOLS, Lynne (1954–)
Studies: diploma in fine arts (Cape Technikon, 1976–1978)
Profile/career: gallery group exhibitions of botanical art, USA (1981, 1983)
Artwork published: *Veld & Flora* (1979–1984); *Flowering Plants of Africa* (1985); flora posters for Durban Parks dept (1980s–1990); trail guides; M. & J. Mander, comp., *Catchment Action: Growing & Knowing Muthi Plants* (Howick: 1995)
Artwork represented: private collections, SA & worldwide
Reference: NBI files

NOALL, Penelope Mary (1952–)
Studies: Camps Bay High School, CT; graphic design (Cape Technikon, CT)
Profile/career: textile designer, CT (14 years); began painting birds & plants (1989); freelance designing & book illustrating for eco-tourism publications
Artwork published: several educational publications (Jhb, Jacana)
Artwork represented: private collections
Reference: –

NORTH, Marianne (1830–1890)
Studies: educated privately; no formal education in botany; knew botanists Francis Galton & Sir William Hooker; lessons in flower painting from Valentine Bartholomew (flower painter to Queen Victoria) & Magdalen von Fowinkel; oil painting lessons from Australian artist, Robert Dowling
Profile/career: travelled the world painting plants in landscapes (from 1871); visited Cape & Natal (Aug.1882–Sept.1883)
Artwork published: *A Vision of Eden* (London, 1980); L. Ponsonby, *Marianne North at Kew Gardens* (Exeter, Devon,1990)
Artworks represented: Marianne North Gallery, Kew
References: M. Arnold, *Women & Art in SA* (CT, 1996); M. North, *A Vision of Eden* (London, 1980); M. North, *Recollections of a Happy Life*, 2 vols (London, 1892); L. Ponsonby, *Marianne North at Kew Gardens* (Exeter, Devon, 1990)

NOWERS, Sheila Marion (1942–)
Studies: Michaelis School of Fine Art, CT (Diploma in Graphic Design, 1962–1965)
Profile/career: lived in England (1966–1970); member of SAAA (1982–); designed several sets stamps (Philatelic Services of the SA Post Office, Pret.); Carpendale Trophy for stamp design (1982); specializes in miniatures, working in gouache, watercolour & pencil, producing variety of general subjects including botanical & landscape; exhibited widely, group & solo, SA & overseas
Artwork published: numerous illustrations in SA magazines (1970–)
Artwork represented: DAG; PAM; SANG; Standard Bank Collection, Jhb; private collections
References: E. Bedford, ed., *Contemporary SA Art 1885–1995 from the SANG Permanent Collection* (CT, 1997); *DSAPS*

OERDER, Frans David (1867–1944)
Studies: Rotterdam Academy of Art, The Netherlands – gold medal & bursary (1880–1885); also Belgium & Italy
Profile/career: lived in SA (1890–1908 & 1938–1944); house painter, Pret. (1890); art teacher, various schools, Pret. (1894–1902); war artist, ZAR (1899); member, SA Society of Artists (1905); lived in East Africa & Europe (1909–1937); married Gerda Pitlo (1910); painted various subjects including flowers (mainly landscape, still-life, horses & portraits), working mostly in oil; winner of international art awards (1909, 1920); exhibited widely, group & solo, SA & Europe
Artwork published: cartoons for newspapers; sets of prints, e.g., 'Magnolias', 'Blossom Time' (E. Schweikerdt, Pretoria & New York Graphic Society); *A&A*
Artwork represented: Albany Museum, Grahamstown; DAG; Den Bosch, N. Brabant, The Netherlands; JAG; KG, PE; MuseumAfrica, Jhb; PAM; Potchefstroom Museum; Pret. City Council; SANG; Tatham Art Gallery, Pmbg; UCT; UNISA; UPret.; WU; War Museum of the Boer Republics, Bloemfontein; WH, Kimb.; numerous private collections, SA & worldwide
References: *A&A*; Arnold; E. de Swardt, *Die Lewe en Werke van die Nederlandse Kunstenaar Frans David Oerder* (n.p., 1973); *DSAB; DSAPS; SESA*

OERDER, Gerda (née Pitlo) (1887–1961)
Studies: unknown
Profile/career: born The Netherlands; married Frans Oerder (1910); moved to SA (1938); curator, PAM (*c.*1953); painted flowers, landscapes, working in oil; group exhibitions, SA
Artwork published: *A&A*
Artwork represented: PAM
References: *A&A*; Arnold; *DSAPS*

O'GRADY, Mona (Mrs W.H.A.) (1899–)
Studies: self-taught as artist
Profile/career: observed & painted flora on her Eastern Cape farm, Port St Johns (from 1966); worked first in watercolour, later in oil applied directly to cloth; participated in several group exhibitions
Artwork published: K.A. Korach, *Catalogue: 3rd International Exhibition of Botanical Art & Illustration* (Pittsburgh, USA, 1972); *Reader's Digest* (SA), Sept 1970 (cover)
Artwork represented: private collections
References: *A&A*; K.A. Korach, *Catalogue: 3rd International Exhibition of Botanical Art & Illustration* (Pittsburgh, USA, 1972); *SA Panorama*, October 1968

OLDENLAND, Hendrik Bernhard (1663–1697)
Studies: medicine, botany (Leiden University, The Netherlands); possibly self-taught as artist
Profile/career: travelled to Cape (1688); collected plants on explorations of Cape interior & assembled *kruidboek* of dried, mounted specimens; a competent artist, thought to have painted originals from which H. Boerhaave's engravings & other examples of Cape flora were made
Artwork published: H. Boerhaave, *Index Alter Plantarum* (Leiden, 1720) – 24 plates of Cape Proteaceae sometimes erroneously attributed to J. Hartog
Artwork represented: Rijksherbarium, Leiden, The Netherlands
References: *DSAB*; Gunn & Codd; M.C. Karsten, *The Old Company's Garden at the Cape & its Superintendents* (CT, 1951)

OLIVER, Edward George Hudson ('Ted') (1938–)
Studies: MSc, PhD (UCT)
Profile/career: Grenfell gold medal, for selection of paintings of genus *Erica* by I. von Below, F. Anderson, C. Letty & E.G.H. Oliver, RHS (1968) – exhibit staged by E.G.H. Oliver
Artwork published: H.A. Baker & E.G.H. Oliver, *Ericas in SthA* (CT, [1967]); *Jl SA*

Bot, 27:1961; 28:1962; 37:1971; *Bothalia*, 11:1973
Artwork represented: NBI, Pret.
Reference: Gunn & Codd

OLIVER, Inge Magdelene (1947–)
Studies: BSc, botany, zoology (UCT); Dipl. Fine Art (UCT)
Profile/career: herbarium botanist, Government Herbarium, Stellenbosch (1973–1975); contract research officer for Ericaceae, Compton Herbarium, NBI, CT (1988–)
Artwork published: *Flowering Plants of Africa;* mainly b/w drawings of Ericaceae in *Bothalia, Novon, Feddes Repertorium, Yearbook of Heather Society;* other families illustrated in *Bothalia* & *Strelitzia*
Artwork represented: NBI, Pret.
Reference: Rourke

PAGE, Mary Maud (1867–1925)
Studies: private schools, England; finishing school, Paris; School of Art (Caldrons)
Profile/career: due to ill health, emigrated to SA (1911); showed flower paintings to Louisa Bolus (1915); botanical artist, Bolus Herbarium, UCT (1915–1925); specialized in Mesembryanthemaceae; participated in group exhibitions
Artwork published: *Flowering Plants of SA;* H. Herre, *The Genera of the Mesembryanthemaceae* (CT, 1971)
Artwork represented: Bolus Herbarium, UCT
References: *A&A; Annals of Bolus Herbarium*, 4(2)1926; Arnold; *DSAB; DSAPS;* Gunn & Codd; Rourke; *Women & Art*

PASCOE, Blythe Dulcie (née Loutit) (1940–)
Studies: Pmbg Girls' High School; Natal Technical College
Profile/career: freelance artist, Natal Herbarium, Pmbg & Angolan Parks Dept (pre 1971); illustrator & technical assistant, NBI, Pret. (1971–1972); freelance artist, State Herbarium, Dept of Nature Conservation, Windhoek, SWA (1974); moved to Skeleton Coast National Park, SWA (1978); devotes life in present-day Namibia to preservation of black rhino & elephant of Damaraland
Artwork published: J.H. Ross, *The Acacia Species of Natal* ([Durban], 1971); O.M. Hilliard, *Compositae in Natal* (Pmbg, 1975); C. Berry, *Trees & Shrubs of the Etosha National Park* (Windhoek, 1982)
Artwork represented: State Herbarium, Windhoek, Namibia; NBI, Pret.; private collections
References: *Fair Lady,* 18.4.1984; NBI files

PATERSON, William (1755–1810)
Studies: horticulture (Syon House, London)
Profile/career: employed as plant collector in SA by Countess of Strathmore (late 1770s); 4 journeys into interior of SA, some of them with Robert Jacob Gordon (1777–1786); many artworks in Paterson Collection (Brenthurst Library, Jhb) are equivalent copies of paintings in Gordon Collection (Rijksprentenkabinet, The Netherlands) – unclear whether Paterson painted himself or if he was given paintings by Gordon or other artists
Artwork published: V.S. Forbes & J. Rourke, eds, *Paterson's Cape Travels, 1777 to 1779* (Jhb, 1980)
Artwork represented: Brenthurst Library, Jhb; Rijksprentenkabinet, The Netherlands
References: *DSAB;* V.S. Forbes & J. Rourke, eds, *Paterson's Cape Travels, 1777 to 1779* (Jhb, 1980)

PEERS, Charles Ernest (1874–1944)
Studies: Liverpool School of Art; Slade School of Art, London
Profile/career: left England to live in SA (1904); member, SA Society of Artists (1905); member, SA Institute of Art (1926); chairman, SA Fine Arts Association (1930); president, Owl Club (1935); colour lithographer for printer, CT (1931); painter of landscapes, scenery, maritime pictures & flower pieces, working in variety of media; winner, *The Cape Times* poster competition (1924); exhibited group & solo, SA & overseas
Artwork published: H. Chilvers, *The Seven Wonders of SthA* (Jhb, 1929); *Our Land: A Series of One Hundred Drawings in Colour* (CT, [1939]); folios of lithographic drawings & other prints
Artwork represented: Albany Museum, Grahamstown; DAG; MuseumAfrica, Jhb; PAM; SANG; Tatham Art Gallery, Pmbg; UCT; WH, Kimb.
References: *A&A; DSAB; DSAPS*

PETRIE, Margaret (active 1830s)
Studies: unknown
Profile/career: daughter of William & Margaret Petrie; painted several species of *Protea*
Artwork published: B. Warner & J. Rourke, *Flora Herscheliana* (Jhb, 1998)
Artwork represented: Brenthurst Library, Jhb; Petrie family descendants
Reference: –

PETRIE, Margaret (née Mitton) (active 1820–1830)
Studies: unknown
Profile/career: wife of William Petrie, deputy commissary general at Cape (1830s); painted local flora
Artwork published: B. Warner & J. Rourke, *Flora Herscheliana* (Jhb, 1998)
Artwork represented: Brenthurst Library, Jhb
Reference: –

PHILLIPS, Nikki (1965–)
Studies: BSc, botany, zoology (UCT, 1985); BSc Hons, botany (UCT, 1986)
Profile/career: research assistant, botany dept, UCT (1987–1990); freelance illustrator – mainly posters (1991–); full-time receptionist for ophthalmologist (1999–)
Artwork published: 3 calendars for UWC (1996, 1997, 1998)
Artwork represented: UWC; private collections, SA & Zim.
Reference: –

PIERNEEF, Jacob Hendrik (1886–1957)
Studies: part-time drawing tuition from architect, Hilversum, The Netherlands (1900); Rotterdam Academy (1901); tuition from Frans Oerder, Pret. (1905–1908) & Erich Mayer, Pret. (1908)
Profile/career: family deported to The Netherlands (1900); returned to SA (1902); librarian, Pretoria State Library, Pret. (1912–1920); art lecturer, Pretoria & Heidelberg Normal Colleges (1920–1923); lived in SWA (now Namibia) (1923–1924); lived in London (1933–1935); lecturer, UNatal (1950); painted cityscapes, landscapes – especially tree studies – working in variety of media; winner of several prestigious art awards; honorary doctorate, UNatal (1951); honorary doctorate of philosophy, UPret. (1957); honorary membership, SA Akademie vir Wetenskap & Kuns (1957); member, SA Society of Artists (1917); founder member, National Academy of Fine Arts (SA) (1932–1933); exhibited widely, group & solo, SA & overseas, including many retrospectives
Artwork published: *A&A;* illustrations in several monographs & journals; designed book & magazine covers; reproductions of artworks published by E. Schweikerdt, Pret.
Artwork represented: Ann Bryant Gallery, East London; Arts Association SWA/Namibia Collection; DAG; Engelenberg House Art Collection, Pret.; Ficksburg High School, Free State; JAG; Jhb Magistrate's Courts; Jhb Railway Station; KG, PE; MuseumAfrica, Jhb; National Cultural History & Open-Air Museum, Pret.; National Museum, Bloemfontein; PAM; Pierneef Museum, Pret.; Pietersburg Collection; Potchefstroom Museum; SS *Pretoria Castle* (1937); RAU; Sandton Town Collection; SANG; SWA/Namibia Administration Collection, Windhoek; UCT; UFS; UNatal; UNISA; UPret.; UStell.; WU; WH, Kimb.
References: *A&A;* N.J. Coetzee, *Pierneef: Land & Landscape* (Jhb, 1992); *DSAB; DSAPS;* J.F.W. Grosskopf, *Hendrik Pierneef* (Pret, 1945); P.G. Nel, ed., *JH Pierneef* (CT, 1990); F.E.G. Nilant, *Die Hout- en Linosneë van JH Pierneef* (CT, 1974); *SESA*

PIKE, Rosalie Barbara (1933–)
Studies: BSc, botany, zoology (WU); Higher Education Diploma (Jhb College of Education)
Profile/career: biochemist & medical artist, dept of surgery & medicine, WU (1953–1955); biochemist, dept of water research, CSIR, Pret. (1956); botanical artist, botany dept, WU (1957–1985); silver medal, Kirstenbosch (2000); exhibited SA & overseas, group & solo
Artwork published: A. Lucas & B. Pike, *Wild Flowers of the Witwatersrand* (CT, 1971); F.D. Hancock, A. Lucas & B. Pike, *Ferns of the Witwatersrand* (Jhb, 1973); Fedsure calendar (Jhb, 2000); S. Sherwood, *A Passion For Plants: Contemporary Botanical Masterpieces* (London, 2001)
Artwork represented: The Blue Train; Fedsure Collection, Jhb; private collections, SA & worldwide, including Shirley Sherwood, England
References: *DSAPS;* Hunt Inst., USA; *Veld & Flora*, 79:1993; 80:1994

POOLEY, Elsa Susanna (1947–)
Studies: Pmbg Girls' High School; self-taught as artist
Profile/career: lived for years in Mkuzi & Ndumu game reserves (Natal Parks

Board); began painting plants of Zululand (1967); became amateur botanist in order to identify plants painted; comprehensive herbarium collection, Maputuland (1967–1983); founding member, BAASA; group & solo exhibitions; bronze medal, Kirstenbosch (2000)
Artwork published: *Flowering Plants of Africa*; 'Art Meets Science: Flowers as Images' (Balgowan, Natal, 1992: original Caversham Press prints first exhibited at Grahamstown Festival, 1992); V. Carruthers, ed., *Wildlife of SthA* (Halfway House, 1997); V. Carruthers, ed., *Trees of SthA* (Halfway House, 1998); V. Carruthers, ed., *Flowers, Grasses, Ferns & Fungi of SthA* (Halfway House, 1998); *Palms of Africa*: limited-edition portfolio (self-published, 1988–1989)
Artwork represented: The Blue Train; Killie Campbell Library, UND; KwaZulu-Natal Nature Conservation Service, Giant's Castle Camp; NBI, Pret.; numerous private collections, SA & worldwide
References: Arnold; *DSAPS*

POORTERMANS, Johannes Cornelius (1786–1870)
Studies: learnt craft of lithography in Amsterdam
Profile/career: moved to Cape (1833); practised as lithographer, CT; also worked as trader & operated drawing school; produced monthly series of lithographs (1844–1850); original artworks naive, but some lithographs (including botanical studies) competent – probably because latter copied from competent originals
Artwork published: *Cape of Gd Hope Lit. Gaz.*, Oct. 1833
Artwork represented: Library of Parliament, CT; MuseumAfrica, Jhb; SA Library, CT
References: *DSAPS; Q. Bull. SA Lib.*, 5:1950; *SESA*

PRICE, Imogen Mary (1932–)
Studies: Diocesan School for Girls, Pret.; St Andrews School, Jhb; BA Fine Art (UNatal, 1952); BA Hons, Literature (UNISA, 1988)
Profile/career: artist, BRI, Pret. (1950s); theatre design & painting scenery (pre 1969); technician, palaeontology dept, creating reconstructions of prehistoric animals, SA Museum, CT (1969); artist, Transvaal Museum, Pret. (1977); curator, displays & art, Transvaal Museum, Pret. (1979)
Artwork published: *Flowering Plants of Africa*; I. Verdoorn, 'The Transvaal Ericaceae', *Jl SA Bot.*, 20:1954; H.A. Baker & E.G.H. Oliver, *Ericas in SthA* (CT, [1967]); C.K. Brain, *Hunters or the Hunted* (Chicago, 1981)
Artwork represented: NBI, Pret.; reconstructions of early man & Karoo reptiles in SA museums
Reference: NBI files

RIPLEY, Lura (Mrs Mason) (1931–)
Studies: Pmbg Girls' High School; BA Fine Art (UNatal, 1954); Central School of Art, London
Profile/career: part-time artist, Kew; part-time artist, British Museum (Natural History) London; freelance artist – commissioned by various publishers in England
Artwork published: several international floras (*Zambesiaca; Tropical East Africa; Iraq; Cyprus; Australia; West Africa; Malagasy*); J.P. Rourke, *The Proteas of SthA* (CT, 1980; 2 ed., 1982); J.S. Beard, *The Proteas of Tropical Africa* (Kenthurst, NSW, Australia, 1993)
Artwork represented: British Museum (Natural History), London; Kew; NBI, Pret.
Reference: NBI files

ROBERTSON, Ann (née Miller) (1944–)
Studies: Girls' High School, Salisbury, Rhodesia (now Harare, Zim.); BA Fine Arts (Rhodes, 1966); Art Teachers' Diploma (Rhodes, 1967)
Profile/career: art teacher, Roosevelt Girls' High School, Salisbury, Rhodesia (1968–1969); botanical artist, NBI, Pret. (1970); as freelance artist worked on SA seaweeds; education officer & guide, Oceanarium, CT
Artwork published: *Flowering Plants of Africa; Jl of SA Bot.* (1978–1990); R.E. Norris, *The Marine Red Algae of Natal, SA* (Pret., 1992)
Artwork represented: NBI, Pret.; private collections
Reference: NBI files

ROUPELL, Arabella Elizabeth (1817–1914)
Studies: apparently self-taught as artist
Profile/career: born in England; lived in India (1840–1843 & 1845–1853); stayed at Cape (1843 & 1844); painter of landscapes, portraits & plants (Indian & Cape)
Artwork published: *Specimens of The Flora of SA by a Lady* (London, 1849); *More Cape Flowers by a Lady* (Jhb, 1964); A. Bird, *The Lady of the Cape Flowers* (Jhb, 1964); A. Bird, *Arabella Roupell: Pioneer Artist of Cape Flowers* (Jhb, 1975)
Artwork represented: MuseumAfrica, Jhb; UCT
References: Arnold; *DSAB; DSAPS*; NBI files; Rourke; *SESA; Women & Art*

ROUX, Jacobus Petrus ('Koos') (1954–)
Studies: MSc (UStell., 1990); PhD (UPret., 1997); self-taught as artist
Profile/career: systematic botanist, specializing in pteridology (NBI, CT); illustrator of own research publications
Artwork published: *Cape Peninsula Ferns* (CT, 1979); 'The fern genus *Elaphoglossum* Schott (Filicales) in SA', *Jl SA Bot.*, 48:1982; 'The genus *Polystichum* in Africa', *Bull. Nat. Hist. Mus., Lond. (Bot.)*, 30:2000
Artwork represented: BRI, CT
Reference: Gunn & Codd

ROUX, Wilma (1947–)
Studies: Menlo Park High School, Pret. (1965); secretarial diploma, Durban Technical College (1966)
Profile/career: agricultural research assistant, NBI, Pret. (1981–1993); freelance artist
Artwork published: G.E. Gibbs, *et al.*, *Grasses of SthA* (Pret., 1990); L. Henderson, *Plant Invaders of SthA* (Pret., 1995)
Artwork represented: NBI, Pret.
Reference: NBI files

SANDERSON, John (1820–1881)
Studies: unknown
Profile/career: journalist, trader & draughtsman; worked for Scottish Press; secretary of Natal Times Co., Durban, which started *The Weekly Times* (1851); traded in Magaliesberg & on Witwatersrand (April 1852); president of Natal Agricultural & Horticultural Society (14 years); designed coats of arms, Durban & Pmbg; corresponded with W. Hooker at Kew & W.H. Harvey in Dublin
Artwork published: *Memoranda of a Trading Trip into the Orange River Free State* (London, 1860); 'Notes on *Flora of Natal*': appendix to J. Chapman, *Travels in the Interior of SA*, 2 vols (London, 1868)
Artwork represented: Kew (sketches accompanying herbarium specimens); Local History Museum, Durban; Natal Museum, Pmbg; NBI, Pret.
References: *DSAB; DSAPS*; Gunn & Codd

SAUNDERS, Katharine (née Wheelwright) (1824–1901)
Studies: music, painting & languages (Belgium)
Profile/career: married James Saunders, Brussels (1851); lived first in London, then left to live in Natal – Tongaat Estate, north of Durban (1854); began painting flowers; developed strong ties with W. Hooker (Kew), W.H. Harvey (Dublin), local Natal botanists & H. Bolus (CT); 5 plants named in her honour; Marianne North stayed with Saunders family (1883)
Artwork published: *Flower Paintings of Katharine Saunders* (Maidstone, Natal, 1979)
Artwork represented: Natal Museum, Pmbg; Tongaat Estate, Maidstone, Natal
References: *A&A; DSAPS; Flower Paintings of Katharine Saunders* (Maidstone, Natal, 1979); Gunn & Codd

SCHUMACHER, Johannes (active 1770–1789)
Studies: unknown
Profile/career: VOC employee who accompanied Hendrik Swellengrebel as draughtsman on 3 journeys into Cape interior (1776–1786); also accompanied Robert Jacob Gordon as draughtsman (1777–1786); may have painted some botanical plates in Gordon Collection or may have made equivalent copies
Artwork published: –
Artwork represented: unknown
References: P. Cullinan, *Robert Jacob Gordon, 1743–1795* (CT, 1992); *DSAB*; A. Gordon-Brown, *Pictorial Africana* (CT, 1975); P. Raper & M. Boucher, eds, *Robert Jacob Gordon: Cape Travels, 1777 to 1786* (Jhb, 1988)

SCHWEIZER, Anne (Rose-Innes) (1930–)
Studies: Eunice High School, Bloemfontein (1947); BA, art, languages (Rhodes, 1950); Continental Art School, CT, under Maurice van Essche (1951); BA Hons, archaeology (UCT, 1961)
Profile/career: Michaelis School of Art, UCT (1952–1956); scene painter, Hofmeyr Theatre, CT (1955–1956); resident artist, SA Natural History Museum, CT (1956–1962); freelance illustrator of botanical, archaeological & ethnographic material, including books & posters (1963–); exhibited group & solo; gold

medal, 'Kirstenbosch Inaugural Exhibition of Botanical Art', CT (2000)
Artwork published: S. Sherwood, *A Passion for Plants: Contemporary Botanical Masterpieces* (London, 2001)
Artwork represented: First National Bank, Jhb; SA Natural History Museum, CT; private collections, SA & worldwide, including Shirley Sherwood, England
Reference: *Catalogue: Kirstenbosch Inaugural Exhibition of Botanical Art* (CT, 2000)

SCOTT, Marguerite (1956–)
Studies: Clapham High, Pret. (1973); self-taught as artist
Profile/career: NBI, Pret. (1974–1976); married (1976); emigrated to Australia (1999)
Artwork published: *Flowering Plants of Africa*; J.H. Ross, *A Conspectus of the African* Acacia *Species* (Pret., 1979); *Flora of SthA*, Vols 10, 16, 22
Artwork represented: NBI, Pret.
Reference: NBI files

SCOTT, Phil (1937–)
Studies: Gipsy Hill Training College, England; part-time, under Anna Vorster (Ruth Prowse School of Art, CT); botanical art master-classes of Katie Lee & Jenny Phillips (CT)
Profile/career: freelance painter, mainly indigenous flora & particularly disa & orchids
Artwork published: –
Artwork represented: Safmarine, Antwerp; several private collections
Reference: –

SEAGRIEF, Stanley Charles (1927–1994)
Studies: MSc (Rhodes); PhD (Cantab)
Profile/career: senior lecturer, botany, Rhodes University, Grahamstown (1957); professor of botany, Rhodes University, Grahamstown (1974–1987); special research interest in marine algae; illustrated own publications; exhibited, SA & overseas; main subject matter marine algae, but Cape flora too
Artwork published: *The Seaweeds of the Tsitsikama Coastal National Park* (Pret., 1967)
Artwork represented: British Museum (Natural History); Hunt Inst., USA; private collections
References: Gunn & Codd; Rourke

SERNEELS, Clément Edmond Théodore Marie (1912–)
Studies: diploma, Académie Royale de Belgique, Brussels (1934)
Profile/career: awarded travelling bursary to visit Belgian Congo (1936); returned to live in Belgian Congo (1938–1958); lived in SA (1961–1980); oil painter of variety of subjects & still-life, particularly flowers; exhibited widely, group & solo, SA & overseas
Artwork published: unknown
Artwork represented: Collection Etat Belge, Belgium; PAM; Willem Annandale Art Gallery, Lichtenburg; WH, Kimb.; private collections, SA & worldwide
References: *A&A; DSAPS*

SETON-TAIT, Adela ('Ada') (active 1900s–1930s)
Studies: South Kensington Art School, London; Slade School of Art, London
Profile/career: art teacher, Jeppe Girls' High School, Jhb (until 1934); art tutor, Art Union, Jhb; painter of portraits, flowers, landscapes, still-life & miniatures, working in oil & watercolour; group & solo exhibitions, SA & overseas
Artwork published: portrait of soldier in *African Art Journal*, 1917
Artwork represented: MuseumAfrica, Jhb
References: *A&A; DSAPS*

SKEEN, Julia Georgina St John (Mrs Mann) (1956–)
Studies: Rhodes University (1976–1978)
Profile/career: born Zim.; moved to SA (1976); member of Grahamstown Group (1976–1978); moved to Botha's Hill, Natal (1980); taught art privately, Westville, Natal (1983–1984); group & solo exhibitions; paints figures, landscapes & botanical subjects, working in various media
Artwork published: C. Mann & J. Skeen, *The Horn of Plenty* (Grahamstown, 1997)
Artwork represented: private collections
Reference: *DSAPS*

SLEIGH, Elsje Catharina (née Uys) (1937–)
Studies: matriculated La Rochelle, Paarl; Mowbray Teachers' Training College, CT; self-taught as artist
Profile/career: began painting local flora (1982); first solo exhibition of botanical art (1984); freelance artist – hundreds of commissioned works to date; silver medal, 'Kirstenbosch Inaugural Exhibition of Botanical Art', CT (2000)
Artwork published: –
Artwork represented: private collections, SA & worldwide
Reference: *Catalogue: Kirstenbosch Inaugural Exhibition of Botanical Art* (CT, 2000)

SMITH, Esther (active 1920s)
Studies: Kensington Art School (London, *c.*1893)
Profile/career: left England for SA; taught art & botany, Collegiate High School for Girls, Port Elizabeth (all 4 Smith sisters taught there); affectionately nicknamed 'Vesuvius' due to temper outbursts; talented dressmaker & gardener – improved school grounds by planting trees & flowers; artist, Bolus Herbarium, UCT (then housed Kirstenbosch Botanic Gardens); worked for H.M.L. Bolus (past pupil, 1893) on *Watsonia* (1924–1925); on retirement, returned to live in London
Artwork published: *Flowering Plants of Africa;* R. Marloth, *The Flora of SA*, 4 vols in 6 (CT & London, 1913–1932)
Artwork represented: Bolus Herbarium, UCT
References: Rourke; School History: Collegiate High School for Girls, Port Elizabeth

SNYMAN, Janet (active from 1990s)
Studies: Commercial & Art High School, Bloemfontein (1979); National Diploma in Graphic Design, Free State Technikon
Profile/career: Provincial Library Services of the Orange Free State, Bloemfontein; National Museum, Bloemfontein; taxonomic illustrator, dept of botany, UFS [then University of OFS], Bloemfontein, for Prof. H.J.T. Venter (1991); moved to Jeffreys Bay, Eastern Cape (1992); freelance artist
Artwork published: *SA Jl of Bot.* (1993–1994)
Artwork represented: dept of botany, UFS; private collections
Reference: NBI files

SPEARS, Frank Sydney (1906–)
Studies: Waverley Road School of Art, Birmingham, England (*c.*1915)
Profile/career: born England; travelled extensively in Europe; lived in SA (1928–*c.*1969); president, SA Society of Artists (1945–1947); taught art privately; painter of variety of subjects (including flower studies) working in variety of media (specializing in stained glass); exhibited widely, group & solo, SA & overseas
Artwork published: *A&A;* several illustrations in magazines & art publications
Artwork represented: Church of the Resurrection, Bonteheuwel, CT; Dutch Reformed Church Centre, CT; Hester Rupert Art Museum, Graaff-Reinet; JAG; National Gallery, Gdansk, Poland; PAM; SANG; UFS; UStell.; WU; WH, Kimb.; private collections, SA & worldwide
References: *A&A; DSAPS*

SPILHAUS, Pauline Augusta Wilhelmina ('Nita') (1878–1967)
Studies: Lübeck School of Art & Kunst Akademie, Munich, Germany (1896–1900); Thanlow Swedish Painting School, Paris
Profile/career: lived in SA (1907–mid 1920s &1938–1967); member, SA Society of Artists (1907); painted mainly in oil, depicting variety of subjects including landscapes, mostly with trees; flower paintings produced during world war years when housebound; exhibited widely, group & solo, SA & Europe
Artwork published: *A&A;* folio of etchings of Lübeck (1900); folios of etchings, 'Trees' & 'Cape Town' published by *The Cape Times*; E. Roworth, *Art of the British Empire* (London, 1917)
Artwork represented: Albany Museum, Grahamstown; DAG; JAG; KG, PE; PAM; SANG; WU; WH, Kimb.
References: *A&A; DSAB; DSAPS*

STADLER, Anne (née Pienaar) (1960–)
Studies: Hoërskool, Fochville, 1978; BA Information Design (UPret., 1982); Higher Teachers' Diploma (UPret., 1984); MA (UPret., 1990)
Profile/career: freelance artist & publisher's assistant (1983–1985); advertisement artist, *Die Laevelder* (1985); production manager, *SA Publishing Today* (1986–1987); graphics artist, advertising consultants (1987–1988); graphics artist & PRO (1988–1990); senior graphics artist, nurseries – National Intelligence (1990–1996); started own graphic art & design company (1996)
Artwork published: *SA Jl Bot.*
Artwork represented: NBI, Pret.; UPret.
Reference: NBI files

STARCKE, Helmut (1935–)
Studies: self-taught as artist
Profile/career: in-service training,

advertising company, West Germany; settled in SA (1958); painter of landscapes, seascapes, plants & interiors, usually large in size; founder member & chairman, Artists' Guild, CT (1982); founder member, Artists Gallery, CT (1982); lecturer, graphic design, Michaelis School of Fine Art, UCT (from 1975); exhibited widely, group & solo, SA & overseas; winner of several prestigious art awards
Artwork published: E. Bedford, ed., *Contemporary SA Art 1885–1995 from the SANG Permanent Collection* (CT, 1997)
Artwork represented: Gencor Collection, Jhb; JAG; PAM; SANG; Sasol Collection, Jhb; private collections, SA & worldwide
References: *A&A;* E. Bedford, ed., *Contemporary SA Art 1885–1995 from the SANG Permanent Collection* (CT, 1997); *DSAPS; Lantern*, 38:1989

STEAD, Jeanette (née MacCallum) (1940–)
Studies: Girls' Collegiate School, Pmbg; commercial art, Natal Technical College, Durban (1955–1956)
Profile/career: freelance artist; exhibited group & solo
Artwork published: designed several sets stamps for Philatelic Services of the SA Post Office, Pret., including 1 set orchids to commemorate Tenth World Orchid Conference (1981)
Artwork represented: SA Philatelic Society, Pret.; private collections, SA & worldwide
References: *DSAPS;* NBI files

STERN, Irma (1894–1966)
Studies: Weimar Academy, Germany (1913); Levin-Funcke Studio, Berlin (1914); Bauhaus, Weimar (1914)
Profile/career: lived in Germany (1899–1903, 1913–1919 & 1922–1926); member, November Group, Germany (1916); member, Neue-Sezession, Germany (1918–1919); member, SA Society of Artists; travelled extensively in SA & central Africa; painted mostly in oil, variety of subjects & often flower studies; influenced by expressionists; home in CT today a museum; exhibited widely, group & solo, SA & overseas, including many retrospectives; winner of several prestigious art awards, SA & overseas
Artwork published: *Visionen: Zehn Steinzeichnungen* (Berlin, 1920); *Dumela Marena: Bilder aus Afrika* (Berlin, 1920); *Congo* (Pret., 1943); *Zanzibar* (Pret., 1948)
Artwork represented: *A&A;* Ann Bryant Gallery, East London; Bielefeld Art Gallery, Germany; Collection of Queen Elizabeth, the Queen Mother, London; DAG; Hester Rupert Art Museum, Graaff-Reinet; Irma Stern Museum, UCT; JAG; Julius Gordon Africana Centre, Riversdale, Cape; KG, PE; Musée de Art Moderne de la Ville de Paris; MuseumAfrica, Jhb; National Museum, Bloemfontein; PAM; SA embassies in Geneva, The Hague, Madrid, Paris, London & Washington DC; SANG; Tatham Art Gallery, Pmbg; UCT; UFS; UPret.; UNISA; UStell.; WU; Willem Annandale Art Gallery, Lichtenburg; WH, Kimb; many private collections, SA & worldwide
References: *A&A;* Arnold; M. Arnold, *Irma Stern: a Feast for the Eye* (CT, 1995); *DSAB; DSAPS;* N. Dubow, *Irma Stern* (CT, 1974); J. Sachs, *Irma Stern & the Spirit of Africa* (CT, 1942); K. Schoeman, *Irma Stern: the Early Years, 1894–1933* (CT, 1994); *SESA; Women & Art*

STEVENSON, Erica Crichton (1926–)
Studies: Diploma in Art & Design (Witwatersrand Technical College, Jhb, 1946)
Profile/career: freelance commercial artist; artworks in variety of media & subjects include indigenous flora; foundation member, Natal panel, Floral Art Judges; founder president, Durban chapter, Ikebana International 170; past chairman, Natal branch, Watercolour Society of SA; honorary life member, Highway Art Group, Cowies Hill, Natal; participant, several group exhibitions
Artwork published: –
Artwork represented: head office, Natal Building Society, Durban
Reference: *DSAPS*

STEYN, Maria Magdalena ('Marietjie') (née van Zyl) (1963–)
Studies: matriculated Durbanville High School, Western Cape (1981); BA (UStell., 1985)
Profile/career: scientific research illustrator, zoology dept, UStell. (1986–1994); moved to Pret. (1994); freelance graphic artist & research illustrator for wide range of disciplines
Artwork published: E.M.A. Steyn & G.F. Smith, *Flora of the World;* Part 3: *Welwitschiaceae* (Canberra, 1999); *Flora of SthA*, vol. 28, part 1, 2000; L. Henderson, *Plant Invaders of SthA* (Pret., 1995)
Artwork represented: NBI, Pret.; Plant Protection Research Institute, Pret.; zoology dept, UStell.; biology dept, Vista University, Pretoria
Reference: NBI files

STOKES, Hazel May (née Hives) (1914–)
Studies: Central School of Art, London
Profile/career: bookjacket designer for London studio; married & moved to SA (1939); freelance artist for book publishers, *Argus* newspaper & various magazines
Artwork published: *What Flower is That? Spring* (CT, 1988); *What Flower is That? Summer* (CT, 1971); *What Flower is That? Autumn & Winter* (CT, 1969); *What Succulent is That?* (CT, 1974); *What Tree is That?* (CT, 1967)
Artwork represented: –
Reference: NBI files

STRACHAN, Lisa (active from *c.*2000)
Studies: textile design, Natal Technikon
Profile/career: designer for fabric manufacturers, Durban, Jhb & London; thereafter, extended period of travel & work in Europe & Australia; currently freelance designer & illustrator; participant, 'Kirstenbosch Inaugural Exhibition of Botanical Art', CT (2000)
Artwork published: –
Artwork represented: –
Reference: *Catalogue: Kirstenbosch Inaugural Exhibition of Botanical Art* (CT, 2000)

STRUBEN, Edith Francis Mary (1868–1936)
Studies: Crystal Palace School of Art, London; further art study in Paris, Rome & CT
Profile/career: exhibited, London & SA; watercolour artist of SA flora for NBG & BSSA (1920s); group exhibitions, SA & England
Artwork published: –
Artwork represented: NBI, CT; SA House, London
References: *DSAB; DSAPS*

TAIT, Cynthia Katie (née Grenfell) (Mrs L. Ussher) (1892–1963)
Studies: self-taught as artist
Profile/career: active in SA as artist (1942–1962); several public exhibitions in CT but painted mainly for personal pleasure; winner of several prestigious art awards (1956, 1961)
Artwork published: –
Artwork represented: NBI, CT; private collections, SA & worldwide
References: Arnold; Rourke; *Veld & Flora*, 85:1999

TENNANT, Alice (née Aubry) (1890–1976)
Studies: Académie de la Grande Chaumière, Paris
Profile/career: lived in SA (*c.*1922–1976); opera singer; painted flowers, landscapes & other subjects, working in variety of media; exhibited group & solo, SA & overseas
Artwork published: illustrated several books for children; wrote & illustrated regular feature for children, *Cape Argus*
Artwork represented: Hester Rupert Art Museum, Graaff-Reinet; KG, PE; PAM; SANG; UPret.; UStell.
References: *A&A; DSAPS*

THOMAS, Wendy Victoria ('Vicki') (née Sholto Douglas) (1951–)
Studies: self-taught as artist; attended botanical art master-classes of Katie Lee & Jenny Phillips (CT)
Profile/career: group exhibitions in SA; winner of several prestigious art awards (1993, 1999); founding member, BAASA; chairman, BAASA, Western Cape (2001)
Artwork published: *Flowering Plants of Africa; Aloe – SA Jl of Succulent Plants*, 30:1993; 31:1994; *American Cactus & Succulent Journal; The Cactus Journal of the UK* (1993); *Bothalia*, 27:1997 & 29:1999; S. Sherwood, *A Passion for Plants: Contemporary Botanical Masterpieces* (London, 2001); calendars, newspaper articles, wine labels
Artwork represented: UWC; Spier Wine Estate, Cape; several private collections, SA & worldwide, including Shirley Sherwood, England
Reference: NBI files

THWAITS, Emily Jane (*c.*1860–1906)
Studies: possibly tutored at art school of Charles Rolando, CT; granddaughter of Thomas Thwaits, art master at Rev. James Beck's school, Roeland Street, CT; sister of Florence Amy Thwaits, also a botanical artist
Profile/career: silver medal for best watercolour paintings, SA Fine Arts Association (1880); married H. Ford (1905); took own life by walking into sea & drowning (Muizenberg, Cape)
Artwork published: 'Art Meets Science:

Flowers as Images' (Balgowan, Natal, 1992: original Caversham Press prints first exhibited at Grahamstown Festival, 1992); *Women & Art*
Artwork represented: NBI, CT
References: *Africana Notes & News,* 22:1977; Arnold; Rourke; *Women & Art*

THWAITS, Florence Amy (active 1880–1890)
Studies: possibly tutored at art school of Charles Rolando, CT; granddaughter of Thomas Thwaits, art master at Rev. James Beck's school, Roeland Street, CT; sister of Emily Jane Thwaits, also a botanical artist
Profile/career: art mistress at Huguenot College, Wellington, Cape; painted a few plates for Marloth's *Flora of SA*; work is unsigned
Artwork published: R. Marloth, *The Flora of SA*, 4 vols in 6 (CT & London, 1913–1932)
Artwork represented: NBI, CT
References: NBI files; Rourke

TONGUE, Mrs E. (active 1890s–1900s)
Studies: unknown
Profile/career: teacher, Collegiate School for Girls, Port Elizabeth (1890s & early 1900s); painted Western & Eastern Cape flora (1890s & early 1900s); moved to Transvaal (1903)
Artwork published: –
Artwork represented: NBI, CT
Reference: Collegiate School for Girls, Port Elizabeth

VAN DE RUIT, Patricia (1932–)
Studies: educated SA; commercial art, Witwatersrand Technical College, Jhb
Profile/career: since marriage, lived mainly in Zim.; illustrations mostly from live material (1972–1976); participated in group exhibitions
Artwork published: J.S. Ball, *SthA Epiphytic Orchids* (Jhb, 1978)
Artwork represented: artist's private collection
Reference: NBI files

VAN DER MERWE, Aleida Hendrika (née de Bruyn) (1933–)
Studies: Afrikaanse Höer Meisiesskool, Pret.; BA Fine Arts (UNP, 1954)
Profile/career: artist, NBI, Pret. (1955–1956); married (1956); artist, NBI, Pret. – worked under Cythna Letty (1965); artist, depts of library services & nature conservation, Transvaal Provincial Administration, Pret. (1968–1972); depts of library services & nature conservation, Transvaal Provincial Administration, Bloemfontein (1973); medical illustrator, faculty of medicine, UFS
Artwork published: *Flowering Plants of Africa;* M. Henderson & J.G. Anderson, *Common Weeds in SA* (Pret.,1966); *Bothalia,* 12(4)1979
Artwork represented: NBI, Pret.; faculty of medicine, UFS; private collections
References: *A&A;* NBI files, Pret.

VAN GOGH, Joan (Mrs Boshier) (1939–)
Studies: stone carving & life drawing, West of England College of Art (1 year); self-taught as botanical artist
Profile/career: freelance artist (1999–2000); designed sets stamps for Philatelic Services of the SA Post Office, Pret.; group & solo exhibitions
Artwork published: *Margaret Roberts' Book of Herbs* (Jhb, 1983); *Growing Herbs With Margaret Roberts* (Jhb, 1985); J. Munday, *Poisonous Plants in SA Gardens & Parks* (Jhb, 1988); J. van Gogh & J. Anderson, *Trees & Shrubs of the Witwatersrand, Magaliesberg & Pilanesberg* (CT, 1988); V. Thomas & R. Grant, *Sappi Tree Spotting: Highveld & the Drakensberg* (Jhb, 1998); R. Grant & V. Thomas, *Sappi Tree Spotting: KwaZulu-Natal* (Jhb, 1998); Jacana & Twisisa, *Sappi Tree Spotting: Lowveld* (Jhb, 1998)
Artwork represented: private collections
References: *DSAPS;* NBI files

VAN HEERDEN, Catherina J.H. ('Katryn') (née du Plessis) (1927–)
Studies: self-taught as artist
Profile/career: began botanical painting (1984); as freelance artist, received commissions from D. Shearing, Beaufort West (1984), Karoo National Park, Beaufort West (1989), P. le Riche, CT (1990), Northern Cape Ministry, Kimberley (1996); Marloth Medal of the BSSA (1995); exhibited in SA
Artwork published: D. Shearing, *SA Wild Flower Guide, 6: Karoo* (CT, 1994)
Artwork represented: Karoo National Park, Beaufort West; Northern Cape Ministry, Kimberley; private collections
Reference: *Veld & Flora,* 80:1994

VAN NOUHUYS, Jan Jozua (1903–1940)
Studies: Landbouw Hogeschool, Wageningen, The Netherlands; BSc (UNISA, 1931)
Profile/career: left The Netherlands for SA (1921); horticulturist, dept of parks, Pret.; gardener in charge, Division of Botany (later BRI, Pret.) (1925); interpreter of aerial photographs for aircraft company, Jhb (from 1933); captain, SAAF (start of Second World War); died in air crash near Nairobi, Kenya; participated in group exhibitions, England; solo retrospective exhibitions, SA
Artwork published: *Flowering Plants of Africa*
Artwork represented: NBI, Pret.; PAM
References: *DSAPS;* NBI files

VAN WIJK, Yvette (née Green) (1941–)
Studies: St Anne's College, Hilton, Natal; BA Fine Art (Rhodes); taxonomy courses I & II (Rhodes)
Profile/career: teacher of art, Diocesan School for Girls, Grahamstown (3 years); operated art gallery (4 years); illustrates mainly own work on herbs & other useful plants
Artwork published: *Eastern Cape Coastal Flora* (Grahamstown, [1984]); *The Practical Book of Herbs: Growing & Using Herbs in SthA* (CT, 1986); *First Aid with Herbs* (Grahamstown, n.d.)
Artwork represented: private collections
Reference: *Fair Lady,* 7.1.1987

VANDERPLANK, Helen Joyce (1919–)
Studies: Clifton High School, Bristol, England; Birkbeck College, London; Froebel Education Institute, Roehampton, England; self-taught as artist
Profile/career: teacher in England; settled in SA (1963); teacher, Diocesan College, Rondebosch, CT (1960s); lecturer, Grahamstown Teachers' Training College (1971–1975); display artist, Albany Museum, Grahamstown; began painting Eastern Cape flora, especially Port Elizabeth area
Artwork published: F.M. Haworth, natural history series, 6 vols (London, 1954–1963); series of cards depicting Eastern Cape flora, Dept Nature Conservation (1976); J.P. Rourke, 'Helen Vanderplank's wild flowers of the Port Elizabeth area', *Brenthurst Archives,* 1:1994; *Wild Flowers of the Port Elizabeth Area: Swartkops to Sundays River* (Hunters Retreat, 1998)
Artwork represented: Albany Museum, Grahamstown; Brenthurst Library, Jhb
References: *Brenthurst Archives,* 1:1994; *DSAPS*

VERSTER, Andrew Clement (1937–)
Studies: National Diploma in Design (Camberwell School of Art, London, 1959); Art Teachers' Diploma (Reading University, 1960)
Profile/career: art lecturer, Witwatersrand Technical College, Jhb (1961); taught art, Michaelis School of Fine Art, CT (1962); art lecturer, University of Durban-Westville (1962–1971); art lecturer, Natal Technical College – later head of division of art (1971–1976); writer & art critic for various newspapers & publications, including *Daily News* (Durban), *Argus* (CT); art advisor & designer, Durban firm of architects (1978); exhibited widely, group & solo, SA & overseas; works in variety of media, subjects including flower studies; winner of several prestigious art awards
Artwork published: *A&A*
Artwork represented: Brenthurst Library, Jhb; DAG; Gencor Collection, Jhb; Hotel Cecil, Bloemfontein; JAG; KG, PE; Mangosuthu Technikon, Umlazi; Natal Technikon; PAM; Pietersburg Collection; Sandton Town Collection; SANG; St John the Divine, Umbogintwini, KwaZulu-Natal; Tatham Art Gallery, Pmbg; Tongaat-Hulett, La Lucia, KwaZulu-Natal; UFS; UNatal; UNISA; UStell.; WH, Kimb.; WU; numerous private collections, SA & overseas
References: *A&A; DSAPS*

VILLET, Carolus Johannes (Jean) (1817–1877)
Studies: apparently self-taught as artist
Profile/career: dealer in natural history specimens & flower paintings, CT; produced latter for sale through father's business (C.M. Villet & Son, CT); work not signed & therefore not easily identifiable
Artwork published: *'Orothamnus zeyheri',* Plate 4357 in *Curtis's Botanical Magazine,* 1848
Artwork represented: Kew
References: *DSAPS;* Gunn & Codd; A. Gordon-Brown, *Pictorial Africana* (CT, 1975) incorrectly attributes unsigned work by E.M. Dixie to Villet

VISAGIE, Marieta (active from 1990s)
Studies: La Rochelle School, Paarl; self-taught as artist

Profile/career: winner of several prestigious art awards (1994, 2000)
Artwork published: *Flowering Plants of Africa;* 4 commemorative fruit notelets – centenary, SA deciduous fruit exportation; Overberg botanical calendar (1990)
Artwork represented: NBI, Pret., private collections
Reference: *Catalogue: Kirstenbosch Inaugural Exhibition of Botanical Art* (CT, 2000)

VOIGT, Leigh (1943–)
Studies: School of Art, Witwatersrand Technical College, Jhb (1961–1962) (daughter of Barbara Jeppe)
Profile/career: artist/designer, advertising agency (1962–1968); active & versatile freelance artist & illustrator thereafter; illustrator of numerous books, including autobiography; bronze medal, 'Kirstenbosch Inaugural Exhibition of Botanical Art', CT (2000); exhibited widely, group & solo, SA & overseas; works in watercolour, ink & pencil
Artwork published: S. Hart, *In the Wild* (Jhb, 1974); S. Hart, *Back in the Wild* (Jhb, 1977); M. Poland, *The Mantis & the Moon* (Jhb, 1978); M. Poland, *Once at KwaFubesi* (CT, 1981); M. Poland, *Shadow of the Wild Hare* (CT, 1986); *Lulu Phezulu* (CT, 1999)
Artwork represented: numerous collections, SA & worldwide
References: *A&A; DSAPS*; NBI files

VOLSCHENK, Jan Ernst Abraham (1853–1936)
Studies: self-taught as artist
Profile/career: member, SA Drawing Club (1894–1902); member, SA Society of Artists (1906); honorary member, SA Akademie vir Wetenskap & Kuns (1931); founder member, National Academy of Arts (SA) (1932); landscape artist, working usually in oil – occasional flower studies; exhibited widely in SA, group & solo, including several retrospectives
Artwork published: *A&A;* landscape reproductions by E. Schweickerdt, Pret.
Artwork represented: Albany Museum, Grahamstown; DAG; JAG; Julius Gordon Africana Centre, Riversdale, Cape; KG, PE; MuseumAfrica, Jhb; National Museum, Bloemfontein; PAM; SANG; UFS; UStell.; Willem Annandale Art Gallery, Lichtenburg; WH, Kimb.
References: *A&A; DSAB; DSAPS;* H. Steyn & M.L. Roux, *Die Volschenks* (Still Bay, *c.*1992)

VON BELOW, Irma (Mrs Kerr) (1920–1984)
Studies: Stella Convent, Durban; Rocklands Girls' School, Cradock; trained as nurse, Jhb General Hospital; few art lessons, local Technical College (early 1950s)
Profile/career: started painting ericas during Cape holiday; gold medal from RHS for exhibit of paintings from *Ericas in SthA* (1968); worked with R. Cowling on dune vegetation studies
Artwork published: *Flowering Plants of Africa;* H.A. Baker & E.G.H. Oliver, *Ericas in SthA* (CT, 1967); R.M. Cowling & S.M. Pierce, 'Southern Cape coastal dunes – an ecosystem lost?', *Veld & Flora*, 71:1985
Artwork represented: NBI, CT; NBI, Pret.
References: *A&A;* Arnold; Rourke

VON KOENEN, Eberhard (1915–)
Studies: Viktoria-Gymnasium, Potsdam (1934); natural science – during war internment (1940–1946); physiotherapy & homeopathy (1960s)
Profile/career: born SWA (later Namibia); internship during Second World War in SA at Andalusia (Jan Kempdorp) (1940–1946); worked in Asia & Germany; botanic research in Kaokoveld in cooperation with Senckenberg Museum, Frankfurt/Main, Germany; founded Gobabeb Desert Research Station, SWA (1963); retired (1986)
Artwork published: *Heil- und Giftpflanzen in Südwestafrika* (Windhoek, 1977); *Medicinal, Poisonous & Edible Plants in Namibia* (Windhoek, 2000)
Artwork represented: –
Reference: *Medicinal, Poisonous & Edible Plants in Namibia* (Windhoek, 2000)

VOORTMAN, Vera Agnes (née Reid) (1903–)
Studies: Rustenburg Girls' High School, CT; part-time study, Michaelis School of Art, UCT
Profile/career: landscape & still-life artist, working in oil & watercolour; under guidance of W.F. Barker & E. du Plessis, Kirstenbosch Herbarium, CT, produced 200 paintings of wild flowers of Queenstown area (1970–1976)
Artwork published: *Flowering Plants of Africa*
Artwork represented: NBI, Pret.
References: NBI files

VORSTER, Elsa Christina (1945–)
Studies: self-taught as artist
Profile/career: unknown; works in variety of media
Artwork published: P. Vorster, ed., *Proceedings of the Third International Conference on Cycad Biology* (Stellenbosch, 1995); C.J. Chen, ed., *Proceedings of the Fourth International Conference on Cycad Biology* (Beijing, 1999); M. Tadesse, S. Demissew & I. Hedberg, eds, *Flora of Ethiopia & Eritrea*, vol. 2, part 1(Addis Ababa & Uppsala, 2000); *Adansonia; Aloe; Excelsa; Jl of SA Bot.; Mitteilungen aus dem Institut für Allgemeine Botanik Hamburg; Nordic Jl Bot.; Novon*
Artwork represented: private collections, SA
Reference: –

VORWERK, Margaret Anne (1949–)
Studies: unknown
Profile/career: botanical artist & herbarium assistant, Durban Herbarium, NBI (1972–1974)
Artwork published: *Flowering Plants of Africa*
Artwork represented: NBI, Pret.
Reference: NBI files

WALGATE, Marion Meason ('Maisie') (Mrs Macnae) (1914–)
Studies: MSc (UCT, 1940); Solly Scholarship (UCT, 1940)
Profile/career: left England to live in SA (1920); lieutenant, artillery (1941–1943); lecturer in botany, UCT (1943–1949); married Dr W. Macnae of WU; herbarium assistant, Moss Herbarium, WU (until 1970)
Artwork published: *Flowering Plants of Africa;* H. Herre, *The Genera of the Mesembryanthemaceae* (CT, 1971)
Artwork represented: NBI, Pret.
Reference: NBI files

WALKER, Joan (1928–)
Studies: unknown
Profile/career: produced over 360 watercolour illustrations, & conducted extensive research on medicinal plant usage, plant habitats, etc. for self-published book on Natal flora
Artwork published: *Wild Flowers of KwaZulu-Natal* (Pinetown, 1996)
Artwork represented: KwaZulu-Natal Nature Conservation Service, Giant's Castle Camp; private collections
Reference: –

WALTERS, Adèle Lennox (1933–)
Studies: St John's High School, Pmbg; Jhb High School for Girls; Alevine High School, Bulawayo, Zim.; BA Fine Art (UNatal, 1955)
Profile/career: scientific illustrator, British Museum (Natural History), London (1957–1959); art teacher, Jeppe Girls' High School, Jhb (1960); botanical artist, NBI, Pret. (1960–1962); settled in Zim.; taught art privately & participated in major Zim. exhibitions; returned to SA
Artwork published: *Flowering Plants of Africa;* M. Henderson & J.G. Anderson, *Common Weeds in SA* (Pret., 1966); *Flora of SthA*, vol. 26, 1963
Artwork represented: NBI, Pret.
Reference: NBI files

WARD-HILHORST, Ellaphie (1920–1994)
Studies: Pret. Girls' High School; studied watercolour with J.P.L. Hilhorst in The Netherlands (1947); self-taught as botanical artist
Profile/career: survey draughtswoman on Witwatersrand gold mines (1940–1946); freelance artist, advertising field (1948–1970); on retirement, developed talent as botanical artist; winner of several prestigious art awards (1988, 1990, 1992); exhibited widely
Artwork published: *Flowering Plants of Africa;* J.J.A. van der Walt & P.J. Vorster, *Pelargoniums of SthA*, 3 vols (CT, 1977–1988); D.A. Snijman, *A Revision of the Genus* Haemanthus L. (CT, 1984); E.J. van Jaarsveld, *Gasterias of SA* (CT, 1994); *The Kew Magazine; Veld & Flora*
Artwork represented: Brenthurst Library, Jhb; Hunt Inst., USA; NBI, Pret.; numerous private collections, including Shirley Sherwood, England
References: Arnold; Rourke; Sherwood; *Veld & Flora*, 1994; *Women & Art*

WEBER, Anna Margaretha ('Rita') (née Louw) (1955–)
Studies: FH Odendaal High School; studied languages, UPret. (1 year); secretarial diploma, Pret. Technikon; self-taught as artist
Profile/career: CSIR (2 years); botanical artist, NBI, Pret. (1977–1979); moved to Mpumalanga; active freelance artist, working in oil & watercolour
Artwork published: *Flowering Plants of Africa;* C.H. Stirton, *Plant Invaders – Beautiful but Dangerous* (CT, 1978); *Flora*

of SthA, Bryophyta, 2 parts, 1981, 1987
Artwork represented: NBI, Pret.; private collections
References: NBI files, *Pret. News*, 24.9.1990

WEHDEMANN, Clemenz Heinrich (1762–1835)
Studies: home-educated by clergyman father, Germany
Profile/career: mercenary soldier, sent to Cape (1784); gave art classes, CT (1795); rose to rank of sergeant in military, but career ended with British occupation of Cape (1806); collected Eastern Cape botanical specimens (Plettenberg Bay area); teacher, Grahamstown; sold sets of paintings of plants & trees, with descriptions; also made quaint book-like boxes of indigenous woods, recording & containing aspects of indigenous Eastern Cape trees; died destitute on farm in Bedford district, Eastern Cape
Artwork published: assistance acknowledged in G. Thompson, *Travels & Adventures in SthA* (London, 1827); *Q. Bull. SA Lib.*, June 1962
Artwork represented: British Museum, London; Library of Parliament, CT; Kew; NBI, Pret.; possibly private collections
References: *DSAB; DSAPS;* A. Gordon-Brown, *Pictorial Africana* (CT, 1975); Gunn & Codd; *Q. Bull. SA Lib.*, June 1962

WELZ, Jean Max Friedrich (1900–1975)
Studies: Realschule, Salzburg, Austria (1915); architecture, Academy of Associated Arts, Vienna (1918); mainly self-taught as artist
Profile/career: architect, Paris (1925–1936); moved to SA (1937); dept architecture, WU, under Professor Pearse (1937–1939); moved to W. Cape (1939); began painting (1940); member, New Group (1942); art teacher & principal, Hugo Naudé Art Centre, Worcester (1943–1948); organizer, SAAA, Worcester (1948); wrote several articles on art; painter of variety of subjects including flowers, working in variety of media; winner of several prestigious art awards (1947, 1969); exhibited widely, group & solo, SA & overseas, including several retrospectives
Artwork published: *A&A;* E. Miles, *The World of Jean Welz* (CT, 1997); portfolio of prints
Artwork represented: Albertina Museum, Vienna; Ann Bryant Gallery, East London; Hester Rupert Art Museum, Graaff-Reinet; Hugo Naudé House, Worcester; JAG; KG, PE; National Museum, Bloemfontein; PAM; Pietersburg Collection; RAU; SA Akademie vir Wetenskap & Kuns; Sandton Town Collection; SANG; Tatham Art Gallery, Pmbg; UCT; UFS; UNISA; UStell.; WH, Kimb.; WU; private collections, SA & worldwide
References: *A&A; DSAPS;* E. Miles, *The World of Jean Welz* (CT, 1997); *SESA*

WENNING, Pieter Willem Frederik (1893–1921)
Studies: son of art master, Leeuwarden, The Netherlands; woodcut technique under Erich Mayer
Profile/career: left The Netherlands to live in SA (1905); member, The Individualists (1911); member, SA Society of Artists (1917); bookseller's & artshop assistant, Pret. & Jhb (1905–1917); travelled extensively in SA; painted in variety of media, mainly landscape, sometimes flower studies; exhibited widely, group & solo, SA & overseas, including several retrospectives; author of 'Kunst in Zuid-Afrika', *Neerlandia*, Sept. 1913; printed etchings, cards & stationery on own press (1913–1915)
Artwork published: *A&A; Johannesburg Impressions* (Amsterdam, n.d.); E. Roworth, *Art of the British Empire* (London, 1917)
Artwork represented: DAG; JAG; Julius Gordon Africana Centre, Riversdale, Cape; KG, PE; PAM; Pret. City Hall; RAU; Roman Catholic Church, Durban; SANG; The Hague Museum, The Netherlands; Transvaalsche Vereeniging, Amsterdam, The Netherlands; UNISA; UPret.; WH, Kimb.; WU; many private collections, SA & worldwide
References: *A&A;* G. Boonzaier & I. Lippy Lipshitz, *Wenning* (CT, 1949); *DSAB; DSAPS; SESA;* H. Wenning, *My Father* (CT, 1976)

WHEELER, Judith Ellen (née FULLER) (1944–)
Studies: largely self-taught as botanical artist; received private tuition & attended specialized courses in various disciplines
Profile/career: numerous group exhibitions, Durban (20 years); painter of landscapes, abstract works, flower studies & detailed botanical works
Artwork published: –
Artwork represented: KwaZulu-Natal Nature Conservation Service, Giant's Castle Camp
Reference: –

WILES, Lucy Mary (née Townsend) (1920–)
Studies: Natal Technical College, Durban
Profile/career: founder member, Queenstown Art Society; lived mainly in Transkei & Eastern Cape; early works signed 'Lucy Mullins'; painter & graphic artist of various subjects, including flowers, working in oil, watercolour & graphic media; exhibited SA & overseas, group & solo
Artwork published: W. Tapson, *Old Timer* (CT, 1957)
Artwork represented: Queenstown Municipal Collection; WH, Kimb.; private collections
References: *A&A; DSAPS*

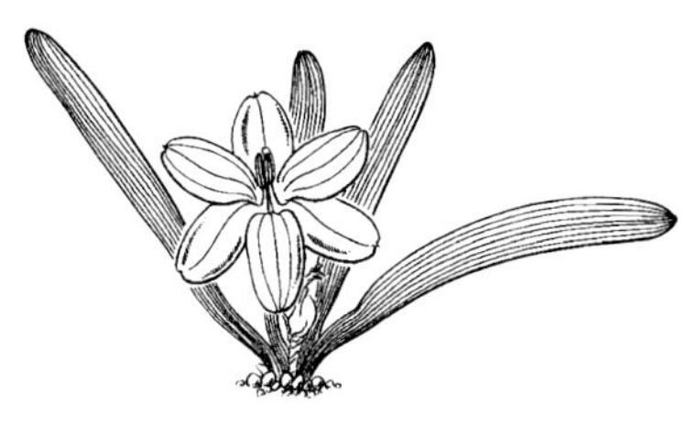

WILLE (later ZAHN), Marjorie (1900–1981)
Studies: Oranje School, Bloemfontein; teaching diploma, College of Arts, CT (1919–1921); teaching diploma, School of Art, Regent Street Polytechnic, London
Profile/career: art teacher, St Andrew's School, Bloemfontein; art teacher, Michaelhouse, Nottingham Road, KwaZulu-Natal; art teacher, Clifton Prep. School, Nottingham Road, KwaZulu-Natal; participated in group exhibitions
Artwork published: H.B. Rycroft, *Trees of the Tsitsikama National Parks* (Pret., 1980); folios of prints of Cape flora; Southern Life calendar (1975)
Artwork represented: private collections
References: *A&A; DSAPS;* NBI files

WILLIAMSON, Graham (1932–)
Studies: BDS; MSc, botany (WU)
Profile/career: practised as dental surgeon (1957–1993); involved in Central African & SA botanical research (since 1962); paintings of Richtersveld flora (from 1977); plates of Zambian endemic Euphorbiaceae soon to be published
Artwork published: *The Orchids of South Central Africa* (London, 1977); *Swartkop Nature Reserve* ([Oranjemund, 1993]); *Richtersveld National Park* (Pret., 1995); *Richtersveld: The Enchanted Wilderness* (Pret., 2000); plates & drawings in many scientific papers
Artwork represented: artist's personal collection.
References: *Aloe*, 34:1977; *Cactus & Succulent Journal*, 71:1999

ZERFFI, Florence Louise Josephine (Mrs Caldecott) (1882–1962)
Studies: Technical School, Stockport, England; Königliche Kunstgewerbeschule, Berlin – silver medal for art studies (1906–1916); private art tuition from Emil Doepler
Profile/career: born England; moved to SA (1916); member, SA Society of Artists (1916); founder member, K Group (1922); curator, Michaelis Collection, CT (1921–1924); married Harry Stratford Caldecott (1924); taught painting privately & at various high schools, CT (1929–1956); founder member, New Group (1938); member, Eastern Province Society of Artists; returned to England (1956); used various media to paint variety of subjects, including flowers; exhibited widely, group & solo, SA & overseas
Artwork published: *A&A;* C.J. Langenhoven, *Die Eensame Hoop* (CT, 1922); various newspapers & journals, including *Die Burger* & *Huisgenoot*
Artwork represented: DAG; JAG; PAM; SANG; UFS; UPret.; UStell.; WH, Kimb.
References: *A&A; DSAB; DSAPS;* J. du P. Scholtz, *Strat Caldecott* (CT, 1970); *SESA*

SELECT BIBLIOGRAPHY

Arber, Agnes, *Herbals: their Origin and Evolution*. Cambridge University Press, 1912.

Arnold, Marion, *Art Meets Science: Flowers as Images*. Johannesburg: Standard Bank, 1992. (Catalogue, Standard Bank National Arts Festival Exhibition, 1992)

Arnold, Marion, 'Thomas Baines and Southern African Flora: "My Small Skill in Botany"', in Michael Stevenson, ed., *Thomas Baines: an Artist in the Service of Science in Southern Africa*. London: Christies, 1999. (Exhibition Catalogue, 1999)

Arnold, Marion, *Women and Art in South Africa*. Cape Town: David Philip, 1996.

Batten, Auriol, *Flowers of Southern Africa*. Sandton: Frandsen, 1986.

Bayer, A., *Flower Paintings of Katharine Saunders*. Maidstone, Natal: Tongaat Group, 1979.

Berman, Esmé, *Art & Artists of South Africa*. Cape Town: Balkema, 1970.

Blunt, Wilfrid and Raphael, Sandra, *The Illustrated Herbal*. London: Francis Lincoln, 1994, 2 ed.

Blunt, Wilfrid and Stearn, William T., *The Art of Botanical Illustration*. Woodbridge, Suffolk: Antique Collectors' Club and The Royal Botanic Gardens, Kew, 1994, 2 ed.

Coats, Alice M., *The Book of Flowers: Four Centuries of Flower Illustration*. London: Phaidon, 1974, 2 ed.

De Bray, Lys, *The Art of Botanical Illustration: the Classic Illustrators and their Achievements from 1550 to 1900*. Secaucus, New Jersey: Wellfleet Press, 1989.

Evans, D.S., Deeming, T.J., Evans, B. and Goldfarb, S., *Herschel at the Cape, 1834–1838*. Cape Town: Balkema, 1969.

Flowering Plants of Africa, The, correspondence relating to. Archives of the National Botanical Institute, Pretoria, File M163.

Forbes, Vernon S. and Rourke, John, eds, *Paterson's Cape Travels, 1777 to 1779*. Johannesburg: The Brenthurst Press, 1980.

Goldblatt, Peter, 'An analysis of the flora of southern Africa: its characteristics, relationships, and origins', *Annals of the Missouri Botanical Garden*, 65, 1978, pp. 369–436.

Goldblatt, Peter, 'Floristic diversity in the Cape flora of South Africa', *Biodiversity and Conservation*, 6, 1997, pp. 359–377.

Goldblatt, Peter and Manning, John, *Gladiolus in Southern Africa*; illustrated by Fay Anderson, Auriol Batten and John Manning. Cape Town: Fernwood Press, 1998.

Gordon-Brown, Alfred, *Pictorial Africana*. Cape Town: Balkema, 1975.

Gunn, Mary and Codd, L.E., *Botanical Exploration of Southern Africa*. Cape Town: Balkema, 1981.

Hewson, H., *Australia: 300 Years of Botanical Illustration*. Collingwood, Victoria: C.S.I.R.O. Publishing, 1999.

Hilton-Taylor, Craig, *Red Data List of Southern African Plants*. Pretoria: National Botanical Institute, 1996.

Hutchinson, John, *A Botanist in Southern Africa*. London: P.R. Gawthorn, 1946.

Jordaan, P.G., 'Rudolf Marloth, South Africa's first trained botanist', *Lantern*, 13 (1) 1963, pp. 86–93.

Karsten, Mia C., *The Old Company's Garden at the Cape and its Superintendents*. Cape Town: Maskew Miller, 1951.

Kerkham, A.S., comp., *Southern African Botanical Literature 1600–1986*. Cape Town: South African Library, 1988.

Killick, D.J.B. and Du Plessis, E., eds, *History of* The Flowering Plants of Africa *and an Index to Volumes 1 to 49* [Pretoria]: Botanical Research Institute, [1988].

Klonk, Charlotte, *Science and the Perception of Nature: British Landscape Art in the Late Eighteenth and Early Nineteenth Centuries*. New Haven and London: Yale University Press, 1996.

Marloth, Rudolf, *The Flora of South Africa*, 4 vols in 6. Cape Town: Darter; London: W. Wesley, 1913–1932.

McCracken, Donal P., *Gardens of Empire: Botanical Institutions of the Victorian British Empire*. London and Washington: Leicester University Press, 1997.

McKay, H.M., 'William John Burchell, botanist'; Part 2: 'Burchell in South Africa', *Journal of South African Botany*, 7, 1941, pp. 61–76.

North, Marianne, *A Vision of Eden: the Life and Work of Marianne North*. Exeter, England: Webb and Bower in collaboration with The Royal Botanic Gardens, Kew, 1980.

Oliver, E.G.H., 'Some observations on two early Cape florilegia', *Bothalia*, 13 (1&2) 1980, pp. 115–125.

Paterson-Jones, C., *The Cape Floral Kingdom*. London: New Holland, 1997.

Ponsonby, Laura, *Marianne North at Kew Gardens*. Exeter, England: Webb and Bower in association with the Royal Botanic Gardens, Kew, 1990.

Procter, M., Yeo, P. and Lack, A., *The Natural History of Pollination*. Portland, Oregon: Timber Press, 1996.

Raven, Peter H., Berlin, Brent and Breedlove, Dennis E., 'The origins of taxonomy', *Science*, 174, 1971, pp. 1210–1213.

Rix, Martin, *The Art of the Plant World*. New York: Overlook Press, 1981.

Rourke, John P., *The Proteas of Southern Africa*, illustrated by Fay Anderson, Lura Ripley, *et al.* Cape Town: Purnell, 1980.

Saunders, Gill, *Picturing Plants: an Analytical History of Botanical Illustration*. London: Zwemmer, 1995.

Sitwell, Sacheverell and Blunt, Wilfrid, *Great Flower Books, 1700–1900: a Bibliographical Record of Two Centuries of Finely Illustrated Flower Books*. London: Witherby, 1956; 2 ed., 1990.

Smith, C.A., 'Common Names of South African Plants', *Botanical Survey of South Africa*. Pretoria: Department of Agricultural Technical Services, 1966. (Memoir No. 35)

Stafleu, Frans A., *Linnaeus and the Linnaeans: the Spreading of their Ideas in Systematic Botany, 1735–1789*. Utrecht: Oosthoek for the International Association for Plant Taxonomy, 1971.

Stearn, William T., *Botanical Latin*. London: David & Charles, 1973, 2 ed.

Stearn, William T., *Flower Artists of Kew*. London: The Herbert Press, 1990.

Warner, Brian and Rourke, John, *Flora Herscheliana*. Johannesburg: The Brenthurst Press, 1998.

Waterhouse, G., De Wet, G.C. and Pheiffer, R.H., *Simon van der Stel's Journey to Namaqualand in 1685*. Cape Town: Human and Rousseau, 1979.

West, Keith, *How to Draw Plants*, foreword by Wilfrid Blunt. London: The Herbert Press in association with The British Museum (Natural History), 1983.

INDEX

Bold numerals indicate pages with illustrations.

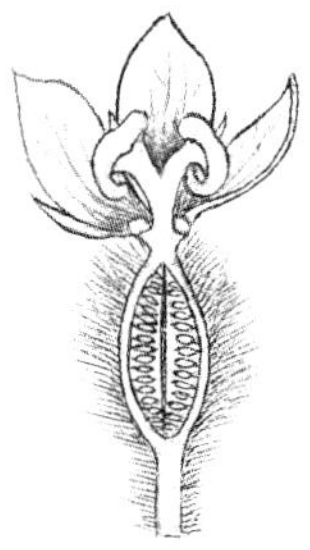

LIST OF SUBSCRIBERS

SPONSORS' EDITION

Arizona-Sonora Desert Museum • Steve Bales
Kenneth S. Birch • Jim Gerard Paul Broekhuysen • M.J. Hyde • Yoshito Iwasa
Jürgen Kögl • Robin Moser • Jan & Elizabeth Nel • Desmond L. Nielsen
Timothy & Marilyn Noakes • Merry Park • Elizabeth Parker • J.P. Rourke
Dr Shirley Sherwood • Jean Turck • Paul van Schalkwyk • Stephan A. Welz
J.A. (Jerry) Windell • Dr Peter R. & Mrs Elvire Young

COLLECTORS' EDITION

William Barlow
Alex A. Barrell
Hanna Boeke-McLachlan
The Brenthurst Library
John E. Bryan, Fellow of the Institute of Horticulture
Irma Burger
Jane Carruthers
Cynthia Cormack
Jalal & Kulsum Dhansay
Mags Dixon
James N.T. Doig, Dundee, Scotland
A.G. Farndell
Fernwood Press (Pty) Limited
Arnold Gain
Rob Gee
Lydia Gorvy
Christopher Greig
Sonia & Rudolf Hanni
Jan Hemmes
R.G. Jeffery
G.E. Jewell, New Zealand
Jean Jordaan
Cynthia Kemp
C.F. Kleiss, Amstelveen, Holland
Dr Harold König
Graham Leslie
R.E. Levitt
Zelma Long & Phillip Freese
Gren Lucas
Peter P.C. Mertens
Mia Karsten Fund, Compton Herbarium, Kirstenbosch
Robin Moser
National Library of South Africa
Ian L. Outram
I.E. Peddle
John Pennefather
B.M. Robson
Marius Stanz
Pieter & Pam Struik
Unisa Library
J.A. Janse van Rensburg
N.P. Janse van Rensburg
Paul F.X. von Stein
Jennie Wessels

STANDARD EDITION

ABC Bookshop
E. Abendanon
J.J. Ackermann
Jill Adams
Jonathan Aldridge
Antonio Federico Alziati
Lia Maria Alziati
Michael & Monica Amm
Dr Ellen Ancker
Anne Aneck-Hahn
Anne & Brian Askew
William & Linda Astley-Jones
Mark Attwood
Hennie Aucamp
Ann Axelson
Peter Badier
David & Linda Baldie
Steve Bales
Marijke & Ken Ball
Mrs R.C.I. Barlow
W. Barsby
David Basckin
Laura Batchelor
Annemarie Baumgartner
Yvonne Becker
Alan & Sonja Begg
Professor Gerhard & Isolde Beukes
Mnr. W.J. en mev. Freda Beukes
Christine Beyers
Richard Bladen
Tamlin Blake
F. de la Rey Blignaut
Agnes Bodenstein

Herbert Bongers
Angela H. Bonning
M.F. Bosazza
Botanical Society of South Africa
Amanda Botha
Hanna Botha
Walter A. Böttger
Dr C. Boucher
Georges & Christine Boucher
Jean Pierre Bouley
J.K. Brinckman
Marilyn Brink
Advocate H. Brits
Shirley Brown
Craig Bruce
John E. Bryan, Fellow of the Institute of Horticulture
Gary Buckley
Boy Burger, Castle Olympus, Citrusdal
Irma Burger
Joyce Burger
Dorothy Byer
Estelle Harrington Byrne
Janet & Mike Cahill
J.E. Carman
Patricia Anne Chemaly
Stephanie Christiané
Clarke's Africana & Rare Books
Glenda Cleaver
P.A.G. Cloete
Susan M.A. Cochrane
Vanessa Cockram
M.J.R. Coetzee
Professor & Mrs Dalvan Coger
'Snooks' Cole
Gael Constable
Sandra Cook & Patricia Rowe
Elizabeth & Ashley Cormack Havens
C.L. Cousins
Mrs D.G.L. Crane
J.M. Therese d'Assonville
Kathé Day
Linda de Luca
Marianne de Villiers
Dr Nic de Wet
Helen & Wal Dean
Tessa Dean
William Dewar
Lisa, Vincent & Marcus Di Bella
In memory of Martin Di Bella
Coba Diederiks
Claire Difford
Maggie Dixon
Dr Philip E. Downs
Gail Dreyer
Enid du Plessis
J.C.M.D. du Plessis
Professor J.E. du Plessis
J.J.R. du Plessis
Niel du Plessis
Deirdre du Preez
John & Pamela Duff
Allan Dunbar
Anne G. Earhart
Ida-Marie Ehlers
Benno M. Eller
Professor G.F.R. Ellis
J.C. Els
R.W.H. Evans
C.P. Everitt
Charles H. Everson
S. Fagan
Pauline Farquhar
Miranda & Alan Feinstein
Peter & Sue Fitt
FNB (A Division of FirstRand Bank Limited)
Kirby W. Fong
Paul Forster
Jutta Fredman
Jonathan A.S. Freedberg
Renée Freedman
Peter V. Gammarano Jnr
Mr G.R. Garlick
Jeremy Gauntlett
Gerrit Germishuizen
David & Gaby Gess
Sarah Gess
H.V. Gibberd
Howard & Monica Gie
Alice Gilbert
Ronald P. Gilbert
Andrew & Moiragh Girdwood
Sue & Martin Goldswain
Dr G.J.M. Gorter
Niki Goulandris, Athens
Teddy Govender
Allison & Clive Graham
Mr & Mrs T.B. Graves
Ryno & Evon Greenwall
Elizabeth Jane Greenwood
Christo en Wilna Grobbelaar
Dr J.P. Grobbelaar
Dr Gottfried Gutte
Gerald Guy
Penny Haig
A.C.J. Hall
Caroline & Benedict Hall
James P. Hanna
Sandor Hanni
Savannah Hanni
Wolf-George Harms
Derek Harraway
Ann J. Harris-Deppe
M.A.C. Harrison
Harry Molteno Library, Kirstenbosch
J.O.C. Hart
Ena Harvey, St. James
R.A. Harvey
Michael W. Hawkes
Geoffrey A. Hedgecock
Doreen & Roger Hemp
Mr & Mrs E.H.O. Hoal
Marÿke Honig
Birgitta Hope
Dr Steve Hopper
Jan & Protea Hubregtse
Erika Huntly
John P. Hurabiell Snr
Dave & Jane Hutson
J.S. Gericke Library, Stellenbosch
Elodie Janovsky
Geoff Jewell
Chris & Ann Job
Jenny & Gavin Johnson
Telita Jorge
Gerald Josman
Paul A.M. Joubert
Sidney H. Kahn
Brian Kalshoven
Andrew Kamiti
Jean Keenan-Smith
Aliki & Clive Kelly
C. Kemp
Donald R. Kendall Jnr
Janet Kennedy
E.W. Kent
Frank Keoghan, Dublin
Giskin Kesterton
Marianne Kien
Mnr. Tommie Kilian
Anna Maria Kirchhoff
Michael Knipe
David & Noleen Knott
Laetitia Knox-Davies
S. Koeltz, Koeltz Scientific Books, Germany
Colleen & Walter Köppe
Shirley Krsic
Peggy Kuschke
Eric & Isobyl La Croix
Dr S.H. Laubscher
James Henry Le Roux
John Lennard
Jean-Marie & Astri Leroy
Ben H. Lim
Claire Linder Smith
Caryl Logie
Don Logie
Elmer J. Lorenz
Johan D. Loubser
Anna en Albertus Louw
Peter J. Louw
Mrs Shëlagh M. Lubbock
Roy A. Lubke
Suzanne Lucas FLS, PRMS, FPSBA
The LuEsther T. Mertz Library, New York Botanical Garden
Patricia Luke
O.J. Mackenzie
Gerhard en Rykie Maritz
M.J. & A. Martinson
Victoria Matthews
Victoria Mayer
Marie McCrae
T.P. McDonogh
Anna Maria McFarlane
Peter McGregor
Candace E. McIntosh
Louise McIntosh
Rhoda & Cameron McMaster
Paul Meaker

Advocate Abri Meiring
Jill Meyer
Julie-Mari Meyer
Lee J. Miller
M. Elton Mills
Dave Minnaar
Buddy & Jenny Mockford
Penny Mockford
P.J. Mockford
Pieter Mohr en Hester Burger
Megan Mooney
Patrick Morant
Milou Morlion
Robin Moser
Ann Mudge
Nigel Wayne Mudge
Andrea Nagel
Bruce & Cilla Nel
Evelyn Nesbitt
Douglas Newton
Sarah Newton
Penny Nicholson
Anne Nicolson
Hans Nieuwmeyer
Mrs G.A. Nikschtat
P.A. Nisbet
William S. O'Brien, Malahide, Ireland
Julian Ogilvie Thompson
Virginia Ogilvie Thompson
Matthew R. Opel
Cyril & Cora Ovens
Owen for Audrey
Valerie Oxley
Raimondo Paladini, Monteroni Di Lecce, Italia
Elizabeth Parker
Jeff & Bev Parsley
Alison Passmore
Wendy Paterson
Rian & Les Pauw
Susan Pearson
Graham S. Peddie
Roydon C. Peden
Peter N. Pentz
Jeffry Perlman
Kathleen Petersen
Alida T. Petrick
Manuel Pita, Madeira Island, Portugal
Pietersburg Art Museum
Pietersburg City Library
Professor A.J.H. Pieterse
Barbara Pike
C.F. Pohl
Elsa & Tony Pooley
Frances Potter
Noel Potter
R.G. Potter
J.D. Pratt
Pieter & Tarisa Pretorius
Yvonne Quénet
Patricia & Adrian Rademeyer
R.F. Raymond
E.A. Reynolds
Dr & Mrs F.D. Richardson
Neil Ritchie
Douglas S. Roberts
Pierre Roberts
Dr George A. Robertson
Charles & Martine Robinson
Cedric Roché
Ian & Viola Ross
Gretel L. Rothe
Neville Rowe
Maraletta Rüde
Dr P. Rush
Ernie & Colleen Sanvido
Stephanie K. Saunders
Rod & Rachel Saunders
Arne & Pixi Schaefer
Robert J. Scholes
Johann Scholtz & Jacqueline Heaton
Brian Schrire
Jonathan Schrire
David M. Schultz
Janine Schulz
Phil & Ian Scott
Françoise Searle
Gary Searle
Patricia Seed
Frederick R. Sellman, London
Rob & Kate Semple
Bo & Anthea Serritslev
Margaret T. Shaw
Peter & Lizzy Simon
William T. Simon
Genard Russel William Sizer
John Reginald Sizer
Stuart Andrew Sizer
Sandy Slater
Dennis & Ansie Slotow
Mrs Barbara Smallwood
Dr J.S. Smiedt
Professor G.F. Smith
Laurie Smith
Miemie Snyman
M.J. Snyman
Burton S. Sperber
Stuart J. Sperber
Mr & Mrs Stainer, Venice, Italy
Marius Stanz, Fairland
Bruce N. Starke
Ines & Basil Stathoulis
George F. Stegmann
Margaret Stevens VPSBA, SGFA
Mrs E.C. Stevenson
Peter Steyn
Charles & Jana Stirton
C. & P. Strand
B. & C. Street
Leonore Strobos
Summerfield's Indigenous Bulbs & Seeds
Andrew & Vanessa Sutherland
Jane Suttle
Isaac M. Sutton
Wayne & Carla Tanner
Norman Tate
C. Taute
J. Terry-Lloyd
Costas A. Thanos
Paule Thiebaut-Zenon
Vicki Thomas
Jennetta Tilney
Sally Townshend
Mrs M.E. Trafford
Dr W.E. Trengove
Derek Tribble
Professor Elizabeth Triegaardt
T.H. Trinder-Smith
Julia Twigg
University of Stellenbosch Museum
University of the Western Cape
Dr Erika van den Berg
Dick & Liz van der Jagt
John & Beth van der Linde
Christo & Ida van der Merwe
François J. van der Merwe
Paul & Gerda van der Reyden
Maria van der Ryst & Siegwalt U. Küsel
B. van der Walt
Jennie van Malsen
Liz van Niekerk
Sarel van Niekerk
Robert & Patricia van Zyl
Paul & Laurie Venter, Honeydew
Daniel & Mabel Visser
Willie & Joan Vlok
Leigh Voigt
M. von Fintel
Dr & Mrs P. Vorster
Vos & McGregor Families
D.J. Wakefield
Robert Waldron
Ethel Walt
Mrs Ann Walters
Dr Schalk & Anita Walters
Brian Warner
Alison Watt
Professor Ian B. Watt
Derick & Ethné Webb
Mr & Mrs William N. Weld
Roland A. Welsbie
Hannarie Wenhold
Donald Wheeler
Dave & Val Whittaker
Colin Whittle
Geraldine Williams
Graham & Françoise Williamson
Nigel Willis
Beverley Wilson
Derric H. Wilson
Julie Wilson
B.C. Winkler
J.H.S. Winter
Roy & Val Wise
Anette Wodrich
B. Nigel Wolstenholme
Rob & Janet Wood
Albina Wortley
Mrs Jean Bianca Wyatt
E. Anne Wyatt-Goodall
Dr Peter R. & Mrs Elvire Young
M. & A. Zonneveld